TR

"Patenaude leaves a vivid impression of the man—his personality, the petty dramas of his life in exile, and the immensely complicated web of relations with his wife, family, patrons, friends, aides, and followers. Patenaude also ably captures the intensity and far-flung nature of the 'Old Man's' ongoing struggle with Stalin."

—*Foreign Affairs*

"Gripping, cinematic. . . . No matter what your political orientation, if you believe—or ever did believe—in the potential betterment of humanity, then you've got something to learn from the strange and tragic story of Leon Trotsky. It's a tale of pride and power and political failure, of genius turned to the service of dogged, dogmatic conviction, of a supremely intelligent man who destroyed others in the name of a cause that then destroyed him."

—Andrew O'Hehir, *Salon*

"Objective and scholarly. . . . Though Patenaude's book concentrates on the period of Trotsky's Mexican exile and provides fascinating pen-portraits of lovers, acolytes, and killers alike (including details of Trotsky's affair with Frida Kahlo that Isaac Deutscher so sweetly veiled), it also encapsulates his earlier life."

—Tariq Ali, *The Guardian* (London)

"A lively and finely detailed description of Trotsky's household."

—Ron Charles, *The Washington Post*

"An extraordinary tale of intrigue and terror leading up to Trotsky's murder in Mexico in 1940, extraordinarily rich in detail and historical background yet as briskly told as a thriller."

—*The Evening Standard* (London)

"Trotsky was no saint, but his last years in Mexico make for a good thriller. Stalin and his henchmen were after him and before successfully killing Trotsky with an ice pick to the head, there were several dramatic attempts. In addition, the author follows Trotsky's friendship with artist Diego Rivera—and their falling out—and his affair with Rivera's artist wife Frida Kahlo." — *New York Post*

"This history, peopled by vibrant figures of the 1930s like Frida Kahlo and John Dewey, recalls with literary flair events that shocked the world. Bertrand Patenaude's Trotsky is an epic character: fiery, vain, contentious, exacting, intellectually lively, ideologically blinded, seductive, even sexually aggressive—and a man keenly aware that the inherent tragedy behind human existence overshadows the petty mishaps of politics, assassination included." —Ken Kalfus

"It is a tribute to Bertrand Patenaude's narrative skill that although we always know how his book is going to end, it is nonetheless readable and utterly gripping. . . . The pace and tension are worthy of a Hollywood thriller. . . . One finishes this splendidly scrupulous and entertaining book struck by the irony that Trotsky—a man soaked in the blood of his countrymen and who sacrificed the lives of millions to his crazed historical logic—became the most famous victim of a regime he had done so much to create."
—Dominic Sandbrook, *The Daily Telegraph* (London)

"What are we to make of Trotsky? Nowadays, young people do not know much about him, and it is not easy to explain to them exactly who he was and why he was important. Bertrand Patenaude's new book will help. It is based on a great deal of research and contains much new material. . . . Patenaude describes Trotsky's exile in great detail." —Paul Johnson, *National Review*

"An absorbing reconstruction of Trotsky's last years in Mexico. . . . Patenaude's hybrid history and detective story grips from start to finish. With rare narrative verve, he chronicles the last years of a revolutionary's life, with its sexual jealousies, paranoia, and finally murder." —Ian Thomson, *The Sunday Times* (London)

"This is an extraordinary, gripping piece of history that gets closer to Trotsky's essential character than any of the vast tomes devoted to him in the past. Patenaude examines the last surreal three years of the revolutionary leader's life, effectively imprisoned in Mexico City, kept afloat by the capricious affection of the painter Diego Rivera and his idealistic supporters from across the United States. . . . Perhaps most extraordinary is the page-turning narrative drive which keeps the reader enthralled despite knowing how the story ends. Don't miss it."   —Misha Glenny

"A haunting and dramatic reconstruction of Trotsky's life and death in exile. The detail is fascinating, almost voyeuristic. . . . No doubt Trotsky's reputation has survived better than Stalin's, but in this painstaking reconstruction of one of the world's great revolutionary thinkers and agitators, Bertrand Patenaude makes it difficult for any reader not to come away with a sense of amazement at the mixture of naïve optimism and moral self-serving that infected the communist vision of the future." —Richard Overy, *Literary Review*

"*Trotsky* reads like a crime thriller. . . . Bertrand Patenaude's magnificent book is a skillful reminder of Trotsky's sheer nastiness and of the brutal world he naturally inhabited."
                    —A. N. Wilson, *Reader's Digest*'s "Books of the Month"

"Patenaude has written a book that is sensitive to the tragedy of Trotsky's end, but alive, too, to the drama of his time in Mexico City. A tale of assassinations and poisonings, Patenaude's account owes more to the films of Francis Ford Coppola and Martin Scorsese than it does to the works of Karl Marx. Trotsky's haunting work on the rise of Hitler is also recalled. As Stalin's posthumous approval ratings continue to rise in Russia, this book provides a melancholic reminder of his greatest enemy's moral clarity."   —*New Statesman*

"Well researched and vividly told."
                              —Robert Service, *The Guardian* (London)

TROTSKY

TROTSKY

DOWNFALL OF A REVOLUTIONARY

BERTRAND M. PATENAUDE

HARPER PERENNIAL

NEW YORK • LONDON • TORONTO • SYDNEY • NEW DELHI • AUCKLAND

HARPER ● PERENNIAL

Grateful acknowledgment for permission to reproduce illustrations is made to the following: Alexander H. Buchman Papers, Hoover Institution Archives: pages 3, 190, 224, 236, 239, 267, 274; Bernard Wolfe Slide Collection, Hoover Institution Archives: pages 29, 42, 60, 169; Bertram D. Wolfe Papers, Hoover Institution Archives: page 80; and Albert Glotzer Papers, Hoover Institution Archives: pages 110, 121.

A hardcover edition of this book was published in 2009 by HarperCollins Publishers.

HarperCollins books may be purchased for educational, business, or sales promotional use. For information please write: Special Markets Department, HarperCollins Publishers, 10 East 53rd Street, New York, NY 10022.

FIRST HARPER PERENNIAL EDITION PUBLISHED 2010.

Designed by William Ruoto

The Library of Congress has catalogued the hardcover edition as follows:

Patenaude, Bertrand M.
    Trotsky: downfall of a revolutionary / Bertrand M. Patenaude.
        p. cm.
    Includes bibliographical references and index.
    ISBN 978-0-06-082068-8
        1. Trotsky, Leon, 1879–1940. 2. Trotsky, Leon, 1879–1940—Assassination. 3. Revolutionaries—Soviet Union—Biography. 4. Statesmen—Soviet Union—Biography. 5. Exiles—Mexico—Biography. 6. Mexico—History—1910–1946. I. Title.
DK254.T6P34   2009
947.084092—dc22   2008052248

ISBN 978-0-06-082069-5 (pbk.)

10 11 12 13 14 OV/RRD 10 9 8 7 6 5 4 3 2 1

CONTENTS

TROTSKY

A Miraculous Escape

n the early-morning hours of May 24, 1940, Leon Trotsky slept soundly inside his villa in Coyoacán, a small town on the southern outskirts of Mexico City. The house was heavily guarded. Five Mexican policemen occupied a brick *casita* on the street just outside the high walls of the property. Inside were Trotsky's private bodyguards, five in all, including four young Americans. One of them, a twenty-five-year-old New Yorker by the name of Robert Sheldon Harte, started his shift that night at 1 a.m. posted inside the barred door to the garage, which was the only entrance to the house. His comrades were asleep in a row of outbuildings set against one of the inside walls of the roughly rectangular patio.

Trotsky had spent most of the previous day dictating a manifesto about the war in Europe and kept at it late into the evening. His major work in progress, a biography of Joseph Stalin commissioned by the New York City publishing house Harper & Brothers, was a year and a half overdue. The war was now a huge distraction, in part because of the bitterly divisive debates it had sparked among his followers in the United States, home to the most formidable of the Trotskyist splinter groups around the world.

Once the most internationally famous leader of the Soviet Union, Trotsky now made his living as a freelance writer. A literary stylist known for his sardonic wit, his most acclaimed work in the West was his panoramic history of the Russian Revolution, published in the early 1930s after he had been exiled by Stalin. He had agreed to write the biography of his archenemy only because he needed the money to support himself and to pay for his security in Mexico. The generous advance from the American publisher was long gone, but the book was nowhere near completion and had become a millstone around his neck. Trotsky often said to his wife, Natalia, that he had become disgusted with it and that he longed to return to writing his biography of Lenin.

Nor were Trotsky's editors in New York especially pleased with the completed chapters. It had been a mistake to expect Trotsky to write an objective biography of the man who had destroyed him politically, wiped out his followers and his family, and transformed his image in the Soviet Union from a dashing hero of the Bolshevik Revolution into its Judas Iscariot. Trotsky's name was readily invoked to account for every accident and failure in the USSR, from a train derailment, to a factory explosion, to a missed production quota. His theatrical appearance—the piercing gaze magnified by the thick lenses of his round glasses, the shock of turbulent hair, the thrusting goatee—and his propensity for striking dramatic poses were a boon to the Soviet caricaturists. He was portrayed as several varieties of barnyard animal, including a pig branded with a swastika feeding at the trough of fascism, and in the title of a cartoon that exploited another favorite motif, "The Little Napoleon of the Gestapo."

It is little wonder then that the Stalin biography had become a slog, and that the Second World War provided Trotsky with a good excuse to procrastinate. The war also gave him the opportunity to earn much-needed income by writing articles for American magazines about the latest diplomatic and military maneuvers. Trotsky's appeal as an analyst of international affairs spiked in August 1939, when the world was stunned by the announcement of the Nazi-Soviet nonaggression pact, a turnabout he had predicted. What did the inscrutable, pipe-smoking Georgian dictator in the Kremlin have in mind when he signed a friendship treaty with his ideological opposite, Adolf Hitler? Trotsky was asked to assess the pact and then its bloody aftermath, as the Wehrmacht and the Red Army swallowed up Poland while the Kremlin asserted its mastery over Latvia, Lithuania, and Estonia, and then invaded Finland. Hitler's preoccupations were France and Great Britain, but it was only a matter of time, Trotsky confidently predicted, before the Führer would turn his armies eastward and invade the Soviet Union.

Stalin's pact with Hitler forced the Soviet cartoonists to expunge the swastikas and the jackboots from their anti-Trotsky propaganda. Communist parties loyal to Moscow had to follow suit, among them the Mexican Communists, who were relentless in their efforts to compromise Trotsky's asylum by portraying him as a meddler in Mexican politics. They had been banging this drum ever since his arrival in Mexico in January 1937, yet the anti-Trotsky campaign they launched in the winter of 1939–40 was more violent and sustained than any that had come before. Its slogan was a point-blank "Death to Trotsky!" And by the time the May Day marchers were shouting in unison for the traitor to be expelled, Trotsky had convened a meeting of his guards to warn them that his enemies were creating the atmosphere for an armed attack on the villa.

These threats put a strain on Trotsky's nerves and his health. He was now sixty years old. He suffered from high blood pressure and insomnia, among other ailments. The best medicine was vigorous outdoor exercise. Trotsky loved to hunt and fish, yet the possibilities were limited in Mexico because of concerns for his safety. A picnic outing required the presence of several armed bodyguards and a detail of Mexican police.

Trotsky in the patio of his "fortress," winter 1939–40.

The Old Man, as his followers affectionately referred to Trotsky, adapted to his more restrictive environment by hunting for various species of cactus, which were transplanted to the patio in Coyoacán. These exhausting expeditions into the countryside were organized once every several weeks. Trotsky's daily exercise these days revolved around his other new hobby, caring for the rabbits and chickens he kept in hutches and a caged yard in the patio. It was prison life, Trotsky often said, and his staff felt the same way. He chafed at his confinement and yearned to find an outlet for his restless energy. Adhering to routine, late in the evening of May 23 he had taken a sedative before going to bed.

At about four o'clock in the morning the nighttime quiet was shattered by the sound of automatic gunfire. Summoned from a deep sleep, Trotsky thought he was hearing fireworks, that the Mexicans were celebrating one of their fiestas. Coming to his senses, he realized that "the explosions were too close, right here within the room, next to me and overhead. The odor of gunpowder became more acrid, more penetrating. Clearly what we had always expected was now happening: we were under attack."

Natalia was quicker to react. She hustled Trotsky off his bed and onto the floor, sliding down on top of him and into a corner of the room. Gunfire came through the two doors facing each other on opposite sides of the room and through the French windows just above the couple, creating a three-way crossfire. As bullets ricocheted off the walls and the ceiling, Natalia hovered protectively over her husband until he communicated to her through whispers and gestures to lie flat next to him. Splinters of glass and plaster flew in all directions in the darkness. "Where are the police?" Trotsky wondered, his mind now racing: "Where are the guards? Tied up? Kidnapped? Killed?" And what had become of Seva? One of the rooms from which the gunfire came was the bedroom of the couple's fourteen-year-old grandson.

The barrage lasted several minutes. For a moment everything went silent, and then they heard the dull thud of an explosion. The door to Seva's room swung open, admitting a fiery glow. Raising her head slightly, Natalia glimpsed a figure in uniform standing at the threshold and silhouetted against the flames, "his helmet, his distorted face, and the metal buttons on his greatcoat glowing red," she recalled afterward. The intruder seemed to be inspecting the Trotskys' bedroom for signs of life. Although there were none, he raised a handgun and fired a round of bullets into the beds, then disappeared.

From the boy's room came a loud, high-pitched shriek: "Dedushka!" It was Seva, calling out in Russian: "Grandfather!" The cry was part warning, part plea for help. For the grandparents, this was the most distressing moment of all. They got up off the floor and went over to his room, which was empty. A small fire was burning the floor beneath a wooden wardrobe, which crackled in the heat. "They've taken him," Trotsky said, fearing that his young American comrades and everyone else in the house had been killed. Sporadic gunfire could still be heard from the patio. Natalia grabbed blankets and a rug to try to smother the fire, as Trotsky reached for his gun.

The American guards had been pinned down in their quarters by an attacker dressed in a police uniform and armed with a Thompson submachine gun. Hearing the rattle of machine guns inside the house, they visualized a massacre. As the gunfire eased up, the chief of the guard, Harold Robins, looked out his door and saw Seva standing in the lighted doorway of the kitchen, crying and speaking gibberish. Robins called to the boy to come to his room and ordered a fellow guard to kill the light. He then aimed his submachine gun across the yard in the direction of the retreating raiders, but the weapon jammed when he tried to shoot. Another guard, Jake Cooper, took aim with his pistol at a man running toward the garage exit, but seeing the stranger's police uniform, he could not bring himself to pull the trigger. Still another guard, Charles Cornell, took a potshot at a different "policeman" retreating toward the garage. These were the only shots the guards managed to fire.

Trotsky, meanwhile, had gone into his bathroom, where he could peer through a window that looked out into the patio toward the guards' quarters. In the semidarkness, he saw a moving figure and called out, "Who is there?" The stranger answered too softly to be understood, so Trotsky fired his gun, missing the target's head—which was fortunate, because the man Trotsky took for an intruder turned out to be Jake Cooper.

Natalia had smothered the fire in Seva's room and returned to her own bedroom. Through the bullet holes in the door leading to Trotsky's study, she observed a peaceful scene: "the papers and books looking immaculate in the calm glow of the shaded lamp on the desk." She tried the door, but the impact of the bullets had jammed the lock. At that moment she heard Seva's voice from somewhere in the patio, this time sounding joyous as he called out the names of friends who were staying at the house. A wave of relief swept over Trotsky and Natalia: The worst had not come to pass after all. They began pounding on the door. Moments later, three of the guards entered the study and forced open the door to the bedroom. Against all expectations, they found Trotsky and Natalia unharmed.

THE MEMBERS OF the household gathered in the patio. Everyone was accounted for, except Bob Harte. Seva had been lightly wounded in the foot. At the sound of gunfire, he had scrambled under his bed and was grazed by a bullet shot through his mattress. Natalia had minor burns from extinguishing the fire, and Trotsky had a few scratches on his face from flying debris. Otherwise, no one was hurt.

From the roof, the guards could see that the five policemen in the *casita* had been tied up. Trotsky ordered his men to go outside and release them, but they hesitated because they could still hear gunfire in the distance and feared an ambush from the nearby cornfield. Trotsky insisted that the assault was over and that either the guards go out and untie the police immediately or he would do it himself.

The freed policemen described how twenty men dressed in police and army uniforms had surprised and overpowered them without firing a shot. Harte, they said, had opened the door for the assailants, apparently unaware of the danger—although it was impossible to say for sure. Nor were the policemen entirely certain whether Harte had been kidnapped or had left with the raiders of his own accord. Both automobiles had been taken, and the garage doors left wide open. The alarm system had been turned off, and the telephone wires were cut.

It was obvious that once the raiders were inside the patio, they knew the precise location of their target. Hundreds of bullets had riddled Trotsky's bedroom, and more than seventy bullet holes were counted in the doors, walls, and windows. Several bullets had sliced diagonally through the pillows and the bolster and the head of the mattress. Three homemade incendiary bombs were found in the patio unexploded. A fourth bomb had ignited the fire in Seva's room.

"We marveled at our unexpected survival," Natalia said later, even though the general sense of relief was tempered by concern for Harte. "It was a sheer miracle that we escaped with our lives." Indeed, Trotsky would be congratulated on his "miraculous escape" by well-wishers near and far in the coming days, although his own view of the matter was more down-to-earth. "The assassination failed because of one of those accidents which enter as an integral element into every war," he observed. He and Natalia had survived only because they had kept still and pretended to be dead, instead of calling for help or using their guns.

The armed attack delivered a shock, but it was not a surprise. Indeed, for a long time Trotsky had been ridiculed by the Mexican Communists for exaggerating the threat to his personal safety. Now he stood vindicated. Or did he? The Mexican detectives who arrived on the scene shortly after the attack were not convinced. The investigation was led by the chief of the Mexican secret police, Colonel Leandro Sánchez Salazar. He found it curious that Trotsky and Natalia and the household members appeared so calm under the circumstances. His suspicions mounted when Trotsky informed him that the perpetrator of the attack was none other than Joseph Stalin, by means of his secret police, the NKVD—although Trotsky persisted in referring to the organization by its former initials, the GPU. And by the time the colonel had finished counting the bullet holes in the bedroom walls and had pondered the spectacular incompetence of the raiders, he strongly suspected that Trotsky's escape was not a miracle but a hoax, a way for him to draw sympathy to himself and to discredit his enemies.

As for the missing American guard, Colonel Salazar quickly arrived at the conclusion that Harte had acted in collusion with the raiders, letting them in the door and then leaving with them of his own free will. Trotsky, refusing to accept that his household had been infiltrated by the GPU, argued strenuously that Harte was a victim, not an accomplice. The unsuspecting guard had been tricked, Trotsky insisted. Prompted by a familiar voice, he opened the door for the raiders, who subdued him and took him as their prisoner. The question was: Who had betrayed Harte?

THE MOOD OF relief at Trotsky's villa soon gave way to a sense of urgency. Everyone assumed that Stalin would not stop until Trotsky had been eliminated. Trotsky was, after all, the last of Stalin's political rivals left alive. In the revolutionary year 1917, when Stalin was a stalwart though obscure Bolshevik, Trotsky was dazzling vast crowds of workers, soldiers, and sailors in Petrograd with his spellbinding oratory. Though a newcomer to the Party, Trotsky proved to be Lenin's most important ally when the Bolsheviks stormed to power in the October Revolution. Then, as the Revolution came under threat in 1918, he created the Red Army and turned it into a disciplined fighting force, which he led to victory against the White armies in the savagely contested civil war.

At Lenin's death in 1924, Trotsky was the heir apparent. Yet he was easily outmaneuvered by Stalin, who expelled him from the Communist Party in 1927, exiled him to Central Asia in 1928, and then cast him out of the Soviet Union altogether in 1929. Stalin would later regret letting Trotsky escape, but it had not yet become acceptable for a Soviet leader, even the general secretary of the Party, to have a fellow Communist arrested and shot.

Trotsky was exiled to Turkey. From there, he requested permission to enter a number of European countries—Germany, Austria, France, Spain, Italy, Czechoslovakia, Norway, the Netherlands, and Great Britain—but each government in turn denied him a visa, in some cases after a contentious debate. During his Turkish exile, he wrote a memoir and his history of the Russian Revolution, while turning out a steady stream of pamphlets and articles. Much of this output appeared in his one-man journal, the *Bulletin of the Opposition*, the political organ of the Trotskyist movement, which was centered in Berlin until the Nazis came to power and then in Paris.

Trotsky lived for four years in Turkey, before receiving permission to enter France, where he spent two precarious years living incognito. The shifting winds of French politics then forced him to move again, this time to Norway. That is where he was living when the first of the sensational Moscow show trials opened, in August 1936. The defendants included several outstanding leaders of the Bolshevik Revolution, notably Grigory Zinoviev and Lev Kamenev, two longtime members of the Politburo. All but one confessed publicly to taking part in a conspiracy, supposedly led from abroad by Trotsky, to assassinate Stalin and other top Soviet leaders and seize power. All were found guilty and were executed for their crimes.

In the wake of the Moscow trial, the Kremlin stepped up pressure on Norway's socialist government to expel Trotsky and, because no country in Europe would accept him, there was a danger he would end up in the hands of the Soviet authorities. Trotsky listened to the menacing voice of Moscow radio fulminating against enemies of the people, while his comrades worked feverishly to find him a safe haven. In early September, he and Natalia were interned in a large house about twenty miles south of Oslo, where their captivity dragged on through the autumn. Deliverance came in mid-December with the news that the government of Mexico, of all places, had offered him asylum, thanks mainly to the efforts of the mural painter Diego Rivera, an avowed Trotskyist, who appealed directly to President Lázaro Cárdenas.

Trotsky was thus able to avoid the fate of the Bolshevik old guard slaughtered in Stalin's Great Terror. Still, in Mexico he lived under a death sentence. Two more Moscow show trials followed, and on each occasion Trotsky was again effectively made the chief defendant in absentia. His comrades and his family were swept up in the Terror and disappeared into the prisons and the camps.

Trotsky knew that Stalin could never forgive the fact that he had openly ridiculed him among the Communist elite as a mediocrity and denounced him in a session of the Politburo as the "gravedigger of the Revolution." Trotsky also understood that Stalin could not allow the alleged mastermind of the grand conspiracies, unmasked in the purge trials, to go unpunished. Yet in Trotsky's mind, Stalin's desire to have him killed was about more than just settling old scores or carrying out the verdict of the Moscow trials. He assumed that Stalin perceived him the way Trotsky perceived himself: as a political force to be reckoned with. As Trotsky said about Stalin shortly after the raid, he "wants to destroy his enemy number one."

Trotsky was predicting that the world war would unleash an

international proletarian uprising that would deal a deathblow to capitalism, already staggering under the effects of the Great Depression. The revolutionary wave would spread to the USSR, where the toiling masses would unite to overthrow the Stalinist bureaucracy that had long maintained a stranglehold on the first socialist state. Trotsky and his followers, rallying under the banner of the Fourth International—the rival to Moscow's Communist International, or Comintern—would be called upon to lead the struggle to restore workers' democracy to the Soviet Union.

If this sounded far-fetched, Trotsky reminded skeptics that the cataclysm of the First World War had created the conditions that enabled the minuscule Bolshevik Party to take power in Russia. Any Marxist-Leninist worth his salt understood that the revolutionary shock waves accompanying the Second World War were bound to be far more destructive. So said Trotsky, who supposed that Stalin feared such a scenario and dared not allow his nemesis to remain at large.

Whatever Stalin may have believed about Trotsky's political prospects, he had motivation enough to want to silence his most prominent critic. And it just so happened that Trotsky's host country had recently welcomed to its shores the kind of men who could help make this happen. When the Soviet Union came to the aid of the Spanish Republic against General Francisco Franco's invading Falangist armies in the civil war that erupted in 1936, Moscow made Spain the international recruiting and training ground of the NKVD. The Republic went down to defeat in 1939, and many hundreds of NKVD recruits and fighters from the International Brigade, which the Comintern had organized, took refuge in Mexico—Madrid's most loyal ally in the Western Hemisphere. Trotsky warned of a gathering danger.

To defend against the threat, the American Trotskyists, headquartered in New York, dispatched reliable comrades to the Coyoacán household to serve as guards, drawing heavily on the Minneapolis Teamsters, a Trotskyist stronghold, for funds and volunteers. Their chief priority was the safety of the Old Man, but they were also worried about his personal archives, which he had been allowed to take with him into exile in 1929. With the help of these voluminous files, Trotsky had exposed the Moscow trials as a sham, and he continued to draw on them to write his biography of Stalin. The purpose of the May 24 commando raid on Trotsky's home, it seemed clear, was not only murder but arson: The bullets were meant for Trotsky, the incendiary bombs for his papers.

The race was now on to prepare for the next assault. The villa must be transformed into a fortress. Turrets must be constructed atop the walls, double iron doors must replace the wooden entrance to the garage, steel shutters must cover the windows, bomb-proof wire netting must be raised, and barbed-wire barriers must be moved into position. But even as these fortifications began to rise up, the NKVD decided to resort to its fallback plan. The assignment of liquidating enemy number one would be entrusted to a lone operative who had managed to penetrate Trotsky's inner circle. The fatal blow would culminate a labyrinthine process that had begun more than three years earlier, as Trotsky sailed for Mexico.

CHAPTER 1

Armored Train

On the night of January 1, 1937, in the middle of the Atlantic Ocean, the Norwegian oil tanker *Ruth* greeted the New Year by blaring its two sirens and twice firing its alarm gun. The tanker carried no oil, only 1,200 tons of seawater for ballast and two very special passengers: Leon Trotsky, the Russian revolutionary exile, and his wife, Natalia. In fact, the Trotskys were the ship's only passengers, strictly speaking, although a Norwegian policeman was on board to escort them. They had sailed from Norway on December 19, after four miserable months of house arrest, which Trotsky said had aged him five years. In spite of this, the couple carried with them warm memories of a marvelous snowy land of forests and fjords, skis and sleighs.

They would sail another week or so before reaching their new home, Mexico—although they were in the dark about what awaited them there, even the port of arrival. The tanker steered an irregular course. The Norwegian government was eager to be rid of Trotsky but anxious to deliver him without mishap—such as what might result from an NKVD bomb—so the ship's departure had been shrouded in secrecy. On board, Trotsky and Natalia were forbidden to use the ship's radio. They were cut off from the outside world.

At the start of the voyage, the seas were rough, and Trotsky found it difficult to write, so instead he avidly read the books about Mexico he had bought just before their departure. Once out on the Atlantic, the seas turned calm, in fact remarkably so for that time of year, and Trotsky began to work intensively, writing an analysis of the Moscow trial that had made him a pariah in Norway and almost everywhere else. Only Mexico had opened its doors to him—"mysterious Mexico," Trotsky called it, wondering to what extent it deserved its reputation for political violence and lawlessness.

The passengers' sense of apprehension rose with the temperature; as the ship entered the Gulf of Mexico on January 6, the cabins grew stiflingly hot. It was early Saturday morning, on January 9, when the tanker finally entered the harbor of Tampico. The oil derricks reminded the couple of Baku, on the Caspian Sea, but otherwise this was terra incognita. They had no idea who or what was waiting for them onshore, and Trotsky warned the captain and the police minder that unless they were met by friends, they would not disembark voluntarily.

Toward 9 a.m. a tugboat approached the *Ruth*, and as it drew up alongside, Trotsky and Natalia caught sight of a familiar face, friendly and smiling, and their worst fears evaporated. The man they recognized was Max Shachtman, an American Trotskyist who had visited them over the years in Turkey, France, and then Norway. He was the first friend Trotsky had laid eyes on in more than two months, and when he stepped aboard the *Ruth*, the two men warmly embraced.

Shachtman was accompanied by the artist Frida Kahlo, introduced as Frida Rivera, wife of the celebrated muralist Diego Rivera. Ill health had kept Rivera off the flight from Mexico City. Frida, darkly beautiful in tightly braided hair and dangling jade earrings and wearing a rebozo and a long black skirt, stood out among the suits and uniforms of the government, military, and police officials there to receive Trotsky. Even the uniformed officers seemed relaxed and friendly, and they made the visitors feel safe and welcome.

A second boat trailed after the tug carrying representatives of the press, who were impatient to interview and photograph the Great Exile. Trotsky was eager to speak, and he answered questions for two hours straight, talking mostly about the Moscow trial. The thumbnail briefing he received from Shachtman, combined with the nature and tone of the reporters' questions, lifted Trotsky's spirits. As Natalia remarked, "the whole New World seemed to have been incensed by the Moscow crimes."

Close to noon, the tugboat brought the Trotskys ashore. Photographers and a newsreel cameraman captured their walk down the wooden pier. Trotsky had performed a number of dramatic entrances and exits over his tumultuous political career, typically adopting a demeanor of stern arrogance. Now, however, as he stepped onto Mexican soil, he looked somewhat tentative and uncertain of himself. Dressed in a tweed suit and knickerbockers, carrying a cane and a briefcase, he projected an image of civilized respectability, looking not at all like a defiant revolutionary. And at five feet eleven inches tall, he hardly resembled the Soviet cartoon image of him as "the little Napoleon." Only when he removed his white cap and exposed his irrepressible white hair did he suggest his old fanatical self. Natalia, conservatively attired in a suit and heels, also looked the part of the harmless bourgeoise, although she seemed frail and uneasy.

At the dock, a Packard was waiting for them. It belonged to the head of the local garrison, General Beltrán, who was the boss of Tampico and had been asked by President Cárdenas to do everything possible to facilitate Trotsky's arrival. Cárdenas had arranged for Trotsky to travel to Mexico City by airplane or by train, whichever he favored. The plane was waiting to take off, but reports of bad weather ruled out flying. The train was still en route from the capital, so the guests were checked into a hotel for the day. From there, Trotsky sent a telegram to President Cárdenas expressing his gratitude and pledging to honor the terms of his asylum. Trotsky and Natalia then retired to their room, reeling from culture shock and frustrated by their ignorance of the Spanish language.

El Hidalgo (The Nobleman), the luxury train that President Cárdenas sent to transport Trotsky to Mexico City, rolled into the Tampico station at eleven o'clock that evening. On board was George Novack, acting secretary of the American Committee for the Defense of Leon Trotsky, the miscellaneous collection of liberals and socialists that had initiated the campaign to find Trotsky a safe haven. Novack arrived in the company of a Mexican lieutenant colonel and a captain of the regular army, a contingent of soldiers from the Presidential Guard, civilian representatives of the Cárdenas administration, and a Russian-language interpreter for Trotsky.

Fifteen minutes later, Trotsky and Natalia, along with Novack, Shachtman, Frida, and the soldiers and officials from Mexico City boarded the train. They were joined by General Beltrán and a number of the most important officials from Tampico, as well as local police officials and detectives. The train, which had once belonged to former President Pascual Ortiz Rubio, was armored with bombproof steel plates and bulletproof windows. President Rubio had had good reason to insist on special protection. On February 5, 1930, his first day in office, as he was leaving the National Palace, a man fired a handgun into his automobile, one of the bullets shattering Rubio's jaw.

Trotsky and Natalia and their friends were placed in the middle car of the train; the car in front of theirs was occupied entirely by soldiers. The train finally pulled out of Tampico at four o'clock in the morning, as the passengers dozed. When daylight came, they looked out on a sunbaked landscape dotted with palm trees and cacti, mountains blazing in the distance. Trotsky's curiosity about the scenery competed with his thirst for information as he huddled in a compartment with Shachtman and Novack, who brought him up to date on what had been happening in the world during his three-week voyage from Norway.

Trotsky's command of English was unsure, so he spoke mainly in German. His comrades described for him how the Moscow trial had sparked a bitter controversy among American liberals and labor leaders. Not long after the Nazis took power in Berlin in 1933, Moscow directed Communists everywhere to support "progressive" governments and antifascist causes. This new Comintern strategy was called the Popular Front. In the United States, the Communist Party had lined up behind the New Deal of President Franklin Delano Roosevelt, while many American liberals, seeing the Soviet Union as a bastion against the rising Nazi tide, reached out to the Communist Party.

The Zinoviev-Kamenev trial of August 1936 troubled many liberals, who suspected the Kremlin of having orchestrated an elaborately staged frame-up. Liberal and socialist skeptics, with the encouragement of the American Trotskyists, formed a committee whose purpose was to lobby democratic governments to grant Trotsky asylum and, once this was achieved, to arrange for him to be given a fair hearing before an international commission of inquiry. The overwhelming majority of the committee's members did not support Trotsky's political views, and some were extremely hostile to them; rather, their fundamental sense of justice told them that he deserved the right of asylum and the chance to defend himself.

Novack showed Trotsky the committee's letterhead, with its roll of seventy names running down the left side of the page. The most prominent politician on the list was the head of the U.S. Socialist Party, Norman Thomas, one of the committee's founding members. The previous June, the small American Trotskyist party had merged with the Socialist Party, hoping to capture its left wing before eventually splitting off with an enlarged group of cadres. The appearance of Thomas's name, therefore, was not unexpected. The identity of another of the committee's initiators, John Dewey, took Trotsky by surprise. He thought it must be a different person with the same name, so when he was assured that it was in fact *the* John Dewey, the famous philosopher, "his whole face was illuminated with satisfaction," according to Novack, "and he said in the most pleased tone with a waggish shake of his head: 'Das ist gut! Sehr gut!' "

Things were indeed looking up, and the mood turned festive in the bright morning sunshine, as the soldiers of the Presidential Guard launched into a series of ballads from the Mexican Revolution. Trotsky asked Shachtman and Novack to perform something from the American radical songbook, so they belted out "Joe Hill," a tribute to the Swedish-American songwriter and labor activist executed by a firing squad in Utah in 1915 after a controversial murder trial. Frida Kahlo then lightened the mood by singing Mexican folk songs. The softness of her voice resonated with the parched panorama of palms, cacti, and agaves rolling by.

DIEGO RIVERA WAS livid that he was unable to accompany Trotsky to Mexico City. He was suffering from kidney trouble, and his doctor had ordered him to bed. Rivera was not only Mexico's most famous artist; he was also its most prominent Trotskyist, and he played the crucial role in arranging Trotsky's new sanctuary. The members of the American committee assumed that the Roosevelt administration would not seriously consider an asylum request, but great hopes were placed on Mexico, a revolutionary country with a radical president. In early December 1936, Anita Brenner, the Mexican-born American writer, art critic, and historian who opened a window onto Mexico's artistic renaissance of the 1920s, sent Rivera a telegram on behalf of the committee, asking him to take up Trotsky's case with President Cárdenas.

At that moment, the president was in the Laguna region, north of the capital, overseeing his government's land redistribution program. Rivera caught up with Cárdenas in the city of Torreón and petitioned him directly in his own name to grant Trotsky asylum in Mexico. To Rivera's great surprise, Cárdenas gave his approval, contingent only on Trotsky's agreement not to involve himself in Mexico's political affairs. In New York, on December 11, the committee cautiously announced the good news, warning that President Cárdenas would now come under tremendous pressure to reverse himself. The committee declared its intention to contact labor and liberal organizations in Spain, France, Britain, and Latin America to urge them to send messages of congratulations to the president on his "splendid decision." Enlightened Americans were encouraged to do the same. But the committee's concerns were unwarranted, because President Cárdenas was not a man easily intimidated.

Lázaro Cárdenas rose to prominence as a military leader, ascending through the ranks of the revolutionary army and then, as General Cárdenas, assuming major commands in the 1920s, when he was a loyal supporter of President Plutarco Elías Calles. He served as governor of Michoacán from 1928 to 1932, where he proved to be a radical social reformer. Despite this, Calles, as Mexico's kingpin, selected him to run for president in 1934 on the assumption that he would be able to control his protégé.

President Cárdenas soon disappointed Calles. His administration claimed to represent "the Revolution" and vowed to make good on the unfulfilled promises of justice and equality spelled out in Mexico's revolutionary constitution of 1917. A top priority was agrarian reform. The president moved to eliminate the large estates, or latifundios, and to distribute their land to collective farms. Very much a hands-on leader, Cárdenas spent a considerable amount of time traveling the country, overseeing his agrarian and other reforms, which is why Rivera had to journey to La Laguna in December 1936 to petition him about Trotsky.

To establish his authority, President Cárdenas had to cultivate left-wing and labor support, starting with the Confederation of Mexican Workers—known as the CTM, its Spanish initials—the largest confederation of unions in the country. The CTM rallied to Cárdenas's side in 1935, when strongman Calles and his supporters challenged the president's authority. In April 1936, Cárdenas had Calles arrested on conspiracy charges and exiled to the United States. The Mexican Communist Party had opposed Cárdenas's candidacy for the presidency, but was drawn into the anti-Calles coalition, and then backed Cárdenas in the name of the Popular Front, as instructed by Moscow.

President Cárdenas invited Trotsky to Mexico because he believed it was the proper thing to do. Yet the gesture also served to demonstrate his independence vis-à-vis the Stalinist left. No one in the Cárdenas administration openly supported Trotsky, but a number of its leading officials were sympathetic to Marxist ideology and were drawn to Trotsky's ideas and stirred by his tragic fate. Shachtman was repeatedly struck by this during his discussions with cabinet officials in the week before Trotsky's arrival. The minister of interior was especially forthright. "We are only too pleased to do this for Comrade Trotsky," he said. "To us he is the revolution itself!" Shachtman responded that the Mexican government had acted nobly. "It was only our duty," the minister replied, prompting another round of handshakes and gracias.

Cárdenas and his ministers anticipated the firestorm of protest that would greet the announcement of Trotsky's asylum. The Communists loudly complained, and vicious anti-Trotsky posters were put up all over the capital. Alongside them appeared Trotskyist counter-proclamations, which featured a pencil-drawn portrait of the exile, although many of these were soon defaced with the Nazi swastika. The Communists, meanwhile, declared open season on the renegade Diego Rivera, himself a former party member. The Trotskyist group in Mexico was insignificant, and unlike in the United States, there was no independent liberal class to take the side of the president.

On New Year's Eve, as political tensions mounted, President Cárdenas summoned Rivera to his residence for a private conference. He assured the beleaguered painter that there was absolutely no reason to fear for Trotsky's safety. Cárdenas was adamant that Trotsky must not land secretly, which would reflect poorly on Mexico and on him as president. A military escort would deliver the distinguished guest safely to his new residence, where he would be provided with full protection. And Trotsky should consider himself a guest and not a prisoner, Cárdenas told Rivera. He would enjoy complete freedom of movement.

The change in the political atmosphere during the next few days provided unmistakable evidence that the president meant business. Communist propaganda backed away from incitement to assassinate Trotsky, instead protesting that his presence in Mexico would divide the labor movement. Rivera assumed that Cárdenas or one of his men had called the secretary of Mexico's Communist Party, Hernán Laborde, and told him to behave himself. The Communists were explicitly warned not to deface the Trotskyist posters or they would be prosecuted for violating the right of free speech.

The attitude of the president left no doubt that he intended to carry out his promises. Yet Mexico's political environment was such that Trotsky's status could be secure only so long as the tough, clever, incorruptible Cárdenas remained in office. And he was now two years into his single, constitutionally permitted six-year term as president. With this in mind, Rivera told Shachtman that Mexico could at best serve Trotsky as a bridge between Europe and the United States. Shachtman, after three weeks of sampling the political culture of the capital, believed that Rivera was "absolutely correct." The American committee must be persuaded to bring its influence to bear on securing Trotsky a U.S. visa. Perhaps the courage shown by President Cárdenas would embolden President Roosevelt to emulate his example.

Shachtman was not optimistic, placing the odds at "one chance in a hundred." But an all-out effort had to be made, because the Old Man's life might depend on it. In briefing Trotsky about what awaited him in Mexico City, Shachtman and Novack thought it wise to leave out some unsettling details, at least for the time being. These included the fact that recently Rivera had narrowly avoided an attempt on his life, that Rivera's caretaker had subsequently been kidnapped and beaten up, and that Shachtman had just put down $100 for a Thompson submachine gun—an outlay to be charged to the American Committee for the Defense of Leon Trotsky.

In the mid-morning on Sunday, January 10, the armored train came to a halt in the town of Cárdenas—no relation to the president—near the foot of the plateau dominating the topography of northern and central Mexico. Trotsky took advantage of the opportunity to get off the train and stretch his legs.

The imposing presence of *El Hidalgo* alerted the local citizens that an important person had arrived. Novack says that as many as four hundred "natives," almost the entire population of the town, crowded around Trotsky and Natalia, the children pushing their way to the front of the crowd. To Trotsky, these brown-skinned, barely dressed little ones were a stark contrast to the children with "china-blue eyes" and "corn-colored hair" he had left behind in Norway. For Novack, the remarkable contrast was right there in front of him: "this striking goateed gentleman with a stick and plus fours thus surrounded by serapeed Indians and barefooted brown-faced gamins."

From the town of Cárdenas, an additional locomotive would be required to haul the heavy train up the plateau. One train, two engines: It was, Trotsky could assure his traveling companions, an arrangement quite familiar to him from his glory days in revolutionary Russia. His own armored train was so formidable that it required two locomotives just to pull it over level ground. That was in the days of the Russian civil war, when Trotsky was the war commissar of the new Soviet regime. For nearly two and a half years, he saw very little of his headquarters, living in a railway carriage that once belonged to the czarist minister of transportation. Trotsky's exploits in his armored train catapulted him to great fame and made "Lenin and Trotsky" synonymous with Russian Bolshevism.

Russia's time of troubles began as a result of the First World War, in which it fought on the side of France, Great Britain, and later the

United States, against Germany and Austria. The Imperial Army's catastrophic defeats at the front, together with the economic hardship at home caused by the prolonged war effort, provoked strikes, mutinies, and riots that led to the fall of the Russian monarchy in the February Revolution of 1917. The Provisional Government that succeeded the autocracy could not halt the country's slide into anarchy. The Bolsheviks seized power in the October Revolution promising to withdraw Russia from the war, a promise they fulfilled in March 1918, when they agreed to the crushingly punitive terms of the Treaty of Brest-Litovsk, the separate peace with Germany that shut down the eastern front. In defending the controversial treaty within the Party's Central Committee, Lenin argued that what revolutionary Russia needed above all was a "breathing spell."

Peace was fleeting, though. A military threat to Soviet power began to materialize within Russia in the spring of 1918. Serious hostilities broke out in the summer, the first shots coming from an unlikely quarter: a legion of 35,000 Czechoslovak soldiers, former Habsburg prisoners of war, who had become trapped inside Russia. The heavily armed Czechoslovaks were attempting to get out of Russia by way of Siberia, strung out along a vast stretch of the Trans-Siberian railroad, when they clashed with Red units and began to topple local Soviet governments. Joining forces with White Guard troops, they seized Samara on the shores of the Volga River, then Simbirsk to the north, then Kazan farther upriver, where the Volga bends west toward Moscow.

Trotsky learned of the fall of Kazan while en route in his hastily organized train, which left Moscow on the night of August 7. The closest his train could get to the ancient Tatar capital was the small town of Svyazhsk, on the opposite bank of the Volga. These Red defeats took place against a backdrop of ominous developments on the periphery. To the west, the Germans had acquired in the peace treaty an enormous swath of territory from the Baltic to the Black Sea; in the north, French and British troops occupied the port cities of Murmansk and Archangel; in the Ural Mountains to the east, and on the Don River to the south, White Russian armies were coalescing. "The civil war front," as Trotsky described it, "was taking more and more the shape of a noose closing ever tighter about Moscow."

Trotsky spent that critical month of August 1918 coordinating operations for the recapture of Kazan from his train, which he called a "flying administrative apparatus." In those early days, the train consisted of twelve cars, all armored, and carried a heavily armed crew of about 250 men, including a squad of Latvian Riflemen and a unit of machine gunners.

The recapture of Kazan was the immediate goal, but the larger challenge Trotsky still faced was to forge a genuine army out of the remnants of the peasant czarist army and the proletarian Red Guard units from Moscow and Petrograd. Desertion from the battlefield remained endemic. Any units retreating without orders, Trotsky now warned, would face the firing squad, starting with the commanding officer. "Cowards, scoundrels, and traitors will not escape the bullet—for this I vouch before the whole Red Army." Such credible death threats produced the desired effect. And after the Reds recaptured Kazan on September 10 and Simbirsk two days later, Trotsky claimed that the significance of the victory far outweighed the liberation of two Russian cities: "A vacillating, unreliable and crumbling mass was transformed into a real army."

This was the first of many turning points in the civil war, as Trotsky coordinated the efforts on innumerable fronts, with more than fifteen armies in the field. They were commanded by former officers of the Russian Imperial Army who were drafted into the ranks of the Red Army because it desperately needed their expertise and experience. In order to ensure the loyal behavior of these often unwilling recruits, Trotsky attached to the ranking officers trustworthy Bolsheviks designated as political commissars. It was an arrangement fraught with tension and controversy.

The Whites would advance to their farthest point in mid-1919, when their armies drove toward Moscow from Siberia in the east and the Ukraine in the south, and threatened Petrograd from the northwest. That summer, Soviet Russia was reduced territorially to the size of ancient Muscovy. Yet its defenders had the advantage of internal lines of operation, enabling the Red Army to shift forces and supplies from one front to another as circumstances required. Trotsky, too, raced from front to front over Russia's run-down railway lines, in disrepair after years of war and revolution. It is reliably estimated that his train traveled more than 125,000 miles during the civil war. "In those years," he later recalled, "I accustomed myself, seemingly forever, to writing and thinking to the accompaniment of Pullman wheels and springs."

One of the train's carriages served as a garage, housing several automobiles and trucks and a large gasoline tank. Where the train could not take him, his car was put into service, transporting him to the front lines on countless excursions covering hundreds of miles. Trotsky and his convoy would set off accompanied by a team of twenty to thirty sharpshooters and machine gunners. "A war of movement is full of surprises," he explained. "On the steppes, we always ran the risk of running into some Cossack band. Automobiles with machine-guns insure one against this."

Trotsky's visits to the front enabled him to ascertain the facts on the ground, but more important, they served to boost the morale of the Red soldiers. The train carried with it experienced fighters and dedicated Communists ready to step into the breach. Armed detachments dressed in black leather uniforms would descend from the train and go into action. "The appearance of a leather-coated detachment in a dangerous place invariably had an overwhelming effect," Trotsky testified. Also effective were the assortment of supplies and gifts distributed from the train: food items, boots, underwear, cigarettes and matches, cigarette cases, medicines, field glasses, maps, watches, and machine guns.

The train was equipped with a telegraph station, so that urgent orders for supplies could be conveyed to Moscow without delay, and news from the outside world could be delivered to the otherwise isolated front-line soldiers. The train carried its own printing press and published its own newspaper, *En Route*. At each stop, the staff would distribute to soldiers and civilians stacks of these newspapers, along with copies of Trotsky's writings. When it came to agitation, however, nothing could surpass Trotsky's rousing whistle-stop speeches. The photographer and the motion picture man who accompanied him on the train captured him in his greatcoat and military cap, his hands in furious motion, his countenance stern, his bearing erect.

Trotsky may have looked the part of the Red warlord, yet he had no military background. In fact, as war commissar he rarely involved himself in questions of strategy or operations, leaving this to the experts. He reserved for himself the role of supreme agitator, and because he was as ruthless as he was ubiquitous, often resorting to bloodcurdling threats to achieve results, he acquired a reputation for brutality, most of all for his merciless treatment of deserters.

Justice was administered by field tribunals. During the battle to retake Kazan, one of these courts martial passed death sentences on every tenth deserter of a dishonored regiment, including the commander and the commissar, causing the executions of at least two dozen men. Trotsky had ordered the punishment, and he defended it without remorse: "to a gangrenous wound a red-hot iron was applied." The Revolution must use all possible means to defend itself, Trotsky believed, although he tended to justify the severity of his regime in traditional terms. "An army cannot be built without reprisals," he declared. "Masses of men cannot be led to death unless the army command has the death penalty in its arsenal."

Trotsky's draconian ways and words made him a lightning rod for anti-Bolshevik propagandists. Especially notorious was his decision to use the wives and children of former czarist officers as hostages in order to inhibit any temptation these officers might have to sabotage the Soviet war effort or to defect to the enemy. White Guard posters and literature made the most of the fact that the Red demagogue was a Jew, thereby tapping into the deepest Russian fears of a mythical "Jewish Bolshevism."

TROTSKY WAS BORN Lev Davidovich Bronstein, in 1879, in the southern Ukraine, then part of the Russian Empire. Like so many of Russia's young Jews of his generation, he was drawn to revolutionary ideas and joined a clandestine circle devoted to propagating those ideas among the lower classes. Arrested at age eighteen, he adopted as his revolutionary pseudonym the family name of one of his jailers, thus becoming Lev, or Leon, Trotsky. In recent decades, Russia had become host to a virulent form of anti-Semitism—its most shocking manifestations were the waves of deadly pogroms in the western and southern borderlands, the former Jewish Pale of Settlement. Czarist Russia was the source of a notorious anti-Semitic forgery known as *The Protocols of the Elders of Zion*, which claimed to reveal a Jewish conspiracy to

dominate the world. White propaganda exploited these festering preju-
dices by portraying Bronstein-Trotsky, his Semitic features cartoonishly
enhanced, as the Jewish-Bolshevik Antichrist.

Yet Trotsky remained something of an outsider among the Bolsheviks. For a decade and a half up to 1917, the year he joined the Party,
he had been a vehement critic of Lenin. But after the collapse of the
Russian autocracy, he hooked his fortunes to the Bolshevik juggernaut,
making history as the organizer of the October coup d'état. Now, as the
Revolution's second most important leader, he appeared too eager to
demonstrate his intellectual superiority and to mug before the mirror of
History. His behavior as war commissar fueled these animosities. Many
Bolsheviks had assumed that the Revolution would put an end to a
centralized regular army, which they considered a vestige of capitalism,
and would rely instead on a volunteer militia to defend itself. Trotsky's
championing of conscription and old-fashioned military command and
discipline ran counter to this spirit. What is more, he seemed to revel in
traditional military culture, instituting awards for bravery and bringing
a military band along on his train journeys. On top of all this, he was
dogged by rumors that he had personally executed Communists.

The backlash against Trotsky was brought on, first and foremost,
by his decision to fill the ranks of the Red Army with tens of thousands of former czarist military officers. This was the core issue at the
start of his running feud with Stalin, who was far more suspicious than
Trotsky of the kinds of treasonous plots these carryovers from the old
regime might decide to hatch. Stalin himself exhibited an aptitude for
scheming insubordination in his capacity as chief political commissar
on the southern front. Sometimes he went over Trotsky's head, directly
to Lenin, in order to get his way. Lenin tried to mediate between his
two headstrong lieutenants, but matters developed to the point where
Trotsky ordered Stalin's removal from the front. Stalin withdrew, but the
problem did not go away.

Trotsky's long absences from Moscow made it easier for his political
enemies to outmaneuver him. His lowest point came in the summer
of 1919, when he suffered a series of setbacks just as the White armies
were closing in. He was overruled by the Communist Party's Central
Committee, its key decision-making body, on questions of strategy and

command appointments at the same time that Stalin's intrigues were undermining his authority in Moscow and an impatient Lenin was reproaching him for the Red Army's reverses on the battlefield. He offered his resignation as war commissar, which the Central Committee rejected.

Trotsky's fortunes turned around in October, when he led a heroic defense of Petrograd. The former capital had come under siege from the Northwestern Army commanded by General Nikolai Yudenich, who was backed by British arms and funds. Lenin concluded that Petrograd ought to be abandoned in order to shorten the front line. Arriving in Moscow, Trotsky argued passionately that the cradle of the Revolution must be saved at any cost, even if it came down to house-to-house combat. If "Yudenich's gang" were to penetrate the city's walls, Trotsky swaggered, they would find themselves trapped in a "stone labyrinth."

Having won the argument, Trotsky hastened to Petrograd, where he found officials demoralized and resigned to defeat. "Exceptional measures were necessary," he decided, "the enemy was at the very gates. As usual in such straits, I turned to my train force—men who could be depended on under any circumstances. They checked up, put on pressure, established connections, removed those who were unfit, and filled in the gaps."

It was in these critical days that Trotsky was presented with his one opportunity to assume the role of regimental commander. He was at division headquarters in Alexandrovka, just outside the city, when he looked up and saw retreating Red soldiers approaching. He reacted instinctively. "I mounted the first horse I could lay my hands on and turned the lines back," he later recalled. It took a few minutes for the commissar of war to make his presence felt among his withdrawing troops. "But I chased one soldier after another, on horseback, and made them all turn back. Only then did I notice that my orderly, Kozlov, a Muscovite peasant and an old soldier himself, was racing at my heels. He was beside himself with excitement. Brandishing a revolver, he ran wildly along the line, repeating my appeals and yelling for all he was worth: 'Courage, boys, Comrade Trotsky is leading you.' "

Petrograd was saved, and Yudenich's army was pushed back into Estonia. The following month, Trotsky was awarded the Order of the

Red Banner. The citation praised the "indefatigability and indestructible energy" he displayed in fulfilling his commission to organize the Red Army and then by leading it so effectively. "In the days when Red Petrograd came under direct threat, Comrade Trotsky, in setting off for the Petrograd front, took the closest part in the organization of the brilliantly executed defense of Petrograd, inspiring with his personal bravery the Red Army units under fire at the front."

By 1920, the civil war was won and the Red Army had five million men in uniform. But Trotsky had also acquired a considerable number of influential enemies. They tended to gravitate toward Stalin, the man who emerged from the war as Trotsky's chief political rival. Stalin would employ the services of these like-minded Bolsheviks to eventually cut Trotsky down to size and to expel him from the Communist Party and then the country. The field was then clear for Stalin's historians to portray their master as a great hero of the civil war, while distorting or omitting Trotsky's revolutionary leadership of the Red Army. By the time the outcast arrived in Mexico, the Party's official history had transformed the organizer of Red Victory into the "despised fascist hireling, Trotsky."

As *El Hidalgo* ascended the Mexican plateau, the air became cooler and it began to rain, which brought relief to all the travelers, but especially to Trotsky and Natalia. Trotsky recorded in his journal: "we soon rid ourselves of the northerner's fear of the tropics which had seized us in the steamy atmosphere of the Gulf of Mexico."

In the evening of January 10, the train stopped briefly in the station at San Luis Potosí, more than 6,000 feet above sea level, then traveled southward, continuing its climb. It was late the following morning when the rail journey came to an end on the northern outskirts of Mexico City, at the tiny station at Lechería. This was how President Cárdenas himself typically arrived in his capital: secretly, and thus safely, in the suburbs.

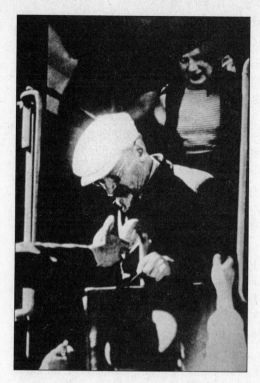

Trotsky and Natalia arriving by train in Lechería, on the outskirts of Mexico City, on the morning of January 11, 1937.

At Lechería, Trotsky was warmly greeted by Diego Rivera, temporarily released from the hospital and "fat and smoldering," according to the *Time* magazine correspondent on the scene. Foreign and Mexican comrades were part of the reception committee, along with government officials, police officers, and more reporters and photographers. In the crush, Trotsky became separated from Natalia, who struggled not to lose sight of Frida, the only face she recognized in the crowd.

Then came what Novack called a "mad dash" by car around the city, southward toward the suburb of Coyoacán. The name Coyoacán meant roughly "place of the coyotes" in the Aztec language, although the only animals visible through the car window on the drive through the neighborhood were donkeys posted outside small adobe houses, cows enjoying an early lunch in the streets, and chickens and dogs patrolling the sidewalks—that is, where sidewalks were present. Coyoacán, the new arrivals discovered, was a village.

It was toward noon when the fleet of automobiles pulled up in front of a squat, one-story house of bright blue stucco arranged in a U-shape around a garden patio. The neighbors called it the *casa azul*, or the Blue House, and Diego and Frida later adopted the name. This was Frida's home growing up, and Diego bought it from her parents after he married her. They had since moved to San Angel, a few miles away, and arrangements had been made for the Trotskys to occupy the Blue House temporarily. They entered a spacious patio filled with plants and flowers, pre-Columbian sculptures, and—what especially caught Trotsky's eye—a fruit-bearing orange tree standing in the middle of the yard. The rooms were fresh and airy, each with its own collection of pre-Columbian artifacts and modern paintings, including works by Diego and Frida.

For the remainder of the day and all of the next, the Blue House was the scene of a celebration. Novack describes the atmosphere as "wild confusion, jubilation, and excitement," with visitors of all kinds coming and going. Trotsky and Diego posed for photographers and gave interviews. Trotsky worked his charm on the Mexican reporters, leaving them with the impression that he intended to retreat into private life and return to work on his biography of Lenin, even though he knew that this was out of the question.

Meanwhile, all Mexico City was talking about the arrival of Trotsky. *La Venida de Trotsky* was the title of a ninety-minute skit featured at the Apollo, a popular burlesque house. The new slogan on the Communist posters was a belligerent "Out with Trotsky, the Assassin." The day after his arrival, the Communists staged a massive demonstration in the city-center Plaza de Santo Domingo, where party secretary Laborde could be heard shouting, "Down with Trotsky who is living in the home of the capitalist painter Rivera!" He called for the exile's expulsion from Mexico on the grounds that he had already violated the terms of his asylum by criticizing the Soviet government. The demonstration was broken up by the police after several speakers attacked the government for harboring an assassin.

Inside the Blue House, Trotsky professed to be indifferent to these fulminations. He had no intention, he said, of "entering into a polemic with flunkeys when ahead lay a struggle with their masters." Moscow was preparing a new show trial. In *Pravda* and *Izvestiia*, Trotsky's former comrades were calling for blood. Even Christian Rakovsky, once Trotsky's best friend and close political ally, added his voice to the bloodthirsty chorus: "No pity for the Trotskyist Gestapo agents! Let them be shot!" American bodyguards would have to be found to supplement the seven Mexican policemen on duty outside the Blue House. And Shachtman urgently appealed to the New York comrades to come up with the money needed to hire a proper secretarial staff so that Trotsky could defend himself against the slanderous charges being prepared against him by Moscow's prosecutors.

At the time of the first trial, Trotsky had been unable to speak out. Now he felt unbound and battle-ready. His emigration to Mexico, he wrote in his journal, had changed the balance of power to the disadvantage of the Kremlin. This may have been true in the short term. But it did not change the harder truth *Time* magazine printed in its blunt assessment of the situation: "Today Trotsky is in Mexico—the ideal country for an assassination."

CHAPTER 2

Mastermind

demand that the mad dogs be shot—every one of them!" This was the cry of a foaming Andrei Vyshinsky, the Soviet chief prosecutor, near the close of the first Moscow show trial, in August 1936. The object of his fury were the sixteen defendants sitting in the dock, accused of plotting to assassinate Stalin and other Soviet leaders. The next day, a screaming headline in *Pravda* echoed his demand: "The Mad Dogs Must Be Shot!" And indeed they were, each of them dispatched with a bullet to the back of the head.

Vyshinsky also prosecuted the second Moscow trial, which opened on January 23, 1937. This time, seventeen defendants were accused of forming a conspiratorial "Anti-Soviet Trotskyite Center" aimed at the dismemberment of the USSR in collusion with Germany and Japan. Trotsky, although not formally a defendant, was once again portrayed as the mastermind of the conspiracy, with his son Lev, nicknamed Lyova, serving as his close accomplice from Paris.

The most prominent defendant in the second trial was Trotsky's former comrade Yuri Pyatakov, whose recent tenure as deputy commissar of heavy industry was exploited by the prosecution to lend credibility to the ancillary charges of sabotage and wrecking. In this way, hundreds of industrial accidents, coal mine explosions, and railway disasters, many of which in fact resulted from the breakneck speed of the five-year plans, could be blamed on the unmasked enemies of the people now on trial for their lives.

This trial, like its predecessor, foreshadowed further purges to come, as the defendants obliged the prosecutor by engaging in a lethal form of name-dropping. Karl Radek, the Bolshevik journalist and Trotsky's onetime ally, put on an inspired performance, which may have saved his life. In his closing statement he remarked that there remained at large "semi-Trotskyites, quarter-Trotskyites, one-eighth-Trotskyites, people who helped us."

This was a strenuous time for Trotsky and his staff at the Blue House. Each day news reports described the latest falsifications out of Moscow, and each day Trotsky issued multiple press releases pointing out the contradictions, improbabilities, and absurdities of the accusations made against him. Trotsky's staff was led by Jean van Heijenoort, a Frenchman who had served as his secretary and bodyguard in Turkey and France. Trotskyist headquarters in New York sent down an American secretary to handle the English-language translations and to deal with the American press; local comrades assisted with Spanish. Each of Trotsky's statements had to be translated immediately from his original French into English and Spanish, then distributed to the international news services and the Mexican newspapers.

Speed was of the essence. As was the case with the first Moscow trial, the entire court proceeding, from the reading of the charges to the final verdicts, took place within a single week. The verdicts were read out to the defendants in the early hours of January 30. All but four were sentenced to death. Later that day, on a wintry Red Square, where the temperature hovered at minus 17 degrees Fahrenheit, Moscow Party secretary Nikita Khrushchev harangued a crowd of 200,000 demonstrators. Some carried banners that read, "The court's verdict is the people's verdict." They heard Khrushchev condemn the defendants as tools of "Judas-Trotsky." "Stalin is our banner," he cried. "Stalin is our will, Stalin is our victory." Across the Soviet Union that day, Soviet citizens gathered at "indignation meetings" to demand that the death sentences be carried out—although the outcome was never in doubt.

In the second trial, as in the first, the confessions of the accused were the only evidence presented in court. These confessions became the subject of endless fascination and speculation in the West. Why would the defendants engage in such acts of self-abasement unless they were in fact guilty as charged? Why did some of them seem to revel in their admissions of guilt? To Trotsky, this was no great mystery. The show trial confessions, he explained, were the result of the prolonged mental torture the victims had endured, which included threats to family members held as hostages. In the end, according to Trotsky, they agreed to confess to the most fantastic of crimes because they knew it was their only chance to save themselves and their families.

While many Western observers found it troubling to think that the confessions were legitimate, it was even harder to imagine that they had been concocted. The defendants were, after all, hardened Old Bolsheviks, veterans of czarist prisons and exile, onetime conspirators in the revolutionary underground willing to take great risks and suffer great hardships. They seemed not only fully capable of conspiring against Stalin, but highly unlikely to fabricate their own confessions of guilt. Besides, Western reporters and diplomats were present inside the courtroom, the October Hall, upstairs in the stately neoclassical House of Unions, not far from the Kremlin. Would Stalin really have risked everything, knowing that one of the doomed men might decide at the last moment to blindside inquisitor Vyshinsky and leave his mark on history by blurting out the truth?

Kingsley Martin, editor of the British magazine *New Statesman*, asked Trotsky why none of the accused chose to go down fighting when they could assume, based on the outcome of the first trial, that they were all going to die. Trotsky became very animated. Even after the first trial, he argued, these men had reason to believe they could escape death. "There is a world of difference between certainty of death and just that much hope of reprieve"—here Trotsky pinched a sliver of space in front of him to indicate the slimmest chance of escape. And in fact, some were spared the executioner's bullet.

Martin persisted, asking Trotsky if perhaps there was something in the Bolsheviks' code of conduct that would "psychologically expose them to serve the Party at the expense of personal honor, by confessing anything that was not the truth." Martin was unaware that Trotsky himself had once offered a striking display of this mentality. It was in 1924, in a moment of political adversity, when he stood before a Party congress and declared: "One can only be right with the Party and through the Party, because history has not created any other way for the realization of one's rightness. The English have the saying, 'My country, right or wrong.' With much greater justification we can say: 'My Party, right or wrong.'"

The idea that the purge trial confessions were a kind of last service to the Party, an act of self-immolation performed by legally innocent true believers, would gain popularity in the years to come, thanks in part to *Darkness at Noon*, Arthur Koestler's 1941 novel dramatizing the trials. Trotsky maintained that the psychology behind the confessions had been best described a century earlier by Edgar Allan Poe, in his short story "The Pit and the Pendulum," in which "the victim is terrorized and psychologically shattered by the slow and systematic descent of death." As he explained to an American reporter, "Human nerves, even the strongest, have a limited capacity to endure moral torture."

Trotsky's explanation of the confessions, and his refutation of the trials generally, failed to sway liberal-to-left opinion in the United States. Part of the problem was that Trotsky's voice could barely be heard above the murderous din orchestrated in Moscow. To remedy this, the American Committee for the Defense of Leon Trotsky arranged for him to deliver a speech about the Moscow trials, live by telephone to an audience in Manhattan. The venue chosen for the event, the Hippodrome, guaranteed Trotsky a large audience and maximum publicity.

The Hippodrome was a monumental Beaux Arts theater on Sixth Avenue between 43rd and 44th streets. In its heyday, after the First World War, it was the scene of extravagant entertainment spectacles, including major circus performances and aquatic shows. Its enormous stage contained an 8,000-gallon glass water tank that could be raised by hydraulic pistons. It was at the Hippodrome in 1918 that escape artist Harry Houdini staged "The Submersible Iron-Bound Box Mystery" and made a 10,000-pound elephant disappear. Now, from this same stage, the incomparable Leon Trotsky would attempt to remove a very large monkey from his back. The New York Trotskyists could barely contain their excitement. "This is going to be one of the most dramatic events of all time in New York," one of them wrote to Trotsky. "It will be an immense newspaper sensation."

On the evening of February 9, Trotsky stood before a microphone in a small room on the premises of the telephone exchange in Mexico

City. He was prepared to speak in English for forty-five minutes, and in Russian for fifteen more. Warned that overtime charges were astronomical, he had rehearsed repeatedly in order to get his timing down. In New York, inside the bright blue walls of the Hippodrome, a capacity crowd of more than six thousand people sat in red-colored seats with gold embroidered crests. Police detectives moved along the aisles, ready to quell any pro-Stalin manifestations, while outside the building 150 officers were on patrol.

Trotsky was scheduled to start speaking at 10:10 p.m., and as the historic moment approached and the audience fell silent, the atmosphere inside the Hippodrome was electric. At exactly ten minutes past the hour, a faint voice could be heard over the sound system speaking in Russian, but a moment later there was a loud click followed by a burst of static. Numerous attempts were made during the next hour to establish a connection, but without success. An act of sabotage, most likely at the Mexico City end, made sure that Trotsky's voice would not be heard.

The audience waited and dozed until, at 11:20, Shachtman executed the backup plan by reading the text of Trotsky's speech, which began with an apology for "my impossible English." Trotsky called Stalin's Russia a "madhouse," and launched into a detailed rebuttal to the Moscow trials, the purpose of which, he believed, was to engineer his deliverance to the Soviet Union and the cellars of the GPU. The speech was laced with Trotsky's usual slashing sarcasm and punctuated with dramatic pauses, as he frequently asked his audience, "Do you hear me?" and "Have you all heard?" They could hear Shachtman perfectly, even as eyes wandered upward from the podium to the huge painting of a prizefighter that towered above the left side of the stage.

Playing to his New York audience, Trotsky took several jabs at some influential American friends of the Soviet regime, in particular Walter Duranty, the Moscow correspondent of *The New York Times*. Duranty, whom Trotsky regarded as the Kremlin's "political prostitute," vouched for the integrity of Stalin and the legitimacy of the trials, explaining the masochistic tone of the defendants' confessions by invoking eternal Russian culture. "No, the Messieurs Duranty tell us, it is not a madhouse, but the 'Russian soul,'" Trotsky explained derisively. "You lie, gentlemen, about the Russian soul. You lie about the human soul in general."

The most memorable moment of the speech—and the one that made headlines the next day—came when Trotsky reiterated his willingness to appear before a neutral commission of inquiry to answer the charges brought against him in Moscow. He capped this promise with a dramatic declaration: "If this commission decides that I am guilty in the slightest degree of the crimes which Stalin imputes to me, I pledge in advance to place myself voluntarily in the hands of the executioners of the G.P.U."

AT THE TIME of Trotsky's Hippodrome speech, there was some uncertainty as to whether the much-discussed commission of inquiry would ever become a reality. The Trotsky defense committee, which sought to lay the groundwork, had recently been buffeted by a series of resignations, nine altogether, which mired it in controversy. The committee's detractors accused it of being a tool of the Trotskyists. They barraged it with letters and telephone calls lobbying against the staging of a counter-trial. Some sixty prominent American journalists and intellectuals signed a petition denouncing the idea.

Those who defended the Moscow trials often took their cue from the two influential liberal magazines of the day, *The New Republic* and *The Nation*, both of which ran editorials asserting there was no reason not to take the trials at face value. They were hurting Moscow's international reputation at a time of mounting international danger, so why would Stalin choose to stage them unless the Trotskyist conspiracy was legitimate? As for Trotsky's claim that the trials had been orchestrated for the purpose of apprehending him, it was impossible to believe that Stalin would go to such lengths and jeopardize the unity of the Popular Front for the sake of personal revenge.

Some friends of the Soviet Union who doubted that Trotsky was guilty as charged reasoned that he was nevertheless morally responsible for the conspiracy uncovered in Moscow. And even if Trotsky were entirely innocent, his personal predicament could not take precedence over the interests of the only socialist country in the world. The trials, in other words, must not be allowed to obscure the Soviet Union's positive

achievements, such as its collective economy and the democratic promise of its 1936 constitution, the most progressive in the world. Besides,
fascism was on the rise, so first things first.

This was not how John Dewey viewed the matter. At a meeting of
the Trotsky defense committee on March 1, he declared that the stakes
involved in the Trotsky case ranked with those of the Dreyfus affair, and
Sacco and Vanzetti. Here Dewey was invoking two landmark cases of
miscarried justice. Alfred Dreyfus was a French army officer of Jewish
extraction who was wrongly convicted in 1894, after an irregular trial,
of spying for Germany and sentenced to life imprisonment. Only a sustained effort by a small number of parliamentary deputies, journalists,
and intellectuals exposed the travesty of justice and eventually led to the
release and reinstatement of Captain Dreyfus.

Sacco and Vanzetti were Italian immigrant anarchists arrested in
1920 for armed robbery and murder in Massachusetts, a case that provoked an international outcry. Although the evidence against them was
compelling, it was widely believed that they had been unjustly tried
because of their political views and their avowed atheism. They went
to the electric chair in 1927, despite worldwide protest demonstrations
and impassioned appeals by men of conscience such as Dewey.

By March 1937, any hope that Trotsky would be given a fair hearing
seemed to hang on Dewey, who was not only an eminent philosopher
and educational reformer, but the most respected public intellectual in
America. Dewey was by reputation a friend of the Soviet Union. Soviet
educators had been influenced by his writings on progressive education,
and he had visited the country in 1928 to observe the results. He came
away endorsing Soviet central planning and social control, although he
sensed that this particular brand of socialism was especially suited to the
Russian national character.

Dewey, now seventy-eight years old, was resisting the effort to draft
him into service on a commission of inquiry. His age was a factor, as
were his family's strong objections to his dirtying his hands in Communist politics. Dewey's reluctance also stemmed from the fact that,
despite his métier and his interest in the Soviet Union, he had not
studied Marxism nor paid much attention to Soviet politics. Yet one
of his most illustrious former doctoral students at Columbia Univer-

sity was Sidney Hook, a leading Marxist philosopher and a professor at New York University. Hook was not a Trotskyist, but after years of tortured relations with the American Communist Party, he had turned against Soviet Communism. He worked assiduously to persuade Dewey to chair the commission of inquiry.

Trotsky himself was enlisted in this cause. On March 15, he addressed a letter to Suzanne La Follette, a radical journalist and a member of the defense committee working closely with Dewey. Trotsky took the position that the sage of Morningside Heights had an obligation to act. "I understand that Mr. Dewey is hesitant about descending from the philosophical heights to the depths of judicial frame-ups. But the current of history has its own exigencies and imperatives." Trotsky pointed out that the philosopher Voltaire had taken it upon himself to redress a gross miscarriage of justice committed in his own day, while the novelist Émile Zola's "J'accuse," his open letter to the French press, had made the Dreyfus affair an international cause célèbre—"and neither lessened his stature by the 'sidestep' in the eyes of history."

Four days later, on March 19, Dewey relented, agreeing to serve as chairman of the newly christened Commission of Inquiry into the Charges Made Against Leon Trotsky in the Moscow Trials. One of the major factors behind his decision, he said, was the campaign of harassment and intimidation aimed at dissuading him. He encountered similar resistance as he attempted to recruit upstanding Americans to serve on what was intended to be a politically neutral commission. Dewey and his colleagues had planned to request U.S. visas for Trotsky and his son so that they could testify before the commission in New York City. But as the prospect of success seemed slight and time was wasting, they decided instead to send a "preliminary commission" to Mexico to take Trotsky's testimony. Dewey agreed to lead it.

Trotsky was overjoyed upon hearing the news. The day he learned that John Dewey was coming to Mexico, he said, was "a great holiday in my life." Friends voiced concern that Stalinists might succeed in infiltrating the commission, as they had the defense committee, and subject him to a hostile interrogation. Trotsky said he had nothing to fear from the questions of the GPU, because truth was on his side: "Dragged into the light of day, the Stalinists are not fearsome." On

April 2, Dewey, accompanied by his fellow commissioners and a team of support staff, boarded the Missouri Pacific "Sunshine Special" bound for Mexico City.

In the days before the hearings, Trotsky and his staff at the Blue House went into overdrive. That staff now included Jan Frankel, who had served as Trotsky's secretary in Turkey and Norway and who arrived in Mexico from Czechoslovakia on February 18. For his defense, Trotsky needed to obtain dozens of affidavits from individuals in numerous countries who could vouch for his innocence of one or another of the Moscow charges. The staff combed through Trotsky's voluminous archives for documents that could be used to rebut Moscow's phony evidence.

As the date of the hearings approached, Mexico City was once again plastered with calls for Trotsky's expulsion. Out of security concerns, the organizers decided to abandon the idea of conducting the hearings in a public hall; instead they arranged for them to take place at the Blue House in Coyoacán, a twenty-minute drive from the city center. Recent rains had flooded the partly paved roads near the house, but on the eve of the hearings there was abundant sunshine and a clear view of the twin snow-capped volcanic peaks rising in the distance to the southeast. The bougainvillea were in bloom, and the houses in the neighborhood were draped in resplendent magenta blossoms, as were the inner walls of the garden patio at the Blue House.

The hearings were to be held in the long dining salon, forty by twenty feet, on the south side of the house facing Avenida Londres. This room had three very tall French windows, which had to be completely covered and then fortified from the inside with six-foot barricades made of adobe brick and sandbags as a protective measure against Stalinist pistoleros. The commission's support staff rushed to complete these defenses the night before the hearings began.

The atmosphere inside the Blue House was tense on the morning of April 10, 1937. The hearings were scheduled to begin at ten o'clock. The police detail outside the house was doubled in strength; inside, an armed secretary verified the credentials of the journalists and the invited guests, and searched them for weapons. Members of the press occupied about twenty of the forty seats set up for the audience, while the rest of the seats were reserved mostly for representatives of Mexican workers' organizations. Each day, would-be spectators had to be turned away at the door.

The five members of the preliminary commission sat at a wooden table at the head of the room. Dewey, tall, bespectacled, with a white mustache and thin white hair parted in the middle, sat in the center. Alongside him were Suzanne La Follette, the commission's secretary; Carleton Beals, a specialist on Latin America; Benjamin Stolberg, a labor journalist; and Otto Rühle, a former Communist member of the German Reichstag and a biographer of Karl Marx.

To the commission's left, at a separate table, the court reporter was flanked by the two legal advisers: John Finerty, counsel for the commission, had argued for the defense in famous radical cases, including Sacco and Vanzetti. Albert Goldman, a Trotskyist lawyer from Chicago, was there to represent Trotsky. At a table across from the lawyers and to the right of the commission sat Trotsky, Natalia, and Trotsky's secretaries.

Klieg lights illuminated the room, creating the feel of a special event. Trotsky was dressed in a gray business suit and a red tie; his hair was neatly groomed. He had decided to testify in English, which would put him at a disadvantage, because although his command of the language had improved since his arrival in Mexico, it was far from fluent. Despite this fact, his American facilitators worried that he would talk too much. Attorney Goldman, coaching him on the workings of Anglo-Saxon jurisprudence, had dissuaded him from opening the hearings with a speech. Goldman also advised him to try to keep his answers short. The preliminary commission hoped to complete its work within a week.

After brief opening statements, Goldman briskly led Trotsky through a series of questions pertaining to his personal background and his career as a Marxist revolutionary. Those hearing the great orator of the Russian Revolution speak for the first time were always struck by the pitch of his voice, which was higher than expected, though vigorous

*Trotsky and Jean van Heijenoort
inside the Blue House during the
Dewey Commission hearings,
April 1937.*

and captivating. Trotsky's English was thickly accented, and he sprinkled it with solecisms, such as "expulsed" instead of "expelled." Except for a brief appearance by secretary Frankel, Trotsky was the only witness during the hearings, which extended over eight days, usually divided into morning and late-afternoon sessions. Trotsky remained seated throughout, frequently turning to ask his secretaries for one or another pertinent document or publication.

Trotsky's testimony about his political career and his relations with the defendants in the Moscow trials ran along smoothly, until Goldman asked him to describe the fate of his children. His two daughters from his first marriage were both dead, one from sickness, the other by suicide in Berlin after the Soviet government stripped her of her Soviet citizenship. The younger of Trotsky and Natalia's two sons had recently been arrested in the USSR. Goldman asked whether under Soviet law, the children of a traitor, or alleged traitor, were also considered guilty. Formally, no, Trotsky replied, but in practice, yes. "All the criminal proceedings, all the trials, and all the confessions are based upon the persecution of the members of the family."

Dewey then asked if this statement would be verified by documentary evidence. "This is simply an opinion," Goldman responded. "It is an opinion of the witness. I will ask him whether there is any documentary evidence—" "Excuse me, it is not an opinion," Trotsky cried in anger. He stuttered, searching for the right words in English, his face twisted in anguish as tears welled up in his eyes. "It is my personal experience," he said at last. "In what way?" Goldman asked. Trotsky replied, "I paid for the experience with my two children."

When Goldman turned to the evidence presented in the two Moscow trials, Trotsky clearly relished the opportunity to expose the Soviet prosecutor's sloppiness in cooking up evidence against him. One of the defendants in the first trial claimed to have had an incriminating meeting with Trotsky's son, Lyova, in the lobby of the Hotel Bristol in Copenhagen in November 1932—but the hotel had burned down in 1917. In the second trial, defendant Pyatakov confessed to having flown from Berlin to Oslo in December 1935 to receive conspiratorial instructions from Trotsky—when in fact no airplane had been able to land at the snowbound Oslo airfield in December, or for the remainder of the winter.

Everyone in the room, and Trotsky more than anyone, wished that a Soviet, or pro-Moscow, attorney could have been present to challenge his testimony and inject some drama into the proceedings. In fact, the commission had invited representatives of the Soviet government, the American Communist Party, the Mexican Communist Party, and the Confederation of Mexican Workers to present evidence and cross-examine Trotsky, but they all declined or ignored these invitations.

After an exhaustive review of the trial evidence lasting three days, Trotsky was questioned at great length about his ideological convictions and his political views. This aspect of his testimony was considered essential to his defense: Would a dedicated Marxist revolutionary like Trotsky, a man who had always repudiated individual terrorism as a political tool, and who even now championed Soviet socialism over

Western capitalism, be remotely likely to conspire with fascists against the USSR, to seek the restoration of capitalism there, or to plot the assassination of Stalin and other Soviet leaders?

Trotsky held forth on a wide variety of topics related to Marxist theory, Bolshevik politics, Soviet history, and Stalin's treachery, sprinting ahead without concern for the hurdles of English vocabulary, grammar, and syntax he toppled along the way. Dewey was riveted, edified by Trotsky's excursions into the theory and practice of communism, and entertained by his flashes of humor and wit. Halfway through the hearings, Dewey wrote to his future second wife in New York, " 'Truth, justice, humanity' and all the rest of the reasons for coming are receding into the background before the bare overpowering interest of the man and what he has to say."

ENLIVENED BY THE occasion, Trotsky was determined to take maximum advantage of the opportunity to set the record straight about his ideas and to dispel the myths propagated in Moscow about a diabolical "Trotskyism." He excoriated Stalin's dictatorial regime, while passionately defending the October Revolution and the actions he took as a Soviet leader to safeguard it.

Trotsky's interrogators showed a special interest in the so-called "dictatorship of the proletariat" established by the Bolshevik Party under Lenin and Trotsky, and how it had evolved into the personal dictatorship of General Secretary Stalin. Finerty asked "if the more correct designation would be dictatorship *for* the proletariat, rather than dictatorship *of* the proletariat," since of course an entire social class could not govern the country. Trotsky maintained, as he had for two decades, that the interests and destiny of the Bolshevik Party and of the Russian proletariat were identical: The Party merely acted as the advance guard of the working class.

Dewey remained skeptical: "I want to ask you what reason there is for thinking that the dictatorship of the proletariat in any country will not degenerate into the dictatorship of the secretariat." "It is a very good formula," Trotsky remarked, though without conceding the point. Stalin's dictatorship, he said, had resulted from Russia's backwardness and isolation. The dictatorship of the proletariat would fare much better in more advanced and less isolated countries.

Dewey was unaware that Trotsky himself, as early as 1904, had warned of the dangers of Bolshevik centralism. His misgivings caused him to turn against Lenin, his mentor, who insisted that only a tightly organized and disciplined group of professional revolutionaries could lead Russia's workers to revolution. Trotsky accused Lenin of engaging in "substitutionism," which was bound to end in authoritarianism. In Trotsky's prophetic formulation: "The party organization substitutes itself for the party, the Central Committee substitutes itself for the organization and, finally, a 'dictator' substitutes himself for the Central Committee." Lenin, Trotsky warned, was threatening to transform Marx's concept of the dictatorship *of* the proletariat into one of dictatorship *over* the proletariat.

Trotsky remained one of Lenin's harshest critics until 1917, when both men rushed to Petrograd from abroad after the fall of the Romanovs. It was then, during the heady days between the February and October revolutions, that Trotsky embraced Bolshevism, recognizing that the Party machinery created by Lenin was the only vehicle capable of carrying out a socialist revolution in Russia. This was his Faustian pact. Lenin's part of the bargain was to endorse Trotsky's concept of the Russian Revolution, which provided the theoretical basis for the Bolshevik seizure of power.

Orthodox Marxism claimed that a socialist revolution could take place only in an advanced capitalist country. Russia in the early twentieth century, although rapidly industrializing, was still a relatively backward country, both economically and politically. It had yet to undergo a bourgeois–democratic revolution to overthrow the autocracy and clear the way for advanced capitalist development. In fact, Russia lagged so far behind the industrialized European countries that its bourgeoisie had grown impotent and was politically unfit to fulfill its historical role. So said Trotsky, who declared that Russia's proletariat, with the support of the peasantry, could make *both* the bourgeois revolution and, close on its heels, the socialist revolution. Trotsky called his theory "permanent revolution."

And the chain reaction would not stop there. A socialist revolution, according to Trotsky's theory, could not be successfully completed within a backward country like Russia. Its ultimate success would depend on its spread to the advanced capitalist countries, starting most likely with Germany. Trotsky and the Bolsheviks thus justified taking power in Russia and establishing a dictatorship of the proletariat by reasoning that their own revolution would serve as the detonator for an international socialist revolution.

The failure of this optimistic scenario became apparent as early as 1920, as the Russian civil war was winding down. The Revolution had triumphed in Russia but had failed to spread. The ruling Bolshevik Party—since 1918 officially called the Communist Party—was forced to retreat from its radical economic program and begin an experiment in limited capitalism known as NEP, the New Economic Policy. Lenin died in 1924 having declared that at some unspecified future date the Party would abandon NEP and resume the socialist offensive. In the power struggle to succeed Lenin, Stalin championed the slogan "socialism in one country," as a nationalistic alternative to Trotsky's "permanent revolution." Trotsky's theory was now turned against him by Stalin, who portrayed his rival as a defeatist, someone who believed that Russia could not proceed to build socialism without assistance from the Western proletariat.

Just the opposite was true, however. Although Trotsky believed that securing the *ultimate* victory of socialism in Russia hinged on the spread of socialist revolution, he did not propose to wait for Europe. In fact, as leader of the opposition in the 1920s, Trotsky urged the Soviet leadership to adopt a faster-paced industrialization and to impose tighter curbs on capitalism in the countryside. Trotsky's enemies, Stalin among them, accused him of being a reckless "super-industrializer" and the enemy of the peasant.

And yet, after Trotsky was defeated and banished from the USSR in 1929, Stalin turned sharply to the left, initiating a crash industrialization drive under the five-year plan and, simultaneously, the forced collectivization of the peasants. This revolution-from-above was far more extreme than anything ever advocated by Trotsky. The human toll was steep. Peasant resisters were branded "kulaks" and slaughtered by the millions, many as a result of the man-made famine in the Ukraine in 1932 and 1933.

Questioned by the Dewey commission about Soviet Russia's great leap to a state-controlled economy, Trotsky explained that while he had opposed the use of "brute force" to achieve collectivization, he never denied its "successes." He also lauded the imposition of state planning in industry, even though he believed it had been carried out recklessly and with unnecessary brutality. Trotsky testified before the commission that the Soviet state's ownership of the means of production made the USSR the most progressive country in the world. Only the Stalinist regime itself was objectionable. Trotsky defined that regime as a parasitic bureaucratic caste, a product of Russia's backwardness and isolation.

Trotsky advocated a revolution to overthrow Stalin's ruling bureaucracy, but he had in mind a narrowly political, as opposed to a social, revolution. The October Revolution created the world's first workers' state, and it remained a workers' state even under Stalin, albeit one that was "degenerated" or "deformed." To Trotsky, the class structure of the USSR made it worth defending against its enemies, despite the purge trials and the terror that were destroying the men and women who had made the revolution and eliminating Trotsky's comrades and loved ones. "Even now under the Iron Heel of the new privileged caste, the U.S.S.R. is not the same as Czarist Russia," he explained to a wealthy American sympathizer who helped finance the Dewey hearings. "And the whole of mankind is, thanks to the October Revolution, incomparably richer in experience and in possibilities."

DEWEY, LIKE FINERTY, probed Trotsky but never seriously challenged him, and the other commission members followed suit—all, that is, except for Carleton Beals, the Latin Americanist. Beals treated Trotsky as a hostile witness, and he provided the hearings with their one moment of contentious drama and scandal. From the beginning, Beals had behaved like the commission's odd man out. He was absent from its pre-hearing meetings held in Mexico City and then missed the opening session. When he spoke, he exhibited a prickliness toward his fellow commission members, especially Dewey.

During the hearings Beals was often seen huddling with *The New York Times* correspondent on the scene, Frank Kluckhohn. Kluckhohn's reporting from Mexico City made it apparent that he had an ax to grind. He wrote a hostile profile of Trotsky and insinuated that the hearings were a whitewash. Even before the commission protested to the *Times*, Kluckhohn's editor had wired him to say that he should do more reporting and less editorializing. This he managed to do for a few days, then he was absent for two more before returning in time for the Beals affair.

On April 16, the penultimate day of the hearings, Beals's questioning veered into provocation when he asked Trotsky whether, as Soviet war commissar in 1919, he had sent a Soviet agent to Mexico to foment revolution. Everyone in the room recognized that the question was intended to jeopardize Trotsky's asylum in Mexico. There was a suspicion that Kluckhohn, who had a habit of posing similarly provocative questions at Trotsky's press conferences, had inspired Beals. His question led to a testy exchange with Trotsky, who bluntly told Beals that his informant was a liar. The next day, Beals informed Dewey by letter of his resignation from the commission. The hearings had proved to be a waste of time, he wrote, and "not a truly serious investigation of the charges."

That same day, April 17, Trotsky delivered his closing statement before the commission. Its text was so long—Dewey called it "a book"—that Trotsky could read only a portion of it at the hearings, the rest being added to the record. He began speaking toward five o'clock in the afternoon and finished close to 8:45.

Most of his presentation was an exhaustive analysis of the Moscow trials, which he called "the greatest frame-up in history." He made the case for his own impeccable Marxist-Leninist credentials and assured his audience of "my faith in the clear, bright future of mankind." He closed with a diplomatic flourish, thanking the committee and its distinguished chairman. "And when he finished," the court reporter testified, "the audience, a singularly diverse one, burst out into applause, which was, believe me, most spontaneous. This moment I shall never forget." Dewey avoided stepping on the moment: "Anything I can say will be an anticlimax." The hearings of the preliminary commission came to a close.

Trotsky and Dewey had thus far been introduced only formally. The organizers had decided that, for appearances' sake, the two men ought to be kept apart, and so they were, even in the Blue House patio during recesses in the hearings. A cartoon in one of the popular Mexican daily papers gave a different impression. It showed Trotsky and Dewey seated side by side in the hearing room. The caption had one audience member remarking to another, "What does Trotsky mean by saying he has been denied liberty when he has been all over the world?" The other man replies, "Yes, so he has, just like a lion [*léon*] in a circus."

Late in the evening after the final session, there was a social gathering for commission members, staff, and journalists at the home of an American well-wisher in Mexico City, an event attended by both Trotsky and Dewey. No longer constrained by protocol, the two men, surrounded by guests, were able to exchange pleasantries. Dewey said to Trotsky, "If all Communists were like you, I would be a Communist." Trotsky replied: "If all liberals were like you, I would be a liberal." The nearby guests erupted in laughter at this good-natured display of mutual diplomacy.

Dewey was disappointed to have to leave for New York without being able to converse privately with Trotsky. He wrote to his former student Max Eastman, who had encouraged him to go to Mexico: "You were right about one thing—If it wasn't exactly a 'good time,' it was the most interesting single intellectual experience of my life."

Dewey canceled his summer vacation plans in Europe in order to direct the work of the full commission in New York. There was other testimony and much documentation to collect, some of it to be supplied by a parallel commission of inquiry set up in Paris. Aside from Dewey and the remaining members of the subcommission—Stolberg, Rühle, and La Follette—there were six other commission members: Wendelin Thomas, a former Communist deputy in the German Reichstag; Alfred Rosmer, former member of the French Communist Party and editor of its newspaper, *L'Humanité*; John R. Chamberlain, former literary critic of *The New York Times*; Carlo Tresca, an Italian-American anarchist leader; Edward Alsworth Ross, a professor of sociology at the

University of Wisconsin; and Francisco Zamora, a Mexican economist and journalist.

When Dewey and the others returned from Mexico, they were surprised to find Trotsky's defenders in such a gloomy state. The American press coverage of the hearings had been less than flattering to the commission. Kluckhohn's reporting in *The New York Times*, including his earnest coverage of the Beals resignation, was reprinted in the Communist and other pro-Moscow publications, which treated the hearings as a sham. With Dewey's encouragement, the American Committee for the Defense of Leon Trotsky decided to go on the offensive. It arranged a public meeting for May 9 at the Mecca Temple in midtown Manhattan, with Dewey as the featured speaker.

Dewey came out fighting. Before a crowd of more than 3,000 people, he upbraided the pro-Moscow liberals for attempting to create the impression that the hearings were a farce. "When did it become a farce in the United States to give a hearing to a man who had been convicted without a hearing?" Dewey accused the liberal apologists for Stalin of suffering from "intellectual and moral confusion." As a past defender of socialism in the USSR, he said he understood that certain liberals were hostile to Trotsky because they wished to protect and preserve the one successful attempt in all history to build a socialist society.

But something more was at work here, Dewey observed. Moscow's defenders believed that Trotsky's theories and views were mistaken. Yet Trotsky had not been convicted for his theories or his views, but rather for the most heinous of crimes, including assassination and treason. To declare Trotsky guilty because of his opposition to the rulers in the Kremlin was "not fair or square," said Dewey. "It is in the name of justice and truth as the end that we ask for your support. We go on in confidence that we shall have it. As Zola said in the Dreyfus case: 'Truth is on the march and nothing will stop it.'"

Dewey had given the best speech of his career, said his friends, who were surprised by the intensity and emotion he displayed, quite unlike his usual professorial manner. Sidney Hook told Dewey that if he wrote his philosophy in the same engaging style in which he delivered that speech, more people would be able to understand it. Dewey replied that he could not get mad writing philosophy.

The full committee went about its business, momentarily interrupted on June 11 by another thunderclap out of Moscow, where the authorities announced they had uncovered a treasonous plot involving the Red Army command in a conspiracy with Nazi Germany, under the banner of Trotsky. Marshal Mikhail Tukhachevsky, the outstanding civil war commander, and seven other top-ranking officers were tried in secret and executed the following day. This was the start of a massive purge of the army's officer corps. Tens of thousands would perish, including a large majority of the civil war commanders.

This time there was no show trial, so there was no call to battle stations at the Blue House. Instead, Trotsky was forced to deal with a challenge from an entirely different quarter. Friends and former comrades in the United States and Europe, without questioning Trotsky's legal innocence in the trials, began to raise doubts about his moral right to challenge Stalin. In doing so, they threatened to erase the thick line Trotsky had drawn between Bolshevism and Stalinism. Had not Lenin and the Bolsheviks, they asked, suppressed the rival socialist parties shortly after the Revolution so that Soviet power quickly came to mean Bolshevik power? Had not Lenin's regime conducted a Red Terror against its declared enemies during the civil war? Had not War Commissar Trotsky, who now condemned Stalin for threatening to execute the wives and children of the trial defendants, seized as hostages the families of former czarist officers serving in the Red Army?

Trotsky had been asked such questions during the hearings in Coyoacán. Typically he invoked the exigencies of the civil war in order to justify Bolshevik violence. Those who confronted him now, however, were far more knowledgeable about these matters. And they believed they could identify Bolshevism's defining moment: the Kronstadt rebellion of 1921. Kronstadt was a fortress city and naval base on an island in the Gulf of Finland, some twenty miles west of Petrograd. The sailors there had played a crucial role in the revolutionary events of 1917. Trotsky, who was their favorite, honored them at the time as "the pride and glory" of the Russian Revolution.

Only a few years later, however, Kronstadt came to symbolize something entirely different. In the frozen winter of 1921, the sailors of Kronstadt, which was the main base of the Baltic fleet, rose up in rebellion against Bolshevik rule. They demanded an end to the Communist monopoly of power, genuine elections to the Soviets, and the cessation of political terror, among other things. A special target of their wrath was "the bloody Field Marshal Trotsky."

The Bolsheviks portrayed the uprising as an act of counterrevolution, in danger of being exploited by Western imperialists and White Guard generals, who perhaps had instigated it. Red Army troops under the command of General Tukhachevsky crossed over the ice to crush the rebellion, which they managed to do only with great difficulty and after suffering heavy losses. Fifty thousand Red Army soldiers made the final assault against nearly 15,000 defenders. Afterward, hundreds if not thousands of rebels were executed without trial.

Trotsky's critics now revived the memory of Kronstadt, making it the centerpiece of their case for an essential continuity between Bolshevism and Stalinism. Trotsky could hardly believe the bad timing of this assault on his reputation, in the middle of his campaign against the Moscow trials. "One would think that the Kronstadt revolt occurred not seventeen years ago but only yesterday," he complained. He accused his critics of romanticizing the Kronstadt sailors, and he claimed he played no role in suppressing the rebellion, although in fact as war commissar his role was central. When he learned of the revolt, he issued a demand for unconditional surrender. The Petrograd authorities then warned the sailors not to put up resistance, or "you will be shot like partridges."

Trotsky answered his critics in a series of short articles and in correspondence, while worrying about the effect this discussion might have on the deliberations of the Dewey Commission. A special source of concern was that one of his most troublesome antagonists on Kronstadt happened to be a member of the commission: Wendelin Thomas, the former German Communist, who had helped lead the Wilhelmshaven sailors' revolt in November 1918. Thomas was still at it, accusing Trotsky of hypocrisy, in December 1937, on the eve of the commission's announcement of its verdict. "That you should seek vindication, I regard as well and proper," he wrote, "that you should deny vindication to your political opponents I regard as good Bolshevism." Trotsky's portrayal of the Kronstadt sailors as political rednecks seeking privileged food rations was a slander, said Thomas. "You call to arms against the calumnies of the Russian state machine of 1937 but at the same time you attempt to excuse and justify the calumnies of the Russian state machine of 1921."

On December 12, simultaneous with the publication of the complete record of the Coyoacán hearings, the Commission of Inquiry into the Charges Made Against Leon Trotsky in the Moscow Trials announced its verdict: the trials were a frame-up; Trotsky and his son were not guilty as charged. Trotsky was jubilant. He told his American secretary, "My boy, we have won our first great victory. Now things will begin to change." The blow to Stalin, he said, was "tremendous." The verdict would deliver a "great moral shock" to public opinion. The initial press coverage of the verdict he described as "the best we could hope for." Even the Mexican press was "extremely favorable."

Yet, unhappily for Trotsky, the truth kept marching on. On December 13, as the first stories about the verdict appeared in the American newspapers, Dewey made a radio broadcast on CBS, warning American liberals away from Soviet Communism. Now more than ever, Dewey said, he disagreed with the "ideas and theories of Trotsky," including his defense of the USSR. "A country that uses all the methods of fascism to suppress opposition can hardly be held up to us as a democracy, as a model to follow against fascism. Next time anybody says to you that we have to choose between fascism and communism, ask him what is the difference between the Hitlerite Gestapo and the Stalinite G.P.U., so that a democracy should have to choose one or the other."

Dewey expanded on this theme in an interview published in *The Washington Post* a few days later. The results of the Soviet experiment were now in, Dewey said, and one of the fundamental things they demonstrated was that democracy could not survive if it was restricted to a single political party. In Russia, the October Revolution had led to the gruesome travesty of justice enacted in the October Hall. Elsewhere, the outcome would differ only in degree. "The dictatorship of the proletariat has led and, I am convinced, always must lead to dictatorship over the proletariat and over the party. I see no reason to believe that something similar would not happen in every country in which an attempt is made to establish a Communist government."

When Trotsky learned about Dewey's statements he was indignant, though of course he could say nothing publicly. Meanwhile, his critics kept nipping at his Achilles heel. Kronstadt would not go away. Well into 1938, Trotsky was forced to defend himself in articles about the rebellion and in one long essay on politics and morality. "Idealists and pacifists have always blamed revolution for 'excesses,' " Trotsky wrote. "The crux of the matter is that the 'excesses' spring from the very nature of the revolution, which is itself an 'excess' of history." In this sense, Trotsky said, "I carry full and complete responsibility for the suppression of the Kronstadt rebellion."

For Trotsky, it came down to the question of whether the end justified the means. "From the Marxist point of view, which expresses the historical interests of the proletariat, the end is justified if it leads to increasing the power of man over nature and to the abolition of the power of man over man." Dewey challenged Trotsky's reasoning in a brief rebuttal article called "Means and Ends." Dewey had no objection to the ultimate end Trotsky put forward, which sounded vaguely like egalitarian socialism. But Dewey detected that Trotsky was deducing acceptable means from an outside source: the Marxist concept of "class struggle," which Dewey classified as "an alleged law of history." Dewey contended that while class struggle might indeed be an appropriate means, it had to be justified and not simply taken on faith.

Dewey, the pragmatist, was alert to the mutual shaping of ends and means. Trotsky, the Marxist, was guided by his belief in an iron law of historical progress. To Dewey, Trotsky was the prisoner of an ideology. "He was tragic," Dewey said in delivering his ultimate verdict on Trotsky a dozen years later. "To see such brilliant native intelligence locked up in absolutes."

CHAPTER 3

Man of October

The departure of the Dewey Commission in mid-April 1937 brought something of a reprieve for Trotsky and his staff at the Blue House after the sweatshop days of the hearings. For several weeks afterward there was more to do to reinforce the case for Trotsky's innocence of the charges brought against him in the Moscow trials: more documents to be collected, more testimonies solicited, and all of it translated into English and sent to New York, where the commission continued its deliberations. Trotsky took a break from these activities in early May, when he headed for the mountains and to Taxco, a silver-mining town of red-tiled roofs and narrow, steeply winding cobblestone roads beneath an imposing Spanish Baroque cathedral, in those days about a four-hour drive from Mexico City.

Yet for Trotsky, a relaxing vacation was out of the question. Mentally he worked overtime trying to anticipate Stalin's next move and those of his provocateurs and assassins. He agonized over the fates of family members left behind in the Soviet Union. And throughout he continued to suffer from the headaches, dizziness, and high blood pressure that had plagued him for years.

Natalia and the staff had hoped that the Taxco getaway would allow Trotsky to rest a bit and enable him to return to work reinvigorated, but this was not the way it turned out. In a letter to Trotskyist headquarters in New York City, one of his American secretaries, Bernard Wolfe, assessed the situation this way on May 26: "The old man relaxed a little bit at Taxco, but suffered, I think, from being shut in and having people around constantly. Now, since his return, he is not feeling well and is trying to rest as much as possible—the last months have been a terrible strain for him." The Blue House should have felt nearly empty after the crush of the Dewey hearings, yet the size of the household—secretaries, guards, and kitchen staff—left Trotsky with little privacy. Nor was escaping the house a simple proposition. Such was the concern for the Old Man's safety that an excursion into the environs of Mexico City required the presence of four bodyguards and all the preparations of a military expedition. And until later that year the automobiles for such outings had to be borrowed, which made planning ahead more difficult and left little room for spontaneity.

The strain on Trotsky took a toll on his secretaries, although as one of the American newcomers, Wolfe was not as vulnerable as the two European veterans, Jan Frankel and Jean van Heijenoort. Frankel, a native of Prague, had joined up with Trotsky in 1930 in Turkey, where he was first exiled after his deportation from the Soviet Union. Van Heijenoort—or Van, as Trotsky called him—came on board in 1932 from his native France. Along with Lyova, Trotsky's elder son and right-hand man now based in Paris, these men were Trotsky's essential *adjutants* during the exile years. More than mere secretaries, Frankel and Van served as translators, political advisers, and bodyguards.

Van also worked as Trotsky's archivist, organizing his papers, both in-house and later at Harvard and Stanford. Afterward, he revealed what it was like to work in close quarters with the Old Man, and in the process he got some things off his chest. "Trotsky displayed all his amiability with visitors and newcomers," Van recalled. "He would talk, explain, gesture, ask questions, and at times be really charming. The presence of a young woman seemed to give him special animation. But the more one worked with him, the more demanding and brusque he became." The situation in Mexico was exacerbated by living in close quarters and the unrelieved security regime. "You treat me like an object," Trotsky once complained to Van.

Van observes of Trotsky that "the three persons toward whom he allowed himself to be the most brusque were Lyova, Jan Frankel, and me." Frankel had taken over as Trotsky's principal secretary in February 1931, after Lyova left Turkey for Berlin. With his dark hair and eyebrows, squarish head, dour mien, and the inevitable cigarette between his fingers, Frankel at thirty-two looked like a somewhat rumpled version of Edward R. Murrow.

It seems that Trotsky's brusqueness—and his explosive temper—may have exceeded the limits of Frankel's tolerance in the days surrounding the Dewey hearings. "One day Trotsky went to Frankel's room to ask him for a document," Van remembered, "and the document was not ready. Trotsky went back to his study and slammed the door. It was a glazed door with many panes, whose putty had long since been eroded by the Mexican rains. Under the impact, the panes fell out, one after another, the crystal din of each fall reverberating throughout the house."

The end of the hearings did nothing to lighten the atmosphere. "Unfortunately," Frankel wrote to a comrade in New York on June 8, "all my predictions have been fulfilled. Our friend, since the departure of the Commission, has been extremely tired, not to say ill. Under these circumstances, all the negative aspects of his external life have become sharper and more critical. The confined life in a small house, without any liberty of movement, surrounded constantly by other people, with no possibility of finding even a corner in the patio where there is no disturbance, creates a terrible tension for him." The tension led to clashes between the two men, until matters reached the point where Trotsky insisted that Frankel move out of the house. That was at the beginning of June. In October Frankel would leave Mexico for New York and the headquarters of the American Trotskyist organization. For the time being, he continued to serve the Old Man loyally in Coyoacán. This included advising the New York comrades about the requirements of the household and appealing for the scarce funds needed to fill them.

One critical need was for an automobile. Diego and Frida often contributed the use of their cars and drivers, but after the hearings Diego went off to the countryside to paint. Frida, who had lately taken ill, could no longer be imposed upon. Her sister, Cristina, who often served as chauffeur in these early months, had recently undergone surgery and was in the hospital. "Thus the old man doesn't leave the house for weeks at a time," Frankel informed New York. "He is a real prisoner." Trotsky's physician, meanwhile, had recommended an extended period of rest for him away from Mexico City, but "here as well we are entirely paralyzed through the lack of money."

An automobile would have gone a long way toward alleviating the problem, but Frankel now put forward a more radical recommendation: Trotsky and Natalia ought to be moved to a different house, one that would allow them greater freedom of movement. He had located what he believed was a far superior residence, one so desirable that "it would be a catastrophe to let this opportunity slip by." Yet the new home would require hundreds of dollars to cover a six-month deposit on rent plus the costs of the move and of security installations—money, Frankel knew, that the cash-poor American Trotskyists were unlikely to be able to raise.

Frankel's use of the word "catastrophe," if not exactly alarmist, seems unwarranted by the circumstances his letter describes. It may have been inspired by an inconceivable turn of events in Coyoacán that he dared not divulge: Trotsky had become romantically involved with Frida Kahlo. The Old Man was having an affair.

Trotsky's liaison with Frida, which got under way sometime after the hearings, did not come as a complete surprise to Frankel or to Van, who had heard stories of his various conquests during his glory days. Based on his observations of Trotsky's behavior with Frida and, shortly afterward, with another Mexican woman, Van figured Trotsky to be an experienced philanderer. Yet this was Trotsky's first romantic adventure since his exile from Russia in 1929. When he lived in Turkey, in France, and then in Norway, his opportunities for an extramarital affair were severely constrained. Now, at fifty-seven, in his final place of exile, Trotsky found that circumstances conspired in his favor.

It is no mystery why Trotsky was attracted to Frida Kahlo. The daughter of a German-Jewish immigrant father and a Mexican mother, at twenty-nine she was a striking and exotic beauty with black hair, audacious almond-shaped eyes beneath batwing eyebrows, and sensuous lips. She was even more attractive than contemporary photographs reveal, to judge by the testimony of the men who made her acquaintance in the late 1930s and were struck by her forceful personality, quick intelligence, and much more. An American friend of the Riveras who was no prude says that Frida could draw from "the richest vocabulary of obscenities I have ever known one of her sex to possess." She had experienced considerable hardship. Polio, contracted at the age of six, had left her with a withered right leg. At age eighteen came a tram accident that nearly killed her, shattering her pelvis, injuring her spine, and crushing her right foot. She was in nearly constant pain as a result of the injuries sustained in the accident and from the multiple surgeries and medical procedures she underwent to treat them—including, in her final decade, a succession of twenty-eight orthopedic corsets.

Frida compensated for her disabilities by transforming her appearance into her best-known work of art. She adopted as her signature costume the colorful apparel of the region of Tehuantepec, most notably the long skirt that hid her deformed right leg. During her extended convalescence after the accident in 1925, when for months she was bedridden, she began to paint seriously. She would use her art to portray her suffering, both physical and psychological, creating shocking imagery and symbolism, much of it intensely personal, as in her breakthrough painting, *Henry Ford Hospital*, a harrowing depiction of the abortion she underwent in Detroit in 1932. Her most compelling paintings indulge in the fantastic and the grotesque, deploying a morbid sense of humor to leaven the profusion of blood and tears. It is little wonder that in the late 1930s the Surrealists wished to claim her as their own.

From 1929, the year she married Diego, to the time she met Trotsky, Frida painted infrequently and in the enormous shadow of her husband's artistic reputation. For now, she was mostly Mrs. Rivera. Her small, finely detailed paintings, depicting herself, sometimes her animals and friends, and occasionally a still life, exhibited none of the political consciousness or commitment that drew Trotsky to Rivera's heavily populated epic frescoes on Mexican history, American industry and technology, and the Russian Revolution.

Shortly after Trotsky's arrival in Mexico, Frida completed a self-portrait called *Fulang-Chang and I*. The artist, flanked by her pet monkey, looks back at the viewer with a self-assured and sensuous gaze. Trotsky and Natalia most likely saw this painting during one of their visits to the Riveras in nearby San Angel, where the painters occupied separate houses connected by a footbridge. Perhaps Trotsky was aware that in

Frida Kahlo, Mexico, 1937.

the Mayan tradition, the monkey is a symbol of lust or promiscuity. In any case, he did not need a monkey to make him aware that Frida was interested. She was a well-practiced flirt and Trotsky knew how to reciprocate. The two spoke in English, which Frida flavored with the slang she had picked up during her three years in the United States. This left Natalia, who spoke no English at all, grasping for the meaning of their knowing exchanges.

"Frida did not hesitate to use the word 'love,' after the American fashion," Van relates. "'All my love,' she would say to Trotsky upon leaving." Trotsky may have been the first to cross the line when he began passing secret notes to Frida. "He would slip a letter into a book and give the book to her, often in the presence of others, including Natalia or Diego, with the recommendation that Frida read it. I knew nothing about this little game; only later did Frida tell me the story."

Trotsky may have been emboldened by an awareness of Diego's own brazen philandering, which included, most devastatingly for Frida, an affair with her sister, Cristina, two years earlier. In the face of this betrayal, Frida became less inclined to contain her own sexual appetite, which seems to have been prodigious. Her "view of life," she told Van, was "Make love, take a bath, make love again."

Frida's promiscuity and her relative youth might give the impression that, of the two of them, she was less invested in the relationship. Yet there is no reason to assume she did not fall hard for *piochitas*—or "little goatee," as she referred to him. In her eyes, Trotsky's reputation as a great revolutionary, Lenin's twin star, remained untarnished by the Moscow frame-up trials. His remarkable performance at the Dewey hearings, which she witnessed as a spectator, demonstrated that he had lost none of his brilliance, courage, and charisma. And perhaps there was an additional source of attraction: an affair with Trotsky, her husband's friend and political hero, would bring revenge for Diego's affair with her sister.

Trotsky and Frida met at Cristina's house on Calle Aguayo, a few blocks away from the Blue House. The members of the household were all aware of these assignations, says Van, including Natalia. "Late in June the situation became such that those close to Trotsky began to get uneasy. Natalia was suffering." Natalia's suffering would have been evident even to a perfect stranger, because she wore it on her face. The writer James Farrell, who attended the Dewey hearings, called Natalia's "one of the saddest faces I have ever seen." She was now fifty-five years old, but her features had aged beyond her years under the burden of the hardships and tragedies of the previous decade, none more heartrending for her and her husband than the uncertain fate of their younger son, Sergei.

By mutual decision of the parents and their son, Seryozha, as they called him, did not accompany the family into exile in 1929. Having rejected the life of politics to which his father and older brother, Lyova, devoted themselves absolutely, the athletic Seryozha joined a traveling circus for two years before pursuing a career in science and technology, becoming an instructor at a higher technical school in Moscow before the age of thirty. His parents believed that bringing Seryozha with them into exile would tear him from his roots and ruin his life.

In Moscow, Seryozha—who like his brother used his mother's family name, Sedov—took precautions not to call attention to his family background. His letters were addressed to his mother only and were devoted exclusively to family news and mundane matters. The hope was that he and his family would be left in peace, and in fact this is the way things worked out for the first few years. But everything changed for Seryozha, as it did for countless other Soviet citizens, after the murder of Leningrad Party chief Sergei Kirov, a rising political star, on December 1, 1934. Kirov's murder, which Stalin may have orchestrated and certainly exploited politically, set off a wave of arrests that launched the Great Terror of the next several years.

Trotsky's younger son was among the victims. His final letter to his parents, then living near Grenoble, France, was written eight days after Kirov's murder. It concluded with an ominous sentence: "The general situation is proving extremely difficult, much more difficult than you can imagine." Trotsky and Natalia now became desperate for news of him. They hoped that his avoidance of politics would spare him, yet they could not help but imagine the worst. Late in May 1935, they learned that Seryozha had been arrested and was being held in a Moscow prison. They imagined him being brutally interrogated in his prison cell, and they blamed themselves for not having insisted that he accompany them into exile. They tried to maintain hope and to support each other, but this was not always possible, and on one occasion Natalia remarked bluntly to her husband: "They will not deport him under any circumstances; they will torture him in order to get something out of him, and after that they will destroy him."

Natalia issued an open letter, published in Trotsky's *Bulletin of the Opposition*, in which she declared her son's innocence and appealed to Romain Rolland, André Gide, George Bernard Shaw, and other European intellectuals sympathetic to the USSR to press Moscow for a commission of inquiry into the repressions following the Kirov murder. Whatever political conspiracy was behind it, she wrote, it could not have involved Seryozha, whose aversion to politics was well known to the GPU, as well as to Stalin, "whose son was a frequent guest in our boys' room."

In the diary he kept in France in the spring of 1935, Trotsky records his wife's "deep sorrow" and constant anxiety: "N. is haunted by the thought of what a heavy heart Seryozha must have in prison (if he is in prison). Perhaps he may think that we have somehow forgotten about him, left him to his fate." She asks her husband if he thinks that Stalin is aware of Seryozha's case. "I answered that he never overlooks such 'cases,' that his specialty actually consists in 'cases' like this."

Seryozha was held in a Moscow prison for several months before being deported to Krasnoyarsk, in Siberia. In January 1937, as his parents landed in Mexico, the Soviet press reported that he had been arrested and charged with attempting, on the instructions of his father, a mass poisoning of workers. Natalia issued another appeal on behalf of Seryozha, addressed "To the Conscience of the World," but by now the situation was beyond hope. In his appearance before the Dewey Commission in April, Trotsky testified that the precise whereabouts in the Soviet Union of Seryozha, whom he referred to sarcastically as the "poisoner," were still unknown. But in a statement to the press he predicted his fate: "Stalin intends to extract a confession from my own son against me. The G.P.U. will not hesitate to drive Sergei to insanity and then they will shoot him."

Living in close quarters with Natalia produced its own particular brand of tension, but Frankel and Van felt a strong sense of loyalty to *"la chère,"* as they referred to her in private. This formed no small part of their concern about the behavior of her husband. Van says he decided to say nothing, but that "Jan Frankel, as I remember, ventured to speak to Trotsky about the dangers inherent in the situation." Over the years this cautious recollection somehow mutated into a dead certainty that Frankel broke with the Old Man over the Frida affair. The real story is less dramatic. Having spoken his mind to Trotsky and withdrawn from the Blue House, Frankel was in a better position to confront him about the Frida matter. Perhaps he also had a clearer sense of the looming catastrophe.

The dangers extended well beyond the health of Trotsky's marriage.

Every nightmare scenario revolved around Diego, who as yet had no notion of what was going on. "Since he was morbidly jealous," Van testifies, "the least suspicion would have caused an explosion." A scandal would compromise Trotsky's reputation while feeding the fury of the Mexican Communists and thereby jeopardizing his security. Diego might feel honor-bound to evict Trotsky from the Blue House; perhaps he would throw his support behind the Communists in their unending pressure campaign against the government to terminate Trotsky's asylum in Mexico.

Or Diego might simply decide to terminate Trotsky. Frida had warned previous lovers that her husband's jealousy could conceivably incite him to murder. One of those lovers was Japanese-American sculptor Isamu Noguchi, with whom Frida had an extended liaison in 1935. He later recounted that when a sock of his turned up at Frida's house, "Diego came by with a gun. He always carried a gun. The second time he displayed his gun to me was in the hospital. Frida was ill for some reason, and I went there, and he showed me his gun and said: 'Next time I see you, I'm going to shoot you!' "

Diego had a habit of threatening people, and he wielded his pistol like an exclamation point, so perhaps there was no reason to fear that he would actually use it as a deadly weapon. But there was no telling what acts of revenge he might be driven to if he were humiliated by the public airing of his wife's affair with the great Trotsky—the man he had helped gain refuge in Mexico.

By the beginning of July, the atmosphere at the Blue House was becoming unbearable, because the discord between Trotsky and Natalia had turned venomous. The only possible remedy was a temporary separation. On July 7, Trotsky left Coyoacán and moved to a hacienda owned by a friend of Diego's near San Miguel Regla, some eighty miles northeast of Mexico City. Here Trotsky would be able to enjoy the outdoors, to fish and ride horseback. Accompanying him were police sergeant Jesús Casas, who commanded the police garrison guarding the Blue House, and Sixto, one of Rivera's chauffeurs.

Four days after Trotsky's arrival at the hacienda, on July 11, Frida paid him a visit. Natalia had wanted to come along on this trip, but Frida had maneuvered to leave her behind. Their meeting was neither secret, nor very private. Frida traveled in the company of Frederico Marin, a brother of Diego's first wife and a medical doctor, whose presence probably provided Frida with the pretext for her visit, just as Trotsky's failing health served as the cover story for his retreat to the hacienda.

Whatever calculations lay behind their rendezvous on that rain-soaked July day, their affair was quickly coming to an end. Van believes that during Frida's visit, the two made a joint decision to call a halt. "Now, in view of the circumstances, it was impossible for them to go further without committing themselves completely. The stakes were too high. The two partners drew back." Frida's alleged remark that she had become "very tired of the old man" might indicate that she alone drew back. Yet she was sufficiently inspired to undertake an arduous nine-hour round-trip by automobile in order to spend a few hours in Trotsky's company.

After Frida and her traveling companions departed, Trotsky wrote to Natalia of his unexpected visitors and about the enjoyable—though not too enjoyable—time they passed together, expressing his regret that she had not felt well enough to undertake the journey. Trotsky tries so hard to make the visit seem uneventful that he ends up sounding like a man with something to hide. "This letter is purely descriptive," he concludes awkwardly, "but it exhausts everything that would interest you."

Hurt and angered at having been left out, and apparently fearing that her husband was part of the conspiracy, Natalia wrote to him demanding an explanation. In his overwrought reply, Trotsky tried to reassure her that her exclusion had been Frida's doing. "Come!" he exclaims, declaring his innocence. His letter sheds light on the nature of the stormy conversations leading up to their separation, which featured a predictable psychological maneuver on his part. Thrown on the defensive about his relationship with Frida, he went on the attack, accusing Natalia of having had an affair of her own almost twenty years earlier, when they occupied an apartment in the Kremlin.

The alleged infidelity was supposed to have occurred in 1918, after Natalia had been appointed director of the museums department of the People's Commissariat of Education. A young comrade on her staff had become infatuated with her, and she had deflected his attentions.

So said Natalia. Trotsky could not be certain. It is unclear whether his suspicion took hold back in the Kremlin days or more recently, as he surveyed the marital landscape with a conscience troubled by his own transgressions and derelictions. Whatever the case, he kept turning the matter over in his mind and confronting Natalia, who called his behavior "recidivism."

The sudden reappearance of the minor functionary from long ago was unquestionably convenient for Trotsky, enabling him to turn the tables on his wife. Yet his display of masochistic jealousy was no mere pose. "I have been reliving our yesterdays, that is, our pangs of memories, pangs of my torment," he said, referring to his suspicions about her extramarital affair. This "insignificant question stands before me with such force, as though our entire life hangs on the answer to it . . . And I run to get a piece of paper and write down the question. Natalochka, I'm writing to you about this with self-hatred."

In this and subsequent letters, Trotsky's emotions bounce back and forth between euphoria and agony, repentance and revenge. "Your letter brought me happiness, tenderness (how I love you, Nata, my only one, my eternal one, my faithful one, my love and my victim!)—but also tears, tears of pity, of repentance and . . . torment. Natalochka, I will burn my stupid, pitiful, self-serving 'questions.' Come!"

In his hour of torment, Trotsky cannot keep still. "After every two or three lines I stand up, walk about the room and weep tears of self-reproach and of gratitude to you, and above all tears for the old age that has taken us by surprise."

They had met in Paris in the autumn of 1902. Trotsky, who turned twenty-three that November, had recently escaped from eastern Siberia; he had spent three years in czarist prisons and exile there, following his arrest for distributing radical texts to the dockworkers of the Ukrainian city of Nikolaev, near the Black Sea. From Siberia Trotsky headed straight to London for his historic first meeting with Lenin, and from there he made his way to the Continent to lecture on Marxist theory to the radical Russian émigré colonies. This brought him to Paris, where he met Natalia Sedova, a smart and attractive young woman who was a member of the radical group associated with the Russian newspaper *Iskra* (*The Spark*), of which Lenin was a leading figure. The daughter of wealthy parents of noble birth, Natalia had been expelled from an exclusive boarding school in Kharkov, in the Ukraine, for reading radical literature and was now a student of art history at the Sorbonne. Like everyone else in the audience that autumn day in Paris, she was impressed by Trotsky's oratory, which "exceeded all expectations." Dialectical materialism had never sounded more appealing, and the two young radicals began seeing each other.

In Natalia, a slight, oval-faced woman with irrepressible wavy brown hair and full lips, Trotsky had a charming and knowledgeable guide to the sights of Paris, especially its art, and the two spent hours touring the Louvre. Trotsky, the Ukrainian farm boy, proved to be a reluctant tourist, however, and generally behaved like a boor, dismissing Paris with the comment: "Resembles Odessa, but Odessa is better." Natalia later recalled that her companion "was utterly absorbed in political life, and could see something else only when it forced itself upon him. He reacted to it as if it were a bother, something avoidable. I did not agree with him in his estimate of Paris, and twitted him a little for this." Trotsky later attested that in Paris "I came face to face with real art" for the first time and that he learned to appreciate it only "with great difficulty" and only thanks to Natalia's persuasive presence. "I had my own world of revolution, and this was very exacting and brooked no rival interests."

The comrades became lovers, which presented Natalia with a dilemma, because she already had a lover. Trotsky was married, but his wife, Alexandra, and their two small children had been left behind in distant Siberia. Natalia felt torn, and she hesitated before committing herself to Trotsky. "He never forgave me," she confided to a friend after Trotsky's death. "It always kept coming up."

Trotsky interrupted the separation with a visit to Coyoacán for three days, July 15 to 18, but his time together with Natalia did nothing to put him at ease. Upon his return to the hacienda, he writes to her that he has decided to confine all his disturbing thoughts and feelings to a personal diary, which they can read together when he rejoins her; this way his letters will not upset her. A moment later, however, he is unable to restrain himself: "I just wanted to say—and it's not a criticism—that my 'recidivism' (as you write) was inspired to a certain extent by your recidivism. You continue (it's hard even to write about it!) to compete, to rival . . . With whom? She is nobody to me. You are everything to me. No need, Nata, no need, no need, I beg of you."

He reassures Natalia that in the diary he has moved on from the question of what actually transpired back in 1918, yet on the following day, July 20, he telephones her from the nearby town of Pachuca and unburdens himself. The scene is utterly preposterous: the hero of Red October shouting at his wife through an uncooperative Mexican telephone about an imagined infidelity of two decades earlier. Natalia protested that it was all in the past, and Trotsky shot back: "The past is the present."

The conversation left Natalia feeling drained and depressed. The next day, she recorded her state of mind in a missive that is part letter, part diary. "Lvionochek doesn't trust me; has lost faith in me," she writes. She was fond of addressing her husband privately as "Lvionochek," a diminutive of Lev that translates as "My Little Lion." "Again he is racked by suspicion, jealousy and speaks about it on the telephone. He tells me to be calm, but how can I be calm when he, Lvionochek, raves, seethes; how can I be calm? . . . It's your pride. I'm stopping to cry."

Once again, Natalia lays out the case for her innocence in 1918, but then subtly, perhaps consciously, shifts the focus to what may have been the real source of her husband's self-torment. She describes their marriage in terms that sound like a rationale for the infidelity she never committed. Her background in art history had helped her land an important position in the upper reaches of the Soviet establishment, but she had none of her husband's experience and confidence, and she looked to him for support. "I remember whenever I wanted to tell you something about my work, something related to personal relations, about some success or failure, looking for your sympathy, or approval, or advice—you avoided me, turned me away, sometimes softly, more often brusquely." Natalia had initially written "sometimes brusquely," but must have decided that under the circumstances honesty ought to trump delicacy.

She won a small victory, she recalls, when her husband agreed to read a report she had written for the Central Committee and then pronounced it "very well written." "This was a moment of great happiness for me. I had wanted to ask you to read it before I sent it, but I couldn't find the right moment, you were always busy." In those days they saw each other fleetingly, usually at lunch and dinner. Risking the disapproval of her colleagues, she would sometimes skip evening meetings in the hope of spending time with her husband. "But usually you arrived home after I had already gone to bed."

Then, without a pause, her tone brightens. "I remember your morning moods. How cheerfully you rose from bed, how quickly you dressed, called the car and with a passing gesture or word encouraged me and . . . Seryozha, who was gloomily getting dressed. How vividly I remember you as such, dear, good, I want to hug you strongly. I would hurry to catch you and together with you drive to work."

Now husband and wife spent almost every day in each other's company and retired together almost every night. Yet, in this moment of despair, Natalia remarks: "Everyone is, in essence, terribly alone." Trotsky underlined these words in red crayon. "This sentence stabbed me in the heart," he told her, repeating the line, as though in disbelief that it came from his wife. Her words seem to have aggravated his growing sense of vulnerability in the face of advancing age, a thread that runs through their correspondence. Lately he has been repeatedly surprised at how old age has crept up on him. A solitary entry in the diary he kept in France in April 1935 reads: "Old age is the most unexpected of all the things that happen to a person." On this the partners were in harmony and offered each other support and encouragement.

Natalia sounds more philosophical about it: "I saw myself in the mirror at Rita's," she tells her husband, "and found I look much older. Our inner state has an enormous importance in old age; it makes us look younger; it makes us look older." Trotsky's response to this observation returned him to their courtship in Paris in 1902: "This was the case with you also when you were young. On the day after our first night together you were very sad and looked ten years older. In happy times you looked like a fawn. You've retained this ability to change all your life."

Confronted with the inevitable, Trotsky strikes defiant poses, twice punctuating one of his missives with the affirmation, "We will live on, Natasha!" In a letter dated July 19, defiance veers toward bravado: "Since I arrived here, not once has my poor cock stood up straight. It's as though it doesn't exist. It is also resting from the stresses of these days. But in spite of it, I myself am thinking tenderly of your old, dear cunt. I want to suck on it, shove my tongue all the way inside it. Natalochka, my dear, I will ever more strongly fuck you with my tongue and with my cock. Forgive me, Natalochka, these lines, it seems it's the first time in my life that I write to you like this."

For Natalia, of course, there was nothing to forgive in this unexpected outburst from "my intimate, old lover." In fact this was just what she needed to hear.

On the following day, Trotsky sets off on a strenuous horseback ride with Casas and Sixto, both veteran cavalrymen. He returns invigorated and writes Natalia an erotic little dissertation on the benefits of a good hard ride. "The shaking is excellent for the organism," he explains. This reminds him of something Leo Tolstoy's wife once wrote in her diary, recounting how her husband, at age seventy, returned from riding full of lust and passion. Trotsky reasons that if Tolstoy at seventy was able to ride and to make love, then the riding itself would have been enough to arouse his passion: "aside from the general shaking, there's the specific friction. . . . A woman who rides a horse like a man, should, in my opinion, experience complete satisfaction."

One week later, Trotsky returned, by automobile, to Natalia in Coyoacán. She did not record whether he behaved like Tolstoy, but it was now clear to Van and the rest of the household that she no longer had to worry about her competitor. "A distance had been established between Trotsky and Frida; the word 'love' was heard no more." Later Frida told Van that Trotsky had asked her to return the letters he had written her during their affair, warning, "They could fall into the hands of the G.P.U." She complied and Trotsky no doubt destroyed this evidence.

The numerous post-affair photographs taken of Trotsky in the company of Frida invite speculation as to what each was thinking as they took part, *en famille*, in picnics, sightseeing, and other excursions. In these images, Natalia is frequently positioned alongside Frida, and each woman seems uncomfortable in the other's presence. Natalia tended to blow hot and cold in her dealings with members of the household, and her behavior toward Frida was no different. On some occasions when Diego and Frida visited the Blue House, Natalia would greet Frida with flowers and warm affection, while at other times she would not come out of her room. Like most everyone else, Frida found Natalia's bursts of affection to be as oppressive as her cold shoulder.

Natalia did emerge from her room on November 7, 1937, her husband's fifty-eighth birthday and the twentieth anniversary of the October Revolution, when Frida presented Trotsky with a self-portrait dedicated to him. The Frida in this painting looks more like a conventional Southern belle, not the folkloric Tehuana Frida one expects. She is tastefully dressed and all done up, with red cheeks, lips, and fingernails, and with a purple carnation and a red ribbon in her hair. She stands between two curtains holding a small bouquet of flowers and a sheet of paper inscribed with the words, "For Leon Trotsky with all love I dedicate this painting on the 7th of November, 1937. Frida Kahlo in San Angel, Mexico." Trotsky hung the painting on a wall in his study in the Blue House.

This proved to be a breakthrough period in Frida's career as an artist. She became more prolific than she had been since her marriage to Diego eight years earlier. She had her first solo gallery exhibition in New York, and soon afterward her paintings were featured in a Paris exhibition devoted to Mexican art and culture. She was emerging from the shadow of her husband. Perhaps this is what she had in mind when she wrote to a friend that Trotsky's coming to Mexico was the best thing that had ever happened in her life.

For Trotsky, Frida was a storm that had passed. He and Natalia were reconciled. It was time to get back to work. But the Little Lion would make one last attempt at escape. The opportunity arose later in the year, when the Mexican Communists launched another of their noisy campaigns of abuse and threats against Trotsky, giving rise to concerns about his safety. Trotsky believed that under the cover of a protest demonstration, the Stalinists might try to storm the Blue House and assassinate him. This was a scenario that Van deemed entirely plausible, and he found Trotsky's plan of escape in such an event to be a clever one. A ladder would be kept in a corner of the second, smaller patio, along the quiet, dimly lit Avenida Berlin. "In case of an attack, Trotsky would put the ladder against the wall, cross over alone and unseen, and quickly walk to the house of a young Mexican woman whom we knew, to take refuge there."

The young woman in question was Frida's sister, Cristina. Before a rehearsal could be arranged, however, she approached Van and explained that during the previous several months Trotsky had on four or five occasions directly and insistently propositioned her. She had managed to deflect these unwanted advances without raising a fuss. She also told Van that Trotsky had divulged to her the escape plan and the anticipated rehearsal. Van was angry that Trotsky would risk compromising his security for a sexual liaison, but he said nothing. There turned out to be no need, because Trotsky stopped pushing for a rehearsal, possibly sensing that its true purpose had been discovered. Yet how many times must the Old Man have raised the ladder and rehearsed the escape plan in his mind.

CHAPTER 4

Day of the Dead

T rotsky turned fifty-eight on November 7, 1937. His birthday co-
incided with the twentieth anniversary of the Bolshevik seizure
of power in Petrograd—known as the Great October Socialist
Revolution because of where the date fell on the old-style cal-
endar used in Russia in 1917. This was Trotsky's first November 7 in
Mexico, and thanks to Diego Rivera and Frida Kahlo, the Blue House
was the scene of a fiesta.

Diego, Frida, her sister Cristina, Antonio Hidalgo, who was Trotsky's
go-between with the Mexican government, and other friends arrived
at the house long before dawn. In fact it was near midnight, according
to Trotsky's American secretary, Joe Hansen, "in the inimitable style of
Mexico." Diego and Frida brought along an enormous collection of
red carnations and they immediately set to work snipping off the flowers and arraying them on the white cloth that covered the long, broad
dining room table. Diego, the master of the large canvas, took the lead
in arranging and rearranging the flowers, before settling on a deco-
rative salute to Trotsky and the Fourth International: "ARRIBA LA
CUARTA INTERNACIONAL/VIVA TROTSKY." As a centerpiece
Diego had ordered an enormous cake covered in red icing and ringed
with candles encircling a hammer and sickle.

Toward 7:30 a.m., two orchestras hired by Diego crowded into the
patio and began serenading outside the French doors to Trotsky and
Natalia's bedroom: sonorous voices accompanied by marimba, accor-
dion, bass viol, and guitars. An hour later, in the bright and warming

winter sunshine, comrades began arriving, most bearing red flowers of one or another variety. The celebrants, about sixty in all, were mostly workers and schoolteachers, the two groups that made up the large majority of Trotsky's Mexican followers. Most were dressed simply, the men in white cotton pants and shirts, the women in the ample dark skirts fashionable at the time. Huarache sandals clustered beside primitive idols, squat rock sculptures engraved with the faces of the gods, each deity surrounded by ferns and cacti. An assortment of foods appeared, Hansen reports, "including a few live chickens, their feet tied. Several five-gallon cans, converted into household utensils, were brought in filled with *atole*, a thin chocolate-flavored gruel which we sipped out of cups."

Atole, a traditional hot beverage with a cornstarch base, is a particular favorite during Mexico's most important fiesta, Day of the Dead, a combination of All Saints' Day and All Souls' Day, November 1 and 2. Despite its name, *Día de los Muertos* is a joyous occasion, a celebration of the lives of the deceased, whose graves become the settings for lively family gatherings. Hansen described it for his wife back in Salt Lake City as a cross between Halloween and Decoration Day, the original name for Memorial Day. On November 2, the cemeteries were crowded with the friends and relatives of the dead, along with vendors selling tacos, candied plums, and cakes, as well as balloons and tinsel propellers. Among the most popular decorations and presents were full-length skeletons, skeleton masks, candy coffins, and sugar skulls with the name of the deceased spelled out across the forehead. Back in those less regimented times, the Day of the Dead usually lasted for an entire week.

All of which means that the revelers who gathered at the Blue House on that November 7 morning were in good practice. At a certain point in the festivities, calls were heard for the guest of honor to make a speech. This was inevitable, of course, yet Hansen noticed that Trotsky seemed hesitant, attributing his reluctance to his limited Spanish, though he could have added that by 1937 the Old Man was seriously out of practice. "He appeared to brace himself, as if he were taking a deep breath." Twenty years earlier, he lit the world on fire as the great orator of the Russian Revolution, racing from one audience to the next, stoking the passions of Petrograd's workers, soldiers, and sailors.

"LIFE WAS A whirl of mass meetings," Trotsky wrote in his auto-biography about his return from exile to the Russian capital in May 1917, ten weeks after the fall of the Romanovs. "Meetings were held in factories, schools, and colleges, in theaters, circuses, streets, and squares." Expand this list to include the Baltic shipyards and various army bar-racks, and one begins to understand why even anti-Bolshevik accounts of 1917 give the impression of Trotsky as a man in perpetual motion. An eyewitness who belonged to one of the political parties vanquished by the Bolsheviks testified that Trotsky "seemed to be speaking simul-taneously in all places. Every Petrograd worker and soldier knew him and heard him personally. His influence, both on the masses and at headquarters, was overwhelming."

One electrifying performance led directly to another and then an-other. "Each time it would seem to me as if I could never get through this new meeting," Trotsky recalled, "but some hidden reserve of ner-vous energy would come to the surface, and I would speak for an hour, sometimes two, while delegations from other plants or districts, sur-rounding me in a close ring, would tell me that thousands of workers in three or perhaps five different places had been waiting for me for hours on end. How patiently that awakening mass was waiting for the new word in those days!"

Trotsky's favorite venue was the *Cirque Moderne*, across the Neva River from the Winter Palace, a dingy hall that became known to his al-lies and enemies alike as his "fortress." In *Ten Days That Shook the World*, John Reed recounts how every night this "bare, gloomy amphitheatre, lit by five tiny lights hanging from a thin wire, was packed from the ring up the steep sweep of grimy benches to the very roof—soldiers, sailors, workmen, women, all listening as if their lives depended upon it." Trotsky appeared there in the evening, sometimes when the hour was late. On each occasion, he remembered, the place was a human tinderbox: "Every square inch was filled, every human body compressed to its limit. Young boys sat on their fathers' shoulders; infants were at their mothers' breasts. No one smoked. The balconies threatened to fall under the excessive weight of human bodies. I made my way to the platform through a narrow human trench, sometimes I was borne overhead. The air, intense with breathing and waiting, fairly exploded with shouts and with passionate yells peculiar to the *Cirque Moderne*."

Landing in the ring, he somehow manages to subdue his audience. The clamor subsiding, he begins to cast his spell. "Above and around me was a press of elbows, chests, and heads. I spoke from out of a warm cavern of human bodies; whenever I stretched out my hands I would touch someone, and a grateful movement in response would give me to understand that I was not to worry about it, not to break off my speech, but keep on." In Trotsky's recollection, faces in the audience theatricalize their emotions, like overwrought actors in a Sergei Eisenstein film montage. Nursing mothers are no less animated, while their passive sucklings serve as conspicuous symbolism. "The infants were peacefully sucking the breasts from which approving or threatening shouts were coming. The whole crowd was like that, like infants clinging with their dry lips to the nipples of the revolution."

As the performance continues, the speaker becomes one with his audience and begins to channel its emotions. Planned remarks give way, as "other words, other arguments, utterly unexpected by the orator but needed by these people, would emerge in full array from my subconsciousness." Trotsky is reliving an out-of-body experience. "On such occasions I felt as if I were listening to the speaker from the outside, trying to keep pace with his ideas, afraid, like a sleepwalker, he might fall off the edge of the roof at the sound of my conscious reasoning."

Having worked his audience into a frenzy, it was now time to seize the moment and administer a revolutionary oath. "If you support our policy to bring the revolution to victory," he exhorted an audience three days before the Bolshevik coup, "if you give the cause all your strength, if you support the Petrograd Soviet in this great cause without hesitation, then let us all swear our allegiance to the revolution. If you support this sacred oath which we are making, raise your hands." Thousands of hands shot up in response.

British intelligence agent Bruce Lockhart, who witnessed Trotsky's heroics in these revolutionary days, wrote in his diary: "He strikes me as a man who would willingly die fighting for Russia provided there was a big enough audience to see him do it."

The oath having been administered, it was time for Trotsky to depart the *Cirque Moderne*. Leaving on foot was out of the question: no passageway could be sliced through that human mass, united in its fervor and in no mood to disperse and go home. There was only one way out: "In a semiconsciousness of exhaustion, I had to float on countless arms above the heads of the people in order to reach the exit." Bobbing along a sea of heads, he would sometimes catch sight of his two teenage daughters from his first marriage, Zina and Nina. "I would barely manage to beckon to them, in answer to their excited glances, or to press their warm hands on the way out, before the crowd would separate us again. When I found myself outside the gate, the *Cirque* followed me. The street became alive with shouts and the tramping of feet. Then some gate would open, suck me in, and close after me."

Now, TWENTY YEARS later, in the bright morning sunshine of Mexico City, the spellbinder of the Russian Revolution was wide awake. As Hansen looked on with anticipation, Trotsky managed to find his voice: "He stepped forward to the balustrade; and he was transformed. He took complete possession and spoke out as if this were completely natural and something he did every day. He pitched his voice so that it soared somewhat and could be heard with complete ease."

For Hansen, the twenty-seven-year-old native of Utah and the truest of the true believers at Trotsky's side in his Mexico years, the moment inspired an epiphany: "It was impossible to think of the great past, the tradition beginning with Marx, continued by Engels, then Lenin, now Trotsky . . . his exile and the genuine appreciation for all this expressed by these comrades without feeling sharp emotion . . . that despite the vicissitudes which revolutionaries as individuals experience, that despite the ingratitude of progress to those who have furthered it, we are part of a great and swelling stream and that we shall inevitably conquer."

One week earlier, on October 29, 1937, Trotsky and Natalia's younger son, Seryozha, was executed in a Soviet prison cell, dispatched with a bullet to the base of the skull. His parents never learned his fate, though they imagined it many times. Seryozha joined the swelling stream of victims that by now included almost all of Trotsky's Bolshevik comrades. They had been shot or were in exile or in the camps, caught up in the blood and fury of the purge and the terror. Both of Trotsky's daughters were dead, victimized by their association with their outcast father; their mother had disappeared into the gulag. Of Trotsky's children, the only one left was Lyova, his favorite and his right-hand man—and next to his father, the chief target of Stalin's secret police.

Just as Trotsky is recognized as the great orator of the Bolshevik Revolution, Diego Rivera was at one time regarded as its outstanding painter. Trotsky once called him October's "greatest interpreter"—though nowadays few would think to bestow this honor on him. Rivera "remained Mexican in the most profound fibers of his genius," Trotsky recognized. "But that which inspired him in these magnificent frescoes, which lifted him above the artistic tradition, above contemporary art, in a certain sense above himself, is the mighty blast of the proletarian revolution. Without October, his power of creative penetration into the epic of work, oppression and insurrection would never have attained such breadth and profundity."

When the Russian autocracy collapsed in 1917, Mexico was in the midst of its tumultuous revolutionary decade, from 1910 to 1920. Its starting point was an uprising against dictator Porfirio Díaz, which ushered in a series of bloody, protracted, and overlapping civil wars led by flamboyant agrarian revolutionaries like Emiliano Zapata and Pancho Villa. The cataclysm left more than a million Mexicans dead. Rivera was absent during Mexico's time of troubles, having left in 1907 at age twenty to go paint in Spain. He later settled in Paris, the art world's metropolis, where under the influence of Picasso, Braque, and Juan Gris he gained a reputation as a credible practitioner of Cubism, then modern art's cutting edge. Yet Rivera knew that his Cubism was essentially imitative, and he felt increasingly constrained by the abstract form, which frustrated his growing desire to express the political and social ideas that had begun to preoccupy him. These he absorbed in the cafés of bohemian Montparnasse, on the left bank of the Seine, where émigrés from Mexico and Russia told firsthand accounts of the epoch-making events under way in their countries.

Mexico's revolution drew to a close in 1920, when the constitutionalist army general Alvaro Obregón was elected president. The multiple civil wars had shattered Mexico's sense of nationhood, and President Obregón and his minister of education, José Vasconcelos, both radical nationalists, sought to unite the country by tapping into the so-called Mexican Renaissance, the movement for cultural renewal that had begun before the revolution. The goal was to forge a fatherland by making Mexicans aware of their common cultural heritage, and public works of art were to play a vital role in this enterprise. Vasconcelos would invite the country's premier painters to use the vast walls of Mexico's government buildings as their canvases.

In Paris, Diego Rivera was eager to answer the call. Before his return to Mexico, and at the urging of Vasconcelos, he spent seventeen months in Italy studying Renaissance frescoes in search of the formula for a genuinely popular art. He arrived in Mexico in 1921 and the muralist movement got under way in earnest the following year. Three major figures would emerge: Rivera, José Clemente Orozco, and David Alfaro Siqueiros, each with his own distinct style, yet like all the muralists, each favoring indigenous over European artistic influences and tending to idealize Mexico's preconquest heritage.

Rivera's artistic breakthrough—the moment he discovered his own style—came in 1923, when he began to paint frescoes at the Ministry of Education building, a recently constructed stone and cement edifice two city blocks long and one block wide done in the style of Spanish convent architecture. In its great inner courtyard, lined with arches on all of its three floors, Rivera painted scenes of Mexico's land and people, labor and festivals. The result was 235 individual fresco panels covering an area of 15,000 square feet. The project would take five years to complete, but already in 1923 the frescoes were a sensation and brought Rivera and the Mexican art movement international renown.

*Diego Rivera,
Mexico, 1920s.*

Planted on the scaffold from dawn to dark, Rivera drew crowds of onlookers. One who had the opportunity to join him up there was Bertram Wolfe, the Brooklyn-born American Communist who became Rivera's comrade, friend, and biographer. Wolfe observed "a bulky, genial, slow-moving, frog-faced man, in weather-worn overalls, huge Stetson hat, cartridge belt, large pistol, vast paint-and-plaster-stained shoes." The frog image recurs. "Frog-toad" was one of Frida's affectionate nicknames for her husband. Wolfe says Diego's eyes "bulged like those of a frog or a housefly, as if made to see a whole crowd, a vast panorama, or a wide mural." In fact, though, they probably bulged as a result of a thyroid condition, which ailed him in later years.

Mexico's painters took the lead among the radical intellectuals who wanted to continue the revolutionary struggle for economic justice, freedom, and democracy, including the liberation of art. In 1922 they formed a union: the Syndicate of Technical Workers, Painters, and Sculptors, as it came to be called. This short-lived and ineffectual body issued a combative founding manifesto, written by Siqueiros, declaring the sympathy of the unionized artists for the oppressed masses and repudiating "so-called easel art and all such art which springs from ultra-intellectual circles, for it is essentially aristocratic. We hail the monumental expression of art because such art is public property." Rivera, Siqueiros, and Xavier Guerrero were elected to the syndicate's executive committee and were coeditors of its newspaper, *El Machete*, whose name reflects the essentially agrarian outlook of Mexico's revolutionaries.

It was at this same time that Rivera joined the Mexican Communist Party, which had been formed by aspiring politicians in 1917 under the influence of events in Russia, but which had since evolved into a party of radical painters with only a few dozen members. Inevitably, Rivera was one of its leading figures. Wolfe, whose radicalism had caused him to flee the United States for Mexico in order to avoid arrest, also joined the party at this time. It did not take him long to figure out that while Rivera was a great painter, in politics he would never be more than "an amateur and a passionate dilettante," and an impulsive one at that. Diego's ideology, Wolfe discovered, was "an undigested mixture of Spanish anarchism, Russian terrorism, Soviet Marxism-Leninism, Mexican agrarianism—the redemption of the poor peasant and the Indian." Nor was the painter sufficiently read in the Marxist classics to pass muster as a credible communist: "All that Diego ever knew of Marx's writings or of Lenin's, as I had ample occasion to verify, was a little handful of commonplace slogans which had attained wide currency."

Rivera made one extended visit to Soviet Russia, beginning in November 1927. He accepted an invitation to attend the tenth-anniversary celebrations of the October Revolution. In Red Square on November 7, Rivera sat on the reviewing stand alongside Lenin's mausoleum beneath the Kremlin wall, observing the daylong passing parade of Red Army soldiers, factory workers, Communist youth, and countless others through the square. Never a mere onlooker, Rivera had by day's end produced forty-five watercolor sketches and filled dozens of small notebook pages with penciled notes and sketches—raw materials for the Russian murals he hoped to paint.

Away from Red Square and undetected by Rivera's panoramic vision, Trotsky and the Left Opposition attempted to stage their own anniversary demonstrations, which were broken up by the GPU. Stalin's thugs stoned Trotsky's automobile as it traveled along a Moscow street, shattering its windows as he ducked down in the backseat. That morning, Stalin had sat down in the Kremlin with Sergei Eisenstein to supervise the editing of his new feature film, *October*, with the result that Lenin's role was considerably reduced, while Trotsky was cut out altogether.

Rivera remained in Moscow for five months. Invited to address public audiences about his art, he used these occasions to encourage Soviet artists to draw inspiration from Russian folk art. "Look at your icon painters," he exhorted them. In his own country such advice would have seemed superfluous; in the birthplace of socialism it had the ring of heresy. Voices from the audience objected that he was glorifying icons and churches and endorsing backward peasant handicraft while minimizing the importance of industry and of economic planning.

Nor was Rivera's artistic vision warmly endorsed by the Soviet establishment. The People's Commissar of Enlightenment had commissioned him to paint a fresco in Moscow's Red Army Club, but the enterprise was undermined by shortages of materials and assistants and by constant delays. He was continually told that something had to be put off until tomorrow—"Zavtra budet"—a Russian variation of the Mexican *mañana*, but in Moscow the sabotage felt deliberate. Perhaps his mood was darkened by his one meeting with Stalin, who sat for the painter. "Judging from the sketch he did," says Wolfe, "he does not seem to have been impressed by his subject." Disillusioned, he stayed long enough to sketch the May Day parade through Red Square, then quietly left the country.

Upon his return to Mexico, Rivera found himself increasingly under fire from fellow Communists and artists, who accused him of political fraud. These attacks intensified as he began to paint what would become one of his most celebrated murals, *The History of Mexico*, in the

National Palace, the seat of the government. This new commission was cited by the Mexican Communist Party as fresh evidence that Rivera was a "millionaire artist for the establishment," a charge that had dogged him throughout the twenties. He had also been criticized by fellow Communists and artists as a "painter for millionaires" because of his private commissions. What really mattered in 1929, however, was the Party line laid down in Moscow, where Stalin, having eliminated Trotsky and the Left Opposition, had turned his sights on Nikolai Bukharin and the right-wing Bolsheviks. The Kremlin now directed the member organizations of the Communist International to unmask the "Right Danger," and in Mexico the Communists decided that Rivera fit the description and expelled him from the party. Appalled at the idea of being classified as a right-wing anything, Rivera declared himself instead to be a Trotskyist. He had little to back up this claim, but it hardly mattered. Rivera was now an ex-Communist.

These political machinations barely registered north of the Rio Grande, where Rivera's reputation as a fashionably radical painter continued to rise. His acceptance of a mural commission in San Francisco inaugurated a three-year sojourn in the United States, the high point of which was a popular one-man show at the new Museum of Modern Art in New York City in December 1931: a retrospective of his drawings and oils, together with movable fresco panels he painted especially for the exhibit. Rivera's exhibit was the second such event at the museum, Matisse being the other artist to have been so honored.

Rivera arrived in San Francisco in November 1930, just as the Great Depression set in. There he completed murals at the Stock Exchange Tower and the California School of Fine Arts—now the San Francisco Art Institute—in 1931. The following year he painted his *Detroit Industry* frescoes at the Detroit Institute of Arts with funds donated by Edsel Ford, president of the Ford Motor Company. These works added luster to Rivera's reputation and they survive as part of his oeuvre. Not so his next project, the mural he was commissioned to paint in the lobby of the new RCA building in New York City's Rockefeller Center in 1933.

The Rockefeller family asked Rivera to produce a work based on the uplifting theme "Man at the Crossroads Looking with Hope and

High Vision to the Choosing of a New and Better Future." His elaborate written proposal for the painting, which reads like an encomium to socialist revolution, leaves little doubt that he saw the Radio City mural as an opportunity to answer his critics on the left by demonstrating his unimpeachable Bolshevik credentials. The plan was approved by the family and would likely have been executed without incident had Rivera not departed from his design by replacing the anonymous face of a prominently placed worker-leader with that of Lenin. The discovery, before the mural was completed, of Lenin's iconic countenance touched off a public controversy, prompting Nelson Rockefeller to request its removal, followed by the artist's refusal, and then an order to suspend work. That was in May 1933. Nine months later, contrary to the reassurances of his patrons, Rivera's unfinished mural was sandblasted into oblivion.

It was in the immediate aftermath of this Battle of Rockefeller Center that Trotsky first established contact with Rivera, in the form of a brief appreciative letter sent from Turkey, dated June 7, 1933. It is curious that this message fails to mention the standoff with the Rockefellers, which must have inspired it. Trotsky expressed his admiration for Rivera's art, which he said he first came upon inadvertently in an American book during his exile in Central Asia in 1928. From the sound of it, he had chanced upon reproductions of the Education Ministry murals. "Your frescoes struck me with their combination of masculinity and gentleness, almost tenderness, their internal dynamic and tranquil equilibrium of form." And such a "magnificent freshness in approach to man and animal!" This generous praise is followed by what would prove to be a telling admission: "I was infinitely far from the thought that the author of these works was a revolutionary, standing under the banner of Marx and Lenin. Only relatively recently did I discover that the master Diego Rivera and that other Diego-Rivera, the close friend of the Left Opposition, were one and the same person." Rivera may have considered this a mixed review, but Trotsky was honored to have him as a comrade. "I am not losing the hope to visit America, see your works in the original, and talk to you in person."

Rivera was paid in full for his unfinished Radio City fresco, and before leaving the United States he was determined to use his Rockefeller earnings—"the money extorted from the workers by the Rockefeller exploiters," as he called it—to paint "the revolution." At the moment Trotsky's letter arrived in New York, he was at work on the next of his epic historical murals, *Portrait of America*, in the New Workers' School on West 14th Street in lower Manhattan. Here Trotsky makes his first appearances in Rivera's art. In the fresco panel titled "World War," Rivera included a passage on the Russian Revolution, represented by a winter scene on Red Square, where Trotsky, his right hand pressed to his forehead in a salute, his left hand clenched in a fist and raised above his head, reviews passing Red Army troops under the banner of the Third International, while behind him Lenin looks on approvingly. Another panel, "Proletarian Unity," re-creates at its center the Lenin of the Radio City mural, here flanked by other revolutionary figures, including the hirsute Marx and Engels, a sinister Stalin lurking in the upper left corner, a frightfully cherubic Bukharin, and a rather harmless-looking Trotsky, with clenched fist raised above a soft-serve hairdo.

Rivera then exhausted his Rockefeller funds at the headquarters of the American Trotskyists, where he painted two minor fresco panels on the Russian Revolution and the Fourth International, which Trotsky had recently begun to proclaim.

After his return to Mexico at the end of 1933, Rivera was able to re-create the Rockefeller Center mural in Mexico City's Palace of Fine Arts. In this second version, the objectionable Lenin returns, now accompanied by Marx and by Trotsky, who helps hold up a banner inscribed with the slogan, Workers of the World/Unite in the IVth International! Like nearly all of Rivera's depictions of communism's high priests, prophets, and contenders, these effigies are unconvincing, "lifeless faces, clichés not men," to quote Wolfe, who observed generally that "Diego's art was poorly served by his attempts at propaganda." Aesthetically, his smokestack-and-tractor utopias cannot compete with his idealized narratives of Mexico's past and present. He was a populist and nationalist. Zapata, not Lenin, was his revolutionary ideologist and hero. His true subject was the Mexican land and people. Labor, in Rivera's art, is not an object of exploitation, as orthodox Marxism instructed, but a "rhythmic dance," in Wolfe's phrase. Industry is not the scene of class struggle, but of intricately beautiful machines, affectionately and, in the Detroit murals, erotically rendered. It is hardly surprising that Trotsky's first impression put him "infinitely far from the thought" that the painter was a Marxist revolutionary.

Having re-created *Man at the Crossroads* in Mexico City, Rivera resumed work on his National Palace mural, completing this and one other mural by the end of 1935, at which point he entered into a kind of exile from the muralist movement that had been inseparable from his name. Nine years would pass before he would be offered another government wall in Mexico. His sense of isolation was magnified by his continued ostracism at the hands of former comrades, led by muralist Siqueiros, a master of political vitriol. No wonder Rivera was increasingly attracted to the figure of Trotsky, a fellow exile, a heroic and tragic figure, and a man with an enlightened view of the arts.

In the autumn of 1936, when Trotsky was imprisoned in Norway as a result of Soviet diplomatic pressure in the wake of the first Moscow trial, and when no other European country would allow him entry, Rivera agreed to approach President Lázaro Cárdenas and ask him to offer Trotsky asylum in Mexico. When Cárdenas granted the request, the Communist attacks on Rivera became an onslaught. Now he really was a Trotskyist.

Diego Rivera's frescoes are far more impressive viewed in the original than as reproductions on the page—a claim not all his fellow muralists could make—and thus Trotsky's estimation of his art was bound to rise accordingly as he was introduced to the murals in Mexico City and Cuernavaca in the beginning of 1937. The sheer physical scale of the murals and their grand narrative sweep could not but inspire awe. Here was class-conscious art accessible to the masses and on permanent public display. "Do you wish to see with your own eyes the hidden springs of the social revolution?" Trotsky wrote in the summer of 1938. "Look at the frescoes of Rivera. Do you wish to know what revolutionary art is like? Look at the frescoes of Rivera."

Trotsky was no less fascinated by the artist himself, his imposing physical presence and outsize personality. He admired the passion and devotion Rivera brought to his work. Naturally, he felt a strong sense of solidarity with him for the unceasing slander campaign he had to endure as a so-called Trotskyist painter. And of course he was delighted to have a great artist associated with the Fourth International.

Trotsky's secretaries shared the Old Man's enthusiasm for Diego and his art, but it was not long before they began to take a more skeptical view of his professed Trotskyism, and even of his Marxism. Diego was a free spirit, and he made no attempt to hide this fact from them. Perhaps he enjoyed scandalizing them. "You know, I am a bit of an anarchist," he liked to say to Van, and he would tell stories about how the political repressions in the Soviet Union had begun before Stalin, in the time of Lenin. "He would say nothing of this, however, to Trotsky, to whom he showed another face." Van recalled that on one occasion when he allowed his skepticism about the painter's politics to peek through, Trotsky reproached him for it: "Diego, you know, is a revolutionary!"

Trotsky and Diego were drawn together by circumstances beyond mutual respect, fascination, and dependence. Diego was fifty years old, only seven years younger than Trotsky, who in his exile years was surrounded mostly by much younger acolytes. Both men were world-famous and they thus experienced that unspoken bond that exists between celebrities. Both were well traveled and had overlapped for several years in Paris, where Diego fraternized with Russian émigrés and even took a Russian as his common-law wife. Among his friends was the Russian writer Ilya Ehrenburg, who based the title character of his 1922 satirical novel, *The Extraordinary Adventures of Julio Jurenito and His Disciples*, on Rivera and his anarchic ideas and temperament. Trotsky and Diego addressed each other in French and in English, and their conversation was enriched by the occasional Russian word or expression.

All of this helps explain why these two apparently mismatched individuals were able to establish a special connection. "Of all the persons whom I knew around Trotsky from 1932 to 1939," Van testifies, "Rivera was the one with whom Trotsky conversed with most warmth and unconstraint. There were indeed limits that could never be crossed in conversation with Trotsky; but his meetings with Rivera had an air of confidence, a naturalness, an ease, that I never saw with anybody else." Trotsky even tolerated Diego's penchant for telling risqué jokes in French, even though they made him squirm.

Diego's friends and acquaintances understood that his artistry extended to the telling of tall tales. Trotsky knew how to tell a story well, how to make the truth compelling—witness his inspired recollection of the *Cirque Moderne*. Diego's imagination, however, often became untethered to any reality. Stuffed with elaborate supporting details, his tales were as wide as they were tall. Often they related to his personal background and experiences. He was of Spanish-Indian and Portuguese-Jewish descent, but apparently this was not exotic enough. He once declared to an audience in Mexico City that his ancestry was "Spanish, Dutch, Portuguese, Italian, Russian and—I am proud to say—Jewish." He told a reporter that he was "three-eighths Jewish." He also claimed that his great-grandmother was Chinese. On one occasion he boasted that his mother "passed on to me the traits of three races: white, red, and black."

Frida used hand signals to alert potential victims when Diego was embroidering or inventing out of whole cloth. She defended her husband's fabrications as products of his "tremendous imagination"; besides, "I have never heard him tell a single lie that was stupid or banal." Trotsky, who was disposed to categorize individuals, was prepared to make generous allowances for the demands of "artistic temperament," but he should have realized early on that there was no one like Diego Rivera.

Diego's precarious politics and infinite imagination aside, those who observed their interactions up close must have wondered whether the two contrasting friends were bound to clash: the rigid, prickly, angular Trotsky and the reckless, riotous, gargantuan Diego. The lion and the elephant. Diego bathed irregularly, dressed carelessly, and seldom arrived on time for anything. Trotsky, meanwhile, was a stickler for neatness, regimen, and routine. Both men had tremendous work ethics, but Diego's self-discipline was restricted almost entirely to his painting. And with brush in hand, he tended to lose track of everything else.

Frida, whose supply of patience had to have been titanic, called her husband the "enemy of clocks and calendars." She once painted him as an overgrown infant in her arms, which must give some indication of what it was like to be married to the man. A British economist, upon meeting Diego for the first time in Mexico in 1938, was struck by the contrast between "the considerable subtlety that his work often displays" and the "childlike simplicity, friendliness, and frankness" of his personality. He loved to make mischief and may have enjoyed flaunting his artistic temperament in Trotsky's presence, as when he greeted him at home in San Angel with a parrot on his head. James Farrell observed of Trotsky that there was an "exactness" about him: "There was not, however, much spontaneity in him—or, rather, his spontaneity was kept in check."

Estimations of Trotsky's personality tend to shade into explanations for his political downfall. His rigidity is seen as of a piece with his haughtiness, which constrained him as a politician. He could inspire the masses, but once Bolshevik rule had been consolidated and the masses had been taken out of the equation, he lacked the personal qualities necessary to organize and lead a political faction in the struggle against Stalin following Lenin's death. The odds were in any case heavily stacked against him, not least because of his late entry into the Bolshevik Party, the extensive paper trail of his often corrosive polemics with Lenin before 1917, his ethnicity, and the vastly superior political instincts and unmatched ruthlessness of his opponent. Hubris alone, therefore, cannot explain Trotsky's downfall, but to the extent that the limitations of his personality are held responsible, it is seen as the tragic flaw.

Some of the most damning evidence comes from comrades who fought by his side, not least Natalia. She later recalled that when he was leader of the Left Opposition in the 1920s, his comrades repeatedly urged him to loosen up a bit, "lest he be thought haughty and arrogant"—which of course was already the case. He dreaded their social gatherings, which required him to engage in that unproductive exercise, small talk. "Nor did he care for the *double entendres*, touched with vulgarity, which were so freely bandied about." The chief perpetrator in those days was the puckish Bolshevik journalist and jester Karl Radek, who had a special talent for telling bawdy jokes but had to clam up whenever Trotsky approached. Natalia explains that "although Trotsky, too, had a sense of humor, it was of a different kind."

As she proceeds with her defense, Natalia begins to protest too much. "The fact was that he used the familiar form of address to hardly anyone, that we neither made nor received visits—in the first place we had no time—and that he went to the theater only very occasionally." Trotsky's circle of friends was restricted to comrades dedicated entirely to the political struggle, she proudly asserts. "But he established the warmest relationships in spite of that." The examples she puts forward, however, describe not genuine friendships but the loyalties of younger comrades.

Natalia's apologetics only serve to reinforce the devastating portrait of her husband drawn by the American writer Max Eastman, who was Trotsky's biographer and the translator of his books into English. Eastman's judgment was that Trotsky lacked "the gift of personal friendship" and that this doomed him as a politician. "Aside from his quiet, thoughtful wife, toward whom his attitude was a model of sustained gallantry and inexhaustible consideration, he had, in my opinion, no real friends," Eastman observed. "He had followers and subalterns who adored him as a god, and to whom his coldness and unreasonable impatience and irascibility were a part of the picture. . . . But in a close and equal relation he managed to get almost everybody 'sore.' One after another, strong men would be drawn to him by his deeds and brilliant conscientious thinking. One after another they would drop away."

Eastman's Trotsky gives the impression of someone who has studied the handbook on how to conduct a friendship. "The part he played was that which a high idea of personal relations demanded of him, but since the whole feeling was not there he fell often and too easily out of the part." Farrell, whose acquaintance with Trotsky was comparatively brief and who, unlike Eastman, remained on good terms with him to the end, found this profile convincing. The man who could move the masses, Farrell agreed, had a proclivity for "seeing individuals as servants to an aim and an idea rather than as personalities in their own right."

Eastman describes the scene at an "anniversary smoker" in the Kremlin in the early twenties, where he discovered Trotsky, who was offended by tobacco smoke and rarely drank alcohol, wandering about "like a lost angel, faultlessly clad as always, with a brand-new shiny manuscript case under his arm, a benign sort of Y.M.C.A. secretary's smile put on for the festivities, but not an offhand word to say to anybody. It seems a funny epithet to use about a commander in chief, but he reminded me of Little Lord Fauntleroy."

Try as she might, Natalia cannot rescue Trotsky here. She recalls his attendance at a New Year's Eve party in 1926 given by his brother-in-law, Lev Kamenev, at his Kremlin apartment, directly above their own. Trotsky's purpose was not entirely pleasure, as he intended to use the occasion to gauge the mood of an oppositionist group from Leningrad. To Natalia's surprise, he returned almost immediately and in a foul mood. "I can't stand it," he fumed. "Liqueurs, long dresses and gossip! It was like a salon."

Trotsky's favorite distractions were hunting and fishing, the essential activities of his exercise regimen up until his departure from Turkey in 1933, after which—in France, Norway, and then Mexico—security concerns restricted his opportunities. Hunting and fishing were not Diego's pastimes, but like Trotsky, he enjoyed escaping Mexico City by automobile. Hansen's letters and notes record the excursions undertaken by Diego and the Old Man—usually identified as "the OM," or as "LD," for Lev Davidovich, his first name and patronymic, which is how the staff addressed him. On November 3, 1937, Hansen reports that, "Last Sunday we drove in the Dodge to the base of Iztaccihuatl, the volcano north of Popocatepetl, and ate some ham sandwiches and tacos with Diego and Frida and Hidalgo. It was a very pleasant drive and a relief for L.D." Later that month the destination is the Huasteca region, 150 miles to the southeast, where the OM spends "a few days relaxing and drinking mineral waters in the company of Diego."

A trip to Guadalajara in July 1938 offered varied diversions, including car trouble: two tires blew out, one when the car was traveling at a speed of 60 miles per hour. "The Old Man enjoyed especially pushing the car out of the mud holes. It made him feel twenty years younger, he declared, as it reminded him of the civil war days. He had not pushed a car out of a mud hole since that time." July falls within the rainy season in Mexico, which explains the ubiquitous presence of mud. "On occasion it was really funny, Diego half up to his knees in mud and so stuck he couldn't move himself without help, and the Old Man one time taking a header into the mud when the car gave a sudden lurch backward."

It begins to sound like the kind of material Hollywood turns into a buddy movie, yet tourism in the company of Diego required a certain amount of patience and stamina. He was never without a tiny scratch pad tucked in the palm of his hand, in which he was constantly sketching whatever came into view: faces, flora, churches, everything, turning over page after page. A typical distraction appeared during the expedition to the volcano in October 1937. "We came through a village where a funeral was in process—black coffin, hexagonal—fitted to the body—ordinary boards—so we had to stop while he sketched that."

These unscheduled stops could get on Trotsky's nerves, as revealed by Hansen's remark about a less enjoyable trip to Guadalajara: "Visiting churches (time *LD lost patience* over *Diego* & and his churches & buying in mkt place & we return alone)."

Trotsky understood that he could not afford to lose all patience with Diego, his sponsor and benefactor, the man who had secured him entry into Mexico and set him up in the Blue House rent-free. The painter also acted as guardian to the vulnerable exile. The first night of Trotsky's stay in Mexico, when security arrangements had to be improvised, Diego went home to retrieve a Thompson submachine gun from his arsenal. He had his own reasons for maintaining vigilance. Several weeks earlier, at the Restaurant Acapulco in Mexico City, four gunmen approached his table and started an argument that was intended to culminate in his assassination. Frida leaped up in front of her husband and made a scene, denouncing the pistoleros as cowards, and they retreated in confusion. Afterward she was sick to her stomach, but her quick action had saved her husband's life.

Diego would end up spending a considerable sum of money on Trotsky's security, despite the fact that his own financial house was in perpetual disorder, another manifestation of his anarchic nature. Contrary to the allegations of his critics, Diego's government-commissioned murals had not made him rich. In fact, the private commissions he accepted and which also brought him derision—oil paintings, watercolors, and drawings sold mostly to Americans—helped to offset the deficits he incurred on his fresco projects and to cover Frida's substantial medical bills. He was able to raise money quickly in an emergency, however, which proved vitally important to Trotsky in February 1938, when the GPU came calling at the Blue House.

It was late in the afternoon on Wednesday, February 2, when a man arrived at the door with several large packages that he claimed had been sent by the minister of communications, Francisco Múgica, Trotsky's most powerful patron inside the Cárdenas government. The packages were said to contain fertilizer for the garden. Why the communications minister would want to donate fertilizer for the Blue House garden is not explained. Neither Trotsky, nor Van, nor Jesús Casas, the chief of the Mexican police garrison assigned to the house, was home when the attempted delivery took place, which might account for its particular timing.

The packages were refused and the deliveryman promised to return the next day with the proper credentials. Later a phone call to the minister revealed that the deliveryman was a fake. Under the circumstances, the only conclusion to be drawn was that the ploy constituted a probe in advance of an assassination attempt. Trotsky was irate that the impostor had been allowed to drive off and that no one could adequately describe his vehicle or had thought to write down its license plate number. Perhaps this is what Van had in mind, in a letter he wrote to Jan Frankel in New York, by the "clumsiness" of those who were present at the house. He added that Natalia "behaved with a criminal lightmindedness" and that Trotsky had not spared her feelings.

The alleged GPU probe now forced a decision to be made. For several weeks, suspicious activities had been observed at the adjacent house on Avenida Londres. A high wall separated the two properties, which made it difficult to monitor developments on the other side. The comings and goings next door were a considerable source of concern for Diego and Van, and now the situation had become intolerable. Diego decided that the surest solution to the problem was for him to purchase the house outright. Financially the timing was especially inconvenient for him. Frida was in the hospital. The previous month, in order to cut expenses he had moved out of his San Angel house-cum-studio and in with Cristina. Now, in order to respond to the emergency and cover the costs—upwards of $2,000—to purchase and renovate the new property and integrate it with the Blue House, Diego mortaged his home.

The purchase would take a few weeks, enough time perhaps for the Stalinists to execute their plan, so it was decided to have Trotsky go live at the home of Antonio Hidalgo in the fashionable Chapultepec neighborhood of Mexico City. On the assumption that the GPU might have an informant, even an unwitting one, among the Blue House staff, Trotsky's absence had to be kept secret. So, on February 13, he quietly slid into the backseat of the Dodge, parked in the rear patio, and lay down on the floor. Van drove through the gate and out onto the street, waving to the guards as he passed. Once the car was safely out of sight, the passenger could sit up for the ride to his temporary sanctuary. Natalia, meanwhile, arranged pillows in Trotsky's bed to make it appear that he was at home but unwell. The servants were told to stay clear of the bedroom, while Natalia pretended to take in tea to her ailing husband.

It would have been an especially unfortunate moment to be bedridden at the Blue House, which was being remodeled in advance of the acquisition of the neighboring property. As Hansen complained on February 14, "the house is in an uproar being changed around with doorways torn in walls, plaster and bricks over everything and everybody on each other's nerves." All the upheaval had put Trotsky on edge, leading to angry clashes with Natalia—reason enough to have him evacuated.

In the early afternoon of February 16, 1938, the third day of his exile from Coyoacán, a telephone call brought the news that transformed the date into what Trotsky later described as "the blackest day in our personal lives." Lyova was dead in a Paris hospital, one week after an emergency appendectomy. Both the Associated Press and the United Press had the story on the wire. Van, who took the call, was thunderstruck. No one in the household was aware that Lyova was even ill. Measures were taken to ensure that Natalia was kept away from the phone and the afternoon newspapers, while Van and Diego tried to call Paris. The shocking news having been confirmed, they immediately drove to Trotsky in Chapultepec, arriving near dusk.

When they entered the room, Diego broke the news. Trotsky's face hardened. "Does Natalia know?" he asked. When Diego replied in the negative, Trotsky said emphatically, "I shall tell her myself!" They left immediately for Coyoacán, Van behind the wheel, Diego beside him, and Trotsky sitting in back, silent and erect.

Natalia was surprised to see her husband enter the house and aghast at his appearance. He was all bent over and his face was ashen; suddenly he was an old man. "What is it?" she asked in alarm. "Are you ill?" "Lyova is ill," Trotsky replied in a low voice, "our little Lyova . . ."

CHAPTER 5

The Trouble with Father

Trotsky and Natalia were stunned by the sudden loss of their elder son, Trotsky's favorite, who was named after his father. "Goodbye, Leon, goodbye, dear and incomparable friend," Trotsky wrote in a moving tribute several days after Lyova's death. "Your mother and I never thought, never expected that destiny would impose on us this terrible task of writing your obituary." This was not the only passage in Trotsky's eulogy where poignancy was allowed to obscure grim reality. The truth is, Trotsky and Natalia had substantial reason to fear that they would outlive Lyova—indeed all of Trotsky's children and grandchildren. Such an outcome was foreshadowed by an incident that took place in Moscow twelve years earlier.

A stormy scene erupted in the Politburo on October 25, 1926, a moment that would prove to be a turning point in the fortunes of the Left Opposition, led by Trotsky, Zinoviev, and Kamenev. For several months their faction had carefully upheld a fragile truce with the Party's majority, led by Stalin and Bukharin, but events now conspired to shatter the accord. Lenin's political testament, suppressed in the Soviet Union since his death in 1924, had just been published in *The New York Times*, including the explosive postscript that sounded the alarm about the danger to the Party posed by Stalin and called for his removal as general secretary. The Opposition leaders, who had heretofore helped suppress circulation of the document under the pressure of Party discipline, decided to endorse the *Times* version of the testament as authentic. This infuriated Stalin, who used the occasion of the Politburo meeting to launch a blistering attack on his rivals, demanding their complete submission.

When Stalin was finished, Trotsky rose to protest against this diatribe, warning that Stalin's malevolence posed a threat to the very existence of the Party. In that moment, Trotsky appears to have been overwhelmed by a feeling of liberation, as though someone had untied his hands. Turning to Stalin, he pointed an accusing finger at him and declared: "The First Secretary poses his candidature to the post of gravedigger of the revolution!" Stalin turned pale and became flustered, then rushed out of the hall, slamming the door behind him. The meeting ended in an uproar. The next morning the Central Committee voted to remove Trotsky from the Politburo.

Trotsky's outburst had dramatically escalated the crisis. His own allies were dismayed that he had needlessly insulted Stalin. Immediately after the Politburo session, several comrades convened at his Kremlin apartment, where Natalia awaited his return. Among them was Yuri Pyatakov, who was especially upset. "You know I have smelled gunpowder, but I have never seen anything like this!" he said, gulping down a glass of water. "This was worse than anything! And why, why did Lev Davidovich say this? Stalin will never forgive him unto the third and fourth generation!" When Trotsky entered, Pyatakov confronted him: "But why, why have you said this?" Exhausted but calm, with a wave of his hand Trotsky brushed the question aside. The damage had been done; the breach with Stalin was irreparable.

Trotsky would recall this episode several years later, as reports of the arrests and deportations of his family members in the USSR reached him in France. "At the time," he wrote in his diary in 1935, "the words about my children and grandchildren seemed remote, rather a mere turn of phrase. But here we are—it has reached my children and my grandchildren . . . what will become of them?"

Now, in February 1938, in shock from their most tragic loss, Trotsky and Natalia once again had occasion to recall Pyatakov's oracle and to contemplate Stalin's vengefulness. Yet the enigmatic circumstances surrounding Lyova's death cast doubt on the culpability of the gravedigger in the Kremlin. Whether Lyova died a natural death or was murdered is a mystery unlikely ever to be resolved.

Earlier that month, Lyova had published a special issue of the *Bulletin of the Opposition* devoted to the recently issued not-guilty verdict of the Dewey Commission. The publication of the *Bulletin* came as a relief both to Lyova and to his father, who had become impatient by its delayed appearance. In his February 4 letter to Trotsky accompanying a copy of the proofs, Lyova gave no hint of his failing health: the sharp abdominal pains, the loss of appetite, the lassitude.

On February 9 Lyova's appendicitis became acute. In part out of mistrust toward the French Trotskyists, he decided to avoid the French hospitals and instead chose to enter a small private clinic owned and run by Russian émigré doctors and staff. The clinic employed both Red and White Russians, spanning the entire spectrum of political enmity toward Trotsky, with the inevitable Stalinist police informants among them. Lyova registered at the clinic under the false identity of a French engineer, using his companion Jeanne's family name, Martin. Evidently he was unconcerned that his illness or the effects of the anesthesia might induce him to speak in his mother tongue.

Emergency surgery took place that same evening and the patient appeared to be recuperating well, until the night of February 13 and 14, when he was seen wandering the unattended corridors, half-naked and raving in Russian. He was discovered in the morning lying on a cot in a nearby office, critically ill. His bed and his room were soiled with excrement. A second operation was performed on the evening of February 15, but after enduring hours of agonizing pain, the patient died the following morning. Lyova was a week shy of turning thirty-two.

According to the doctors, the cause of death was an intestinal blockage, but Trotsky and Natalia could only assume that their son had been poisoned by the GPU. An autopsy turned up no sign of poisoning or any other evidence of foul play, yet Lyova's relapse seemed unaccountable to his parents, who retained an image of their son as a vibrant young man. And if poison was not involved, then why had one of the doctors asked Jeanne, just before Lyova's death, if he had recently spoken of suicide? Then there was the matter of the Russian clinic, a choice that must have seemed perverse, especially considering that one of the family's most trustworthy friends in Paris was an eminent physician who could have arranged for Lyova to have the best medical care.

Such were the perplexities that afflicted the grieving parents, who secluded themselves in their bedroom at the Blue House. Joe Hansen recalled hearing Natalia's "terrible cry"—perhaps at the moment she was told the news. Otherwise silence reigned over the house. For several days, the staff caught only an occasional glimpse of Trotsky or Natalia, and the mere sight of them was heartbreaking. Tea was passed to them through a half-opened door, the same ritual as five years earlier when they learned of the suicide of Trotsky's daughter Zina in Berlin. Yet for Trotsky the loss of Lyova was indeed incomparable. As he explained in a press release on February 18, "He was not only my son but my best friend."

LYOVA WAS ONLY eleven years old at the time of the Bolshevik Revolution. He idolized his father, who once allowed the boy to accompany him to the front on his armored train. Lying about his age, Lyova joined the Komsomol, the Communist Youth League, before reaching the minimum age and later moved out of his parents' Kremlin apartment in order to live in a proletarian student hostel. When Trotsky led the Opposition against Stalin, Lyova plunged headlong into its activities, dropping out of technical school to become his father's closest aide and bodyguard. "Lyova has politics in his blood," Trotsky remarked approvingly. When the Opposition went down to defeat at the end of 1927, Lyova decided to leave behind his wife and son and join his parents in exile.

On the evening of January 16, 1928, Trotsky, Natalia, and Lyova were preparing to depart a wintry Moscow by train for Central Asia. Their baggage had been taken to the station ahead of them, and the family gathered in the apartment Trotsky and Natalia had occupied since moving out of the Kremlin the previous autumn. They were joined by twenty-year-old Seryozha, whose aversion to politics had recently mellowed, in part due to his father's tribulations. As the evening progressed, the family assembled in the dining room to await the police. The train was scheduled to depart at 10:00 p.m. As they nervously watched the clock, the appointed hour passed and they puzzled over this.

Shortly afterward, a GPU official telephoned to inform Trotsky that his departure would be delayed for two days. This produced further puzzlement until friends arrived with the news that a "tremendous demonstration" by Trotsky's supporters at the station had caused the postponement. They described an unruly scene of resistance around the railroad car reserved for Trotsky. His supporters set up a large portrait of their hero on the roof of the car, as people cheered and shouted "Long live Trotsky!" Demonstrators blocked the tracks and clashed with the GPU and the local police, which led to casualties on both sides and arrests.

The mood at Trotsky's apartment was suddenly buoyant. For the next few hours, jubilant supporters kept telephoning with descriptions of what had transpired at the station, and deep into the night family and friends turned over the possibilities. Late the next morning the doorbell rang and two women friends entered. A moment later the doorbell rang again and the apartment filled with GPU agents in civilian clothes— a surprise abduction was under way. Trotsky, still in his pajamas, was handed an arrest order but did not intend to cooperate. He and Natalia and the two guests locked themselves in a room, and tense negotiations ensued through the glazed glass door, until the agents decided to telephone for instructions. The calm was broken by the sound of shattering glass, as an arm reached inside to unlock the door.

One of the GPU men on the scene, a former Red Army officer named Kishkin who had often accompanied Trotsky on his armored train, behaved oddly, as though distressed by the Red commissar's reversal of fortune. As the agents broke through the door he kept repeating, "Shoot me, Comrade Trotsky, shoot me." Trotsky replied coolly: "Don't talk nonsense, Kishkin. No one is going to shoot you. Go ahead with your job." They found Trotsky's slippers and put them on him, then his fur coat and winter hat. Still he refused to move, at which point the policemen lifted him in their arms and began to carry him off.

Natalia hurriedly pulled on her snow boots and fur coat and walked out to the landing. The door slammed behind her and she heard a commotion on the other side of it. A moment later she watched as the door flew open and her two sons burst out, followed by the women guests. "They all forced their way through with the aid of athletic measures on Seryozha's part." Descending the stairs, Lyova frantically tried to rally support, ringing every doorbell and crying out, "They're carrying

Comrade Trotsky away!" His efforts were hopeless. "Frightened faces flashed by us at the doors and on the staircase," Natalia remembered.

Seryozha sounds like a real bruiser. At one point during the drive to the train station the policemen had trouble containing him inside the speeding car. He tried to jump out near the workplace of Lyova's wife in order to alert her to her husband's unscheduled imminent departure. As frigid air rushed in through the open door, the agents struggled to restrain the young athlete and appealed to Trotsky to convince his son to relent.

When they arrived, Trotsky had to be lifted out of the car and carried into the station, which was nearly empty: this time there would be no protesters to obstruct his departure. A desperate Lyova tried to recruit supporters from among the scattering of railway workers, shouting, "Comrades, look! They're carrying Comrade Trotsky away!" A GPU agent named Barychkin, someone who used to accompany Trotsky on his hunting and fishing trips, grabbed Lyova by the collar and tried to cover his mouth with his hand. Natalia says that Seryozha intervened with "a trained athlete's blow in the face," which forced the policeman to retreat. This was no proud mother's idle boast. Several years later Trotsky recorded in his diary that Natalia was tormented by the thought that this "thoroughly corrupted and depraved" GPU man would be allowed to take his revenge on Seryozha in his prison cell. "He will remind Seryozha of that now," she told her husband.

Trotsky, Natalia, and Lyova were placed in the railroad car for the first stretch of the journey to Alma Ata, in Kazakhstan. During the ensuing year of internal exile—Lyova's apprenticeship in the art of conspiracy—he served as his father's liaison with Trotskyists throughout the USSR. Trotsky wrote proudly of his son's contributions in this period, "We called him our minister of foreign affairs, minister of police and minister of communications." When Trotsky was expelled from the Soviet Union the following year, Lyova decided to accompany his parents, although he was not formally exiled himself. On Turkey's Prinkipo island, he assisted Trotsky in the writing of his autobiography and his history of the Russian Revolution, served as editor of the *Bulletin of the Opposition*, and helped direct the assortment of parties and groupings that constituted the incipient international Trotskyist movement.

A few months into his Turkish exile, Lyova became homesick for Moscow and his family, and he decided to attempt to return to the USSR. He applied at the Soviet consulate in Istanbul, but several weeks later he was informed that his request had been rejected. He now had no alternative but to continue as his father's aide-de-camp. Relations between Trotsky and Lyova were never easy, and they became increasingly fraught under the pressures of working together in isolation in a foreign land. Anyone who served as Trotsky's secretary could testify that he was difficult to please, but only Lyova knew how difficult it was to please his father. Sensitive by nature, Lyova was deeply wounded by his father's carping criticisms of his efforts as "slipshod," "slovenly," and worse. Trotsky was aware that his severity could be oppressive, but apparently he did not fully grasp its toll.

Lyova's involvement with a woman placed a further strain on his relationship with his father. She was Jeanne Molinier, the wife of Raymond Molinier, at the time one of Trotsky's most valued French followers and a frequent visitor to Prinkipo. At the end of one of the couple's visits, Raymond returned to Paris alone, and not long afterward Lyova and Jeanne began an affair. Jeanne at first considered it no more than a fling, while Lyova took the matter so seriously that he even threatened to commit suicide unless Jeanne agreed to live with him. Trotsky strongly disapproved of the liaison, and a few years later, when the French Trotskyists split into two rival groups, he had even more reason to do so, as Jeanne sided with the renegade faction led by her former husband against the orthodox group under Trotsky and Lyova.

There is no telling how things might have developed had Lyova not gone to live in Berlin in February 1931. The rationale behind this move was to facilitate Trotsky's leadership of the movement by having Lyova represent him at its organizational nerve center in the German capital. There Lyova would take full control of the *Bulletin of the Opposition*, whose publication would be transferred from Paris. In order to secure a German visa, he enrolled as a student at Berlin's Higher Technical School. This was no mere ruse, however, as he was intent on resuming his education toward an engineering degree cut short in Moscow. The family's sadness must have been tempered with relief.

Perhaps the separation would make it easier for Lyova to serve as his father's indispensable comrade.

As the years passed, the family would have plenty of occasions to consider what Lyova's fate would have been had he managed to return to the Soviet Union in 1929. The decision to deny him a visa was made at the highest echelon. When informed of Lyova's application, Stalin said with a sneer, "For him it's all over. And the same for his family. Reject it."

Before Lyova's departure for Berlin, the family was joined on Prinkipo by Zina, the elder of Trotsky's two daughters by his first marriage. She arrived from Moscow with her son, five-year-old Seva, a blond-haired boy with plumpish cheeks who spoke beautiful Russian, "with the singsong Moscow accent," in Trotsky's words. Many years later, Albert Glotzer, a young American Trotskyist who came to Turkey in this period, still remembered Seva's high-pitched voice calling out to his grandfather, "Lev Davidovich!" The boy was embarked on an extended period of upheaval, during which time he would lose, among others, his mother, his uncle Lyova, and then his grandfather, while he himself would barely escape death in the commando raid of May 1940.

It seems certain that Zina was already mentally unstable by the time she moved into her father's house in Prinkipo. Her younger sister, Nina, had died of tuberculosis in 1928, a victim of the privations and persecution she was forced to endure because of her association with her father, who was in exile in Alma Ata during the final stage of her illness. Nina's husband had been arrested and exiled, and she had lost her job. As Trotsky's daughter, she had difficulty getting proper medical care. She died at twenty-six. Her two children were taken in by Trotsky's first wife, Alexandra, in Leningrad.

Zina was also tubercular, and she received permission to go abroad for treatment. She was allowed to take along only one child, leaving behind her daughter from a previous marriage. Her husband, Seva's father, an outspoken Trotskyist, had been arrested in 1929 and deported to Arkhangelsk, near the White Sea. Zina suffered from chronic depression and seemed to believe that close contact with her father could provide a cure.

Of Trotsky's four children, Zina most resembled him, both physically and in her emotional intensity. She worshipped her father, and yet they barely knew each other. He had left his daughters as infants when he made his first escape from Siberia in 1902, and had had little contact with them over the years. Now father and daughter were to live together under the same roof. This arrangement would last nearly ten months, during which time the tension between them mounted until the lid almost blew off.

In Moscow Zina had been active in Opposition politics and twice had been detained by the police. Arriving in Turkey, she hoped to be welcomed as one of her father's trusted disciples. Trotsky, however, refused to entertain the idea, not least because Zina's increasingly evident instability made it impossible to trust her with confidential information. Zina was told that as she intended to return to Moscow after her convalescence this arrangement was for her own protection, though she took it instead as a form of rejection. She was intensely jealous of Lyova for his close collaboration with Trotsky, and during the brief period they overlapped in Turkey the two half-siblings clashed. Even more perilously, Zina competed with Natalia for Trotsky's affections, which led to angry scenes between father and daughter, and when Trotsky raised his voice, Zina fell apart. "To Papa," she often said, "I am a good-for-nothing."

Zina's lungs responded to treatment, but her mental health deteriorated. She was prone to fits of anger and delirium. Trotsky began to encourage her to go to Berlin for psychoanalysis, a proposal she resisted until he prevailed. She departed from Turkey near the end of 1931, leaving young Seva behind. According to Zina, in their final conversation her father said to her: "You are an astonishing person. I have never met anyone like you." "He said that," she told Lyova, "in an expressive and severe voice."

In Berlin, she continued to slide—with some assistance from the Kremlin. On February 20, 1932, the Soviet government deprived

Trotsky and all his family members abroad of their Soviet citizenship. For Zina, this meant that she would never again be able to see her daughter, her husband, or her mother. At the same time, she sensed her father drifting away from her emotionally, and increasingly she blamed her condition on the growing distance between them. Lyova saw her occasionally, and one such encounter left him shaken. "Zina is terribly oppressed, depressed, she looks utterly destroyed," he wrote to Trotsky. "I pity her, Papochka, very, very much. It's painful to look at her." Lyova urged his father to write to Zina, but Trotsky was incapable of sending his daughter the kind of letter she was increasingly desperate to receive.

Trotsky, meanwhile, was angry that his daughter had left Seva in his care. "Mama is tied down both hands and feet by Seva," he complained to Lyova in June 1932. "We must settle the question of Seva as quickly as possible." Yet Van asserts that when he arrived in Prinkipo in October 1932 to take up his secretarial duties, he found a "gentle, quiet little boy, who went to school in the morning and made himself scarce in the house. Natalia was far from being 'tied hands and feet' by him." Inconvenience aside, one can only speculate on the source of Trotsky's discomfort. Glotzer recalls that at school Seva "suffered the usual little cruelties inflicted by children" because he was different. One day, he remembers, Trotsky asked him if he could teach Seva how to box. Their first lesson broke down almost immediately, never to be resumed.

Seva was reunited with his mother in Berlin in the final days of 1932, but his presence may have aggravated her condition, perhaps by exacerbating the feeling that her father had rejected her. Two weeks earlier, on December 14, she had written in her final letter to him: "Dear Papa, I expect a letter from you, if only a few lines." On January 5, 1933, she barricaded herself in her apartment and turned on the gas taps. She had taken steps beforehand to arrange for Seva to be with friends and left instructions to explain to the boy that she was confined in a hospital for infectious patients. "Poor, poor, poor child. But nothing could be more horrible for him than a psychologically deranged mother." The barricade she constructed inside the door of her apartment ensured that rescue was impossible. She was thirty-one years old.

In Prinkipo, when the news arrived by telegram Trotsky and Natalia immediately isolated themselves in their room. The household understood that something terrible had happened, but the nature of the tragedy was revealed only with the arrival of the afternoon newspapers. A few days passed before Trotsky emerged from his room and returned to work. "Two deep wrinkles had formed on either side of his nose and ran down both sides of his mouth," Van observed. His first act was to compose an open letter to the Central Committee of the Soviet Communist Party in which he placed the blame for his daughter's death on Stalin: by forever cutting off Zina from her family in Moscow, the dictator had driven her to madness and suicide.

This was how Trotsky explained Zina's death to her distraught mother, Alexandra, then in Leningrad, but she refused to believe it. "I will go mad myself if I do not learn everything," she wrote to him after receiving the news. As a radical young activist in the southern Ukraine in the 1890s, Alexandra Sokolovskaya had introduced Trotsky, then known as Lev Bronstein, to Marxism. The lovers had married in a Moscow transit prison in 1900, submitting to this bourgeois ceremony in order to be sent jointly into Siberian exile. There, two years later, convinced that her husband was destined for greatness, she encouraged him to escape and pursue his ambitions among the Russian Marxist émigrés in Europe. She remained a loyal Trotskyist through the 1920s and raised two fervent Oppositionist daughters. Now she had lost them both.

"Where is my bright radiant darling," Alexandra grieved, "where is my Zina?" She quoted a letter Zina sent her a few weeks before her death in which she blamed her illness on her father's indifference toward her. "Papa never writes to me," she complained repeatedly. "He will never write to me again." Nor did she believe she would ever be able to see him again.

"I wrote to her that it was not so tragic as it seemed to her," Alexandra recounted to Trotsky, "that much is explained by your character, by the difficulty you have expressing your feelings, although often you understand that this has to be done." Unlike Trotsky, Alexandra was not inclined to attach any special significance to Zina's loss of her Soviet citizenship. Zina was a "public person" whose life was never focused on her husband or her children. She prized above all her father's "tender solicitude, but she did not get enough of it." Psychoanalysis, it was obvious to Alexandra, was hardly appropriate for someone like Zina. "She was by nature very reserved, and getting her to talk was very difficult. This was a quality she acquired from both of us. And yet here she was forced to talk about things that she didn't want to talk about."

Of course, had Zina remained in Russia, Alexandra understood, tuberculosis would have killed her. "Our daughters were doomed," she declared, and she feared that the same was true of their grandchildren. "I look at them in horror. I no longer believe in life, I don't believe that they will get to grow up. All the time I expect some new catastrophe." She wondered whether Trotsky would assume care for Seva.

"It was hard for me to write this letter and it is hard to send it off," she concluded. "Forgive me this cruelty towards you, but you should know everything about our kinfolk."

Trotsky's response to this letter sought to explain and console. He wrote it in his own hand and gave it to Van in a sealed envelope, which he himself had addressed "in his fine handwriting." The letter was sent by registered mail, according to Van, with proof of delivery requested. "The return receipt never came back."

Van says that Trotsky's appearance changed markedly in the first half of 1933. "The two furrows that had appeared on his face after Zina's death did not disappear and, with time, grew deeper." His hair grayed considerably, and he began to comb it to the side "instead of wearing it proudly brushed back." Within those few months, "his features became what they would remain till his death."

It was also at this time that Trotsky shed the habit of casually remarking about his adversaries, "You know, they should be shot"—a practice he probably adopted during the Revolution, when it was more than just a manner of speaking. "After the spring of 1933," Van claims, "the word vanished from his vocabulary. He would no longer allow himself this kind of irony."

Two years later, in the spring of 1935, with the Great Terror under way, Alexandra was arrested and exiled to Siberia. She was shot in 1938.

Trotsky's sons-in-law, already in exile, were rearrested in 1935 and sent farther on. Both were later shot, Seva's father in 1936.

That same year, Trotsky's sister, Olga Kameneva, four years his junior, who was at one time an official in the Soviet foreign ministry, was arrested and imprisoned shortly after her former husband, Lev Kamenev, was executed in the wake of the first Moscow trial. Her two sons were shot in 1936; she was shot in 1941.

Trotsky had an older sister, Elizaveta, who died of natural causes in the Kremlin in 1924. He had an older brother, Alexander, a former director of a sugar factory with whom he had maintained only distant relations since the early 1920s. The elder Bronstein was "unmasked" in February 1938 as an agent of his brother, "the bandit chieftain Trotsky." He was shot in April. Natalia's brother, Sergei Sedov, was arrested in 1937 and sentenced to five years in a camp, where he died the following year.

After the arrest of Alexandra, her daughter Nina's orphaned children were placed in the care of Alexandra's ailing sister in the Ukraine. They disappeared without a trace.

Zina's death in Berlin in January 1933 occurred as the National Socialists were storming their way to power in Germany. On January 30, Adolf Hitler was appointed chancellor. A month later came the Reichstag fire which the Nazis used as the pretext to arrest communists and socialists and suspend civil liberties, as they moved toward the establishment of a one-party state. Lyova and Jeanne barely managed to escape Germany, fleeing to Paris, which became the new headquarters of the sparse and struggling Trotskyist movement.

As Trotsky's chief lieutenant in Berlin and then Paris, Lyova carried a tremendous load. He served as his father's liaison with the many and fractious Trotskyist groups, edited and published the *Bulletin of the Opposition*, and acted as Trotsky's literary agent in Europe. Himself penniless, he constantly worried about his parents' lack of funds. From Moscow he received despairing letters from his wife, Anna, a daughter of the proletariat, who wrote of the hardships suffered by her and their boy, Leon, and who threatened to commit suicide. She would be arrested and shot one month before Lyova's death. Their son would vanish completely. Lyova's relationship with Jeanne was volatile and often contentious. In early 1935 they became surrogate parents to Seva, who arrived from Vienna where he had been sent to live with friends of the family after his mother's death. Now nine years old, he had forgotten his Russian and French; at school in Paris he was taunted as "le Boche" (the German).

Devotion to his father kept Lyova going, yet his father often made his life miserable. "I think that all Papa's deficiencies have not diminished as he has grown older," Lyova wrote to his mother, "but under the influence of his isolation . . . have gotten worse. His lack of tolerance, hot temper, inconsistency, even rudeness, his desire to humiliate, offend and even destroy have increased." Having thus unburdened himself, Lyova decided not to mail this indictment, but in any case his mother would not have disagreed with it. "The trouble with father, as you know, is never over the great issues, but over the tiny ones," she observed, resigned to her husband's arbitrary ways. "What is to be done—nothing can be done," was a refrain of her letters to her son, and on one occasion she commiserated, "I am writing as you are, with my feelings and my eyes closed."

During Trotsky and Natalia's two-year sojourn in France, which began in the summer of 1933, father and son were able to confer directly on political and other matters, especially during the winter of 1933–34, when Trotsky lived in Barbizon, thirty miles outside Paris. Trotsky was later forced to move to Norway, which left Lyova alone to absorb the shock of the Moscow trial in August 1936. Lyova remembered these Old Bolsheviks as family friends from his childhood days. Lev Kamenev had been an uncle to him. Now prosecutor Vyshinsky railed against them as "scum" and "vermin." The very fact of such a trial was astonishing; its outcome was inconceivable. On a Paris street, when Lyova read the news that all sixteen of the accused had been executed, he became hysterical, bawling uncontrollably, with no concern to hide his face. People stopped and stared as he walked by, crying like a child.

That autumn, when the Norwegian government, under pressure from Moscow, interned Trotsky and Natalia, the somewhat shy and in-

Trotsky and Natalia in the patio of the Blue House, 1937.

secure Lyova was forced to emerge from his father's shadow. The trial itself made this inevitable by charging both Trotsky and his son with having masterminded an elaborate conspiracy to bring down the Soviet regime. This frame-up needed to be exposed, and with Trotsky in confinement the responsibility fell on Lyova. With vital assistance from Van, he produced a careful refutation of the purge trial evidence under the title *The Red Book on the Moscow Trial*, published in French, German, and Russian. In Norway, Trotsky was allowed to read a copy. "I became completely engrossed," he later wrote in admiration of his dead son. "Each succeeding chapter seemed to me better than the last. 'Good boy, Levusyatka!' my wife and I said. 'We have a defender!' How his eyes must have glowed with pleasure as he read our warm praise!"

When the second Moscow trial began, Lyova's parents were in distant Mexico. Once again, old comrades and family friends were arrested as "enemies of the people" and confessed the most fantastic of crimes, carried out with inspiration and assistance from abroad by Trotsky and Lyova. Victor Serge, the Brussels-born Russian Trotskyist writer who won his release from Soviet captivity in 1936 as a result of a campaign by left-wing writers and activists—a rare bit of good fortune—came to Paris and sought out Lyova. "More than once, lingering until dawn in the streets of Montparnasse," he recalled, "we tried together to unravel the tangle of the Moscow trials. Every now and then, stopping under a street lamp, one of us would exclaim: 'We are in a labyrinth of sheer madness!'"

It was around this time, in the first weeks of 1937, that Lyova began to have concerns about his safety. He sensed the presence of a spy in his midst, and he wondered if the GPU might try either to kidnap him or to kill him and make his death look like something other than murder. He published a statement in a Paris newspaper declaring that he was of sound health, both physically and mentally, and that were he to die suddenly, it would likely be as a victim of Stalin's secret police.

Lyova's relations with his father came under increasing strain as Trotsky and his staff prepared for the Dewey Commission hearings that April. Documents were urgently required, chiefly depositions from European witnesses in order to expose the many internal contradictions of the evidence presented in Moscow. Much of this burden fell on Lyova, who initially questioned the value of such a counter-trial. As Trotsky's demands proliferated, so did his reproaches of his son for delays and incompetence. The success of the hearings in Coyoacán brought only partial relief, because the commission continued its investigations from New York City.

"I am a beast of burden, nothing else," Lyova complained to his mother that summer. "I do not learn, I do not read." He had precious little time and energy to devote to his studies of math and physics at the Sorbonne, his third attempt—after Moscow and Berlin—to complete an engineering degree. Money was a constant source of worry, both in Paris and in Coyoacán. To cover expenses he proposed to take a job in a factory, dismissing his mother's suggestion that he earn a living as a writer. "I cannot aspire to do any literary work; I do not have the light touch and the talent that can partly replace knowledge."

Lyova was depressed, and perhaps, as speculation has it, his condition impaired his judgment, which led him to enter a Russian clinic pretending to be a French engineer. Van, whose French upbringing and acquaintance with Lyova make his testimony authoritative, calls this disguise "a ridiculously transparent pose," even had Lyova been in perfect health. "In two minutes, other Russians could not have failed to realise that he was Russian." It turns out, however, that the transparency of Lyova's deception, and the fact that he ended up roaming the corridors speaking delirious Russian, probably made no difference. The fact is, the moment he set out from his apartment on rue Lacretelle for the clinic, his closest comrade picked up the telephone and alerted the GPU.

His real name was Mark Zborowski. In the movement he went under the pseudonym Étienne. His GPU code names were "Mack" and "Tulip." He was born in Russia, near Kiev, in 1908 and later emigrated with his parents to Poland, where at some point in the 1920s he took up radical politics, joined the Communist Party, and served time in prison for organizing a strike. In 1928, he and his wife moved to France, where he seems to have become something of a professional student, first in Rouen, then in Grenoble, then in Paris at the Sorbonne, where he decided to specialize in ethnology, eventually obtaining a degree at the École des Hautes Études. Perennially hard up for money, he was easily recruited by the GPU. By befriending Jeanne he was able to infiltrate Lyova's circle, making himself useful and, before long, indispensable as Lyova's personal assistant. One quality that set him apart from the other Trotskyists in Paris served to ingratiate him with Lyova: the two men could converse together in Russian.

Looking back on Zborowski long after his identity as a spy had been exposed, Van remembered a man with a "sullen, frowning face" and a "colorless manner," who behaved "rather like a mouse. He did not make himself conspicuous in any way. . . . There was nothing you could grapple with in him, except his insignificance." As it happens, Lyova suspected that the mouse might be a rat. His doubts about Zborowski would come in waves lasting five or six days, but after each wave receded, their friendship was restored. Lyova never told Trotsky about his suspicions, which over time seem to have evaporated entirely.

So complete did Lyova's trust become that he granted Zborowski access to his mailbox and to Trotsky's most confidential files, some of which were stored in Zborowski's apartment. Before leaving Paris for a few weeks in August 1937, Lyova wrote to his father: "In my absence, my place will be taken by Étienne, who is on the closest terms with me here, so the address stays the same and your missions can be carried out as if I were in Paris myself. Étienne can be trusted absolutely in every respect." Based on this and similar evidence, Van judged Lyova's blindness toward Zborowski to be "astonishing"—although Van himself had to confess, "I never had any special suspicions about him."

It seems that Zborowski's role was limited to that of informant and finger man, as in November 1936 when thieves made off with a portion of Trotsky's archives stored in an apartment on rue Michelet—files that ended up in the Kremlin. Zborowski kept Moscow thoroughly acquainted with the activities of Trotsky and Lyova, who were assigned the unimaginative code names "Old Man" and "Sonny." Lyova came to confide in Zborowski and these confidences were passed along to Moscow in top secret reports. One item in particular must have come under special scrutiny. At the end of the second Moscow trial, in January 1937, Lyova is supposed to have remarked to Zborowski: "There's no reason to hesitate any longer, Stalin must be killed." This sounds like one of Lyova's emotional outbursts, but Zborowski framed it as an endorsement of assassination.

To judge from these police files, in the month of November 1937 Lyova was experiencing some kind of mental crisis. One explanation for this was a letter he had received from his father rejecting a proposal put forward by Lyova's comrades that he leave Paris for Mexico. Lyova learned of Trotsky's decision at a time when he had begun to fear for his life. The source of his alarm was the circumstances surrounding the recent defection of Ignace Reiss, a Soviet spy based in Europe. Reiss went into hiding after being ordered to return to Moscow, the first of several such desertions by men who had good reason to believe that their next trip home would be their last. Having orchestrated the murders of many of Lenin's closest comrades, Stalin was now intent on

exterminating those who knew about his crimes, starting with the in-
quisitors behind the show trials. Reports began reaching agents abroad
about a bloodbath of their old comrades in the Cheka—the original
name for the Soviet secret police—a purge that extended to the upper
reaches of the GPU.

Reiss made contact with Trotsky's allies and informed them that
Stalin had decided to "liquidate Trotskyism" outside the USSR. He described the methods of torture and blackmail used to extract the purge
trial confessions and portrayed doomed Trotskyists facing death with
cries of "Long live Trotsky!" Reiss declared his loyalty to the Fourth
International.

The liquidation of Trotskyism abroad, it turned out, began with
Reiss himself. On September 4, 1937, his bullet-ridden body was found
on a Swiss road near Lausanne where he had been lured to his death.
The police ascertained that the gang that killed him had been tailing
Lyova and had earlier in the year laid a trap for him near the Swiss
border, an appointment with death that his ill health prevented him
from keeping. The investigation revealed that the GPU was receiving
detailed information about Lyova's movements and activities, including
the fact that he had arranged a rendezvous with Reiss two days later
in the northern French city of Reims. Had Reiss eluded his killers in
Switzerland, a GPU hit squad was waiting for him in Reims.

Against this sinister background, Lyova's friends in Paris wrote to
Trotsky and Natalia in early November 1937 urging them to persuade
Lyova to get out of France and join them in Mexico. Lyova was sick,
exhausted, and in danger, yet convinced he was "irreplaceable" in Paris
and must "remain at his post." This was not the case, they insisted. "He
is able, brave, and energetic; and we must save him."

Trotsky probably withheld these admonitions from Natalia. He replied to these friends that Lyova would be forced to lead the life of a
"*demi-prisonnier*" in Mexico, which should serve him only as "*le dernier
refuge.*" Trotsky also wrote directly to Lyova to quash the idea of his
withdrawal from Paris. In the event that the French government decided to expel him, "Mexico always remains a possibility," he assured
his son; in the meantime, however, Lyova could count on the protection
of the French police, which had assigned him a special guard after the

Reiss murder. Whatever his feelings were about the prospect of moving in with his father, Lyova may have been discouraged by his preemptive settlement of the matter. "*Voilà, mon petit*, this is what I can tell you," Trotsky concluded on an eerily fatalistic note. "It isn't much. But . . . it's all. Naturally, you should keep whatever money you can collect from the publishers. You will need it." He signed off in his own hand: *"Je t'embrasse. Ton Vieux."*

A GPU report from this period depicts "Sonny" as alcoholic and depressed. On one occasion, after several continuous hours of boozing, he apologized to "Mack" and "almost in tears begged his forgiveness for having suspected him of being an agent of the GPU when they first met." He also confessed that he had "lost all faith" in his father's cause as early as 1927, and that wine and women were now more important to him than anything else. Secret police reports of this type, written to satisfy the spymasters back in Moscow, must be approached with a great deal of skepticism. But if what Zborowski told his GPU handlers approximated the truth, it indicates that Lyova was in a downward spiral when appendicitis struck.

The Soviet intelligence officer who later organized Trotsky's murder claimed that he and his colleagues were stumped when they learned of Lyova's death. No one stepped forward to take the credit; no decorations were given out in secret ceremonies in the Kremlin. Nor, decades later, did the partial opening of the KGB archives help to clarify matters, though perhaps the orders to liquidate "Sonny" were issued verbally or the written evidence was destroyed once his case was closed. If so, this raises the question of what incentive there was to eliminate Trotsky's son at a time when his most trusted comrade was a Soviet police informant. Or was Lyova's murder—if murder it was—simply a way to lash out at Trotsky?

Mourning their loss, Lyova's parents grappled with the deeper mysteries. "Both of them have aged terribly in this one day," Hansen wrote on

February 17, the day after that blackest day. "Poor poor Natalia. She is absolutely prostrated. When the news was broken to the OM, he said, 'This is the finish of Natalia.' They have remained in their room without coming out, windows and doors closed, the room in darkness. They are utterly agonized." As a precaution, the staff decided to remove the small automatic pistol that served as a paperweight on Trotsky's desk.

Trotsky's grief was compounded by guilt for his censorious treatment of Lyova. How could it be otherwise, after the long stream of invective? "Slovenliness bordering on treachery" was how he characterized his son's performance in a letter from February 1937. "It is difficult to say which are the worst blows, those from Moscow or those from Paris. . . . [A]lthough in recent months I have had to endure a lot, I have not experienced such a dark day as today, after receiving your letter. I opened the envelope confident that I would find the affidavits, but instead found excuses and promises."

Expressions of remorse would follow and help mollify his incomparable friend, but only until the next dressing-down. On one occasion, long-suffering Lyova struck back, reminding Trotsky of his limited resources—"sometimes I do not even have the money to buy postage stamps"—and objecting to the way his father belittled him to other comrades. "I thought that I could count on your support. Instead you are making me your butt and are telling all and sundry about my 'criminal carelessness.' "

These fulminations recurred to the end. Trotsky's penultimate letter to his son, dated January 21, 1938, perhaps the last one Lyova read, conveyed his exasperation that the issue of the *Bulletin* that was to feature the Dewey Commission's verdict had yet to appear in print. He called this an "outright crime" and once again threatened to move the journal to New York. Lyova's last letter to his father, sent on February 4, accompanied proofs of this special issue of the *Bulletin*—though this package would arrive too late for Trotsky to be able to convey his relief and appreciation.

Now, behind the blackened windows, Trotsky wrestled with his guilty conscience. Work on remodeling the Blue House had been halted, leaving the patio littered with brick, plaster, lime, and sand. On the morning of February 20, while inspecting the grounds, Hansen noticed that the French doors to Trotsky and Natalia's bedroom were open. The OM was seated at a small table he had placed in the doorway for the light. He was writing Lyova's obituary. "In the evening he was still there working under a lamp he had set up. . . . He worked very late."

The result was his affecting tribute, "Leon Sedov—Son, Friend, Fighter," which sketched scenes from his son's life, beginning with a pregnant Natalia confined in a St. Petersburg jail during the Revolution of 1905. "His mother—who was closer to him than any other person in the world—and I are living through these terrible hours recalling his image, feature by feature, unable to believe that he is no more and weeping because it is impossible not to believe. . . . Together with our boy has died everything that still remained young within us."

Trotsky's testimonial was also a statement of partial contrition and self-exculpation. His intimate collaboration with his son had sometimes led to fierce clashes between them, he revealed, and not only over political matters. He admitted that Lyova had borne the brunt of the pedantic and exacting behavior that often made him insufferable to family and friends. "To a superficial eye it might even have seemed that our relationship was permeated with severity and aloofness," he elucidated, laying down a marker for his future biographers. "But beneath the surface there glowed a deep mutual attachment based on something immeasurably greater than ties of blood—a solidarity of views and judgments, of sympathies and antipathies, of joys and sorrows experienced together, of great hopes we had in common. And this mutual attachment blazed up from time to time so warmly as to reward us three-hundred-fold for the petty friction in daily work."

Lyova was hunted by the GPU, who intercepted his mail, stole his papers, and listened in on his telephone conversations, Trotsky affirmed. "His closest friends wrote us three months ago that he was subject to a danger too direct in Paris and insisted on his going to Mexico." But Lyova objected that while the danger was undeniable, it would be criminal of him to abandon his post in the midst of the battle. "Nothing remained except to bow to this argument." This was how Trotsky preferred to remember the episode, from the perspective of a mere bystander.

As to the cause of Lyova's death, Trotsky held the GPU responsible. "This young and profoundly sensitive and tender being had had far too much to bear. Whether the Moscow masters resorted to chemistry, or whether everything they had previously done proved sufficient, the conclusion remains one and the same: *It was they who killed him.*"

Trotsky and Natalia's isolation came to an end after several days, though it would be several weeks before they rejoined the rest of the household at the dinner table. The period of mourning was convoluted by the onset of a protracted struggle with Jeanne over custody of Seva and of Trotsky's archives. At first, relations were warm, with Trotsky inviting Jeanne to come live in Mexico as "our beloved daughter." Before long, however, a chill set in. It emerged that Lyova, in a testament he produced in great haste before his departure for the clinic, had left Jeanne as the executor of Trotsky's papers. As an adherent of the breakaway group of French Trotskyists, Jeanne was not inclined to relinquish these archives. As surrogate mother to Seva, moreover, she had become attached to the boy and did not wish to give him up. In the face of Jeanne's intransigence, Trotsky, who at the outset seemed open to the idea that Seva should be raised in Paris, became determined to gain custody of his grandson.

An agreement on the transfer of the archives was arranged within a few months, but the custody battle dragged on in court and was the subject of sensational press coverage. When the law eventually ruled in Trotsky's favor, an overwrought Jeanne abducted Seva and hid him in a religious institution in the Vosges region of eastern France. He was rescued by Trotsky's allies then nearly kidnapped by Jeanne's confederates. It was not until August 1939 that the boy, then thirteen years old, was brought to Mexico.

Throughout this ordeal, Natalia retained feelings of compassion for Jeanne, and the two women consoled each other in a tearful correspondence. Trotsky could not abide such divided loyalties, and the resulting marital strife was registered by Hansen in his telegraphic penciled notes: "*The only times of anger between OM & N*—his banging window shattering glass—The translation from Russian—Why don't you (get a divorce) go & marry/live with some one else"—an outburst, Hansen seemed to recall, that had to do with Seva.

One afternoon during those tempestuous days, while Trotsky was having a siesta, Natalia came to Van's room very upset, with tears running down her cheeks. "Van, Van, you know what he told me?" she cried out. "You are with my enemies." She quoted her husband in French then repeated his remark in the original Russian. Of course, Trotsky had his own reasons for being upset, Van allowed. "But the harsh fact remains that six weeks after Liova's death, while Natalia was still devastated with grief, Trotsky had spoken to her in the most cutting and brutal terms possible."

Meanwhile, Trotsky's most treacherous enemy continued to elude detection, as the French Trotskyists nominated the reliable Étienne—agent provocateur Mark Zborowski—to take Lyova's place in Paris. From Moscow, Zborowski was given a new assignment: to penetrate Trotsky's household in Coyoacán.

CHAPTER 6

Prisoners and Provocateurs

n the course of their secret meeting on Prinkipo island in 1929, GPU agent Yakov Blumkin warned Trotsky that it would take at least twenty trained men to guarantee his safety. This number was well beyond Trotsky's means. In Turkey he never had more than three or four comrades by his side—men selected, moreover, on the basis of their political fealty and secretarial abilities; guarding was their night job. The Turkish government supplied essential protection by stationing a half dozen policemen outside Trotsky's home. Later, in France, where a clean-shaven Trotsky, with the collusion of the French government, lived mostly incognito, security was left to a few secretaries who doubled as guards.

Mexico presented a more dangerous proposition. In Mexico, anyone could tell you, political disputes were apt to be settled at the point of a gun. Lyova was distressed when he learned that his parents' next destination was, as he called it, "the country where one can hire an assassin for a few dollars." Since the first Moscow trial in August 1936, the Mexican Communist Party had been loud in its condemnation of the renegade Trotsky. His alleged co-conspirators had all been executed for their crimes. Lyova's nightmare was that a Mexican pistolero would finish Stalin's work.

As Trotsky sailed from Norway in December, Lyova dispatched longtime secretaries Jean van Heijenoort and Jan Frankel to Mexico, proposing that Frankel devote himself exclusively to matters of security. His first letter to his parents in Coyoacán reveals Lyova's obsession with the topic: "The question of life and death has never as yet been posed so directly as today." He urged his father to organize a rigorous security regime. One of his recommendations was that Trotsky live incognito and reside at Diego Rivera's home only officially. He also proposed an electric alarm system "like the ones they use in American banks"—a bit of wisdom he probably picked up watching gangster movies.

President Cárdenas, having granted Trotsky asylum, had a strong interest in keeping him alive. His government stationed a police garrison of a half-dozen men on the street outside the Blue House. It also authorized selected local Trotskyists to carry firearms, and in the early days these Mexican comrades were drafted two or three at a time to guard the house from inside the high blue walls. The Trotskyist leadership in New York planned to replace these recruits entirely with trustworthy Americans who could double as secretaries.

Trotsky's top priority during the initial months in Mexico was to expose the Moscow trials as a frame-up. Brains being in greater demand

Trotsky and his secretaries in the patio of the Blue House, 1937. From left: Bernard Wolfe, Jan Frankel, Trotsky, Jean van Heijenoort.

than brawn, the New York office selected Bernard Wolfe, a twenty-two-year-old New Haven native and Yale graduate, to serve as his American secretary. The future author of numerous pornographic, science fiction, and other novels, Wolfe was regarded among the comrades as an effective stylist in English. He could also read German and French. In Coyoacán he would put these abilities to considerable use, notably during the Dewey Commission hearings in April.

The three secretaries—Van, Frankel, and Wolfe—divided the night shift into three watches, with the middle shift, from midnight to four, considered the most demanding. The Blue House was drafty and without heat, and it could get chilly at night 7,500 feet up on the plateau, with temperatures dipping into the forties. Seated at the twenty-foot-long lemon yellow dining room table cluttered at one end with typewriters, books, and papers, Wolfe preoccupied himself nightly by disassembling and then reassembling the Luger he had been issued. One night, Trotsky entered the dining room on his way to the bathroom and noticed the guts of Wolfe's handgun spread out on the table in front of him. He spoke softly but his tone was grave. "You know, in the Revolution we lost more people than the enemy could claim credit for. Many young comrades killed themselves with their guns and suicide was very far from their minds." Wolfe tried to put him at ease: "You don't have to worry about me, L.D., I always take the clip out and make sure the chamber's empty, there's no danger here."

By the time the summer rains began to fall, the search was under way for a sturdier man to replace Wolfe. By then the question of Trotsky's security had begun to loom larger. This partly had to do with the unfolding events in Spain, where the civil war between the Nationalist forces under General Franco and the Loyalist militias of the Republic was entering its second year. Franco's armies received the military backing of Germany and Italy, while the Republican government—a Popular Front coalition of liberals, socialists, communists, and anarchists—relied on weapons and advisers from the Soviet Union, including a sizable contingent from the GPU. Otherwise, the Republic looked to the support of tens of thousands of volunteers from numerous countries, mostly Communists, who made up the International Brigades, including nearly 2,800 Americans who fought with the Abraham Lincoln

Battalion. Britain and France, wary of a confrontation with Hitler and Mussolini, chose to stand aside from the conflict.

Stalin had a complicated political agenda in Spain. He wanted to get credit for Soviet military support to the Republic, yet not go so far as to get drawn into a direct clash with Hitler. He was also intent on preventing the emergence in Spain of an alternative model to Soviet Communism. Toward this end, he set out to neutralize the non-Communist forces fighting on behalf of the Republic and hoping to transform the civil war into a socialist revolution. These included anarchists, syndicalists, Trotskyists, and others heavily concentrated in the northeast, in Catalonia, whose government in Barcelona was affiliated with the central government in Madrid.

Catalonia was the home of the POUM, the Spanish initials for the Workers' Party of Marxist Unification, led by Andrés Nin. Nin, one of the founders of the Spanish Communist Party, had worked for the Comintern in Moscow in the 1920s, when for a time he served on Trotsky's staff and cast his lot with the Left Opposition. Trotsky had since broken with Nin, accusing him of centrism and "treachery" for taking the POUM into Spain's Popular Front government and then joining Catalonia's ruling coalition. This did not prevent Moscow from branding the POUM a Trotskyist organization.

In the first week of May 1937, the GPU made its move in Barcelona, spurring the Communists and the local police to launch an attack against the anarchists and the POUM, even as Franco's armies were attacking the city. During these May Days, Barcelona was the scene of intense street fighting that left hundreds dead. It was a civil war within the civil war. On June 16, Nin and most of the POUM leadership were arrested. Nin subsequently disappeared, as did many others, including Erwin Wolf, a citizen of Czechoslovakia who had served as Trotsky's secretary in Norway for a year. Like Nin, Wolf was presumed to have been kidnapped and killed by the GPU.

The fighting in Spain had a special resonance in Mexico, the only country other than the USSR to declare its support for the Republic. Mexico provided a haven for political refugees from the conflict, who began to arrive on its shores in the summer of 1937. This ongoing emigration was viewed warily at the Blue House, where the new arrivals, not all of them Spaniards, were perceived as potential recruits to the GPU.

One individual who attracted particular interest among the Trotsky-ists both north and south of the Rio Grande was an American by the name of George Mink. Although he claimed to be American-born, Mink was believed to have come from Russia. As a Communist union organizer and goon squad enforcer among the longshoremen on the East Coast, he reportedly traveled back and forth between Moscow and the New York waterfront. For several years he had been employed by the Yellow Cab Co. in Philadelphia. In 1935 he was arrested in Co-penhagen for the attempted rape of a hotel maid, at which time he was discovered to be carrying forged passports for purposes of espionage. Sentenced to prison for eighteen months, he served out most of his term before being deported to the Soviet Union. From Moscow he was sent to Spain.

Comrades escaping Spain told bloodcurdling tales of Mink's monstrous deeds, although few claimed to have actually seen him there. By the summer of 1937 Mink had acquired a considerable notoriety as a GPU executioner. In Europe he was known as the Butcher. It was believed that after the Communist purge in Catalonia, he had left Spain for America. By autumn, rumor had him heading to Mexico.

By then the Blue House was on high alert because of an attack cam-paign directed at Trotsky by the Mexican Communists and the trade union organizations. Posters and newspaper articles accused the exile of plotting to overthrow President Cárdenas and install a fascist dictator-ship. The goal of this slanderous campaign was to compromise Trotsky's asylum so that he would be turned over to Moscow, but Trotsky had to assume that an additional motive was to prepare the atmosphere for his assassination.

At this same time, unsettling reports began to reach Coyoacán about the police investigation of the murder in Switzerland of GPU defector Ignace Reiss. Although the two trigger men had escaped, accomplices were arrested, and it emerged that one of the killers had been in Mexico City earlier in the year and the other had been living in an apartment building in Paris next door to Lyova.

This confluence of ominous developments convinced Trotsky that it was time to enlist a full-time American guard for the Blue House. A leading candidate for the job was quickly identified in Harry Milton, born Wolf Kupinsky, an American Trotskyist who had just returned from Spain. Milton had fought with the POUM militia in northeastern Spain, where he got to know George Orwell. While at the front in Huesca, Orwell took a sniper's bullet in the throat. In his civil-war classic, *Homage to Catalonia*, he recounts how an American sentry came to his aid: "Gosh! Are you hit?" That American was Milton, who called for a knife to cut open Orwell's shirt, a procedure that led to the discovery that the bullet had passed through his neck.

Milton landed in prison during Barcelona's May Days, and was able to win his release only after organizing a hunger strike among the prisoners. Arriving in New York, he made a generous contribution to the legend of George Mink's exploits in Spain. Milton had not seen Mink in Barcelona, but he had known him for years in New York and would have no trouble recognizing the stocky, thuggish former taxi driver.

Milton possessed no secretarial abilities and could speak no foreign languages—even his Spanish, he confessed, was wretched. Nonetheless, Trotsky was persuaded of his suitability as a guard. On November 6 he wrote to Milton that in view of the recent "gangsterist activity of Stalin's agents abroad," his services were needed in Coyoacán. Just after Trotsky had reached his decision, however, Jan Frankel, arriving in New York from Mexico, preempted Milton's appointment. Frankel endorsed an alternative candidate, Hank Stone, arguing that his engineering background would be an invaluable asset in organizing the defense of the Blue House.

Frankel was persuasive, but the hesitation gave Trotsky an opportunity to reconsider the entire proposition. Even as he authorized the hiring of Milton, Trotsky was worried that a man sent down to serve as a pure guard, with no secretarial duties to distract him, would become bored and begin to feel more like a prisoner than a guard. The presence of a full-time guard, moreover, would mean an additional resident at the Blue House. A worker in the New York office, briefed by Frankel, warned colleagues that "the O.M. is extremely restive about being constantly surrounded and has been balking at security arrangements." Which is precisely what Trotsky now proceeded to do. On December 1, he suggested a postponement of the "Milton, Stone matter" for two months, proposing in the meantime to rely on the Mexican comrades.

One factor that influenced Trotsky's decision was the performance of Joe Hansen, who had recently replaced Bernard Wolfe as his American secretary. The search for Wolfe's successor had turned contentious because of Trotsky's insistence that the new man be an experienced driver. He had become paranoid about ending up in a car wreck on Mexico's hazardous roads. Reliance on Diego's driver and on Cristina Kahlo was no longer practical. Frankel did not drive, and although Van had a French driver's license, he was timid about using it. What the household needed was a driver able to negotiate the broken terrain in the countryside, elude pursuers, and take evasive action in the event of an ambush along the narrow, congested streets of Mexico City.

The New York office was slow to appreciate Trotsky's obsessiveness on this subject. James Cannon, the leader of the Trotskyist party, was miffed when the person he recommended to replace Wolfe was rejected: "Fifty million Americans drive autos and it is reasonable to assume that comrade Gordon can learn in a week or two." This made it clear to Trotsky that Cannon failed to understand his situation. In order to get the point across, the job was redefined as "secretary-driver," and New York was instructed that the Old Man would settle for nothing less than "a chauffeur-comrade, sure, reliable and experienced." This led to the selection of Hansen, who had learned to drive in the mountains out west and who could pass Cannon's political litmus test.

Hansen was born in the farming town of Richfield, Utah, and studied at the state university in Salt Lake City, where under the influence of one of his teachers, the Canadian poet Earle Birney, he was drawn to Trotskyism, joining the Communist League of America in 1934. Two years later he moved to San Francisco, where he reported on that city's maritime strikes for *The Voice of the Federation*, the organ of the West Coast maritime unions, and for *Labor Action*, a paper put out by Cannon. Hansen knew some French, the language of the Coyoacán household. He could type and was trained in shorthand.

He was married, but willing to endure an extended separation for the cause. As for Hansen's ability to handle a car, New York demonstrated its confidence in him by arranging for him to deliver Trotsky a new Dodge Sedan, which he drove down from Ohio at the end of September 1937.

Eager to make a good first impression, Hansen had the car washed in Mexico City before driving it up to the Blue House in the bright morning sunshine. The Dodge got a hearty reception, as did the driver. Trotsky threw his arms around Hansen in a warm embrace, which helped put the new secretary at ease. He had been warned that the Old Man could be very difficult to get along with. Three weeks later Hansen still could not get beyond his image of Trotsky as a towering historical figure. "This puts me on an uneasy edge," he wrote to the comrades up north, "as if I were trying to get on friendly terms with a volcano. But he is kindness, courteousness, itself, and goes out of his way to make things easier."

One day shortly after Hansen's arrival, Trotsky needed to pay a visit to the home of the Fernández family in the suburb of Tacuba, northwest of Mexico City and about a twenty-minute drive from the Blue House. This would be Trotsky's first opportunity to assess the new car as well as the new driver. Hansen took the wheel, Van gave directions from the passenger seat, and Trotsky as usual sat in back. Hansen drove slowly because he was unfamiliar with the route, and he might have been distracted by the volcano in his rearview mirror. At each intersection, Van prompted him—"Left," "Right," "Straight ahead"—a performance that was repeated on the drive home.

The next day, for some reason Trotsky found it necessary to return to the Fernández home, and once again Hansen relied on Van to guide him. Upon their return to the Blue House, Trotsky asked Van to follow him into his study. "Don't you think we ought to send Hansen back to the United States?" Trotsky asked. Van did not hide his surprise. "He will never learn!" Trotsky exclaimed. Van pleaded Hansen's case, explaining how difficult it was to memorize the complicated route to Tacuba, but Trotsky remained skeptical: "We shall see!"

A few days later, while chauffeuring Natalia in the city, Hansen drove through a red light and was pulled over by a cop, who "gave me hell." Trotsky was not inclined to let him off so easily: "LD much disappointed in my driving and refused to go down town."

All was forgiven, however, in the second week of October, when Hansen drove Trotsky and Natalia to Taxco. As the Dodge made its way along the winding highway down the plateau and then up the steeply winding roads into the mountains, Trotsky grew more and more confident that he had found his chauffeur-comrade. As Hansen boasted to his wife, "my driving through the mountains astonished him." Trotsky praised the driver for his "courage and caution"—meaning that Hansen sped away from a car on his tail and stopped at railroad crossings. On the drive back to Coyoacán, Trotsky kept commenting in his thickly accented English, "The driver is good." He now resolved to learn to say "Joe," a syllable that gave him great difficulty.

Out of all the American comrades who came to live in Coyoacán, Hansen ended up being Trotsky's favorite. Certainly no other chauffeur was in his class. "With me we can have hairskin shaves from death and he thinks it's just fun," Hansen observed after nine months behind the wheel, "as the other day when a truckload of gasoline shot out of a blind alley and nearly plowed into us. I left rubber all over the road, and the Old Man came sliding out of his seat and onto the floor with the cushion and his hat flying into the front seat—and he thought it was more fun than firecrackers."

Natalia also admired Hansen's driving, to the point where she refused to let anyone else take her to the market. On the evening of September 16, 1940, Mexican Independence Day, she asked him to take her for a drive. She had no particular destination in mind. More than three weeks had passed since they accompanied Trotsky on his last ride, in the back of an ambulance, its siren blaring as it raced to the hospital. The house now felt empty and quiet, and Natalia needed to escape.

Hansen drove her through the countryside and then into the city, whose streets were jammed with holiday traffic. He steered only with his left hand because he had broken the right one subduing Trotsky's assassin. The cast had recently been removed, but the hand was still swollen and tender and could not grip the steering wheel. As they drove along, Natalia struggled to hold herself together. She could get through the days somehow, she told Hansen, but the nights were terribly lonely.

The bedroom was cold; there was nothing to do. "She kept crying as we drove along."

Every new recruit to Trotsky's staff of secretaries and bodyguards quickly learned that although the gravest dangers lurked outside the high walls of the Blue House, the threats to domestic tranquility were ever present. Life in the Trotsky household was marked by frequent periods of tension and petty strife which at times had the effect of undermining Trotsky's security. Mealtimes offered the greatest challenge, because of the often objectionable food on the menu, but even more so because of Trotsky's unpredictable behavior at the dinner table. The volcano might explode at any moment.

The way Trotsky saw it, he was in an awkward position. All the most respected Marxists, starting with Marx and Engels, enjoyed wine and cigars, while the revisionists and other backsliders tended to be ascetics. Trotsky behaved like a revisionist. Hansen, sitting through another uninspired Blue House dinner, was consoled by this bit of self-deprecating humor. Trotsky, he reflected, was sensuous about ideas and historical processes rather than food, drink, or tobacco.

Trotsky's indifference to food was by now an old story to Van, who testified that, "During the seven years that I had meals with him three times a day, seated at his right, I never heard him make a remark about the food." This sounds implausible, yet Van also records Trotsky's bitter complaint about life's daily routines: "To dress up, to eat—all these miserable petty things that one has to repeat from day to day!"

There were generally two types of meals at the Trotsky household. The first were those that took place in near total silence, with Trotsky "lost somewhere in the clouds a million miles from the table," as Hansen put it. He seemed to be the only one at the table not conscious of the eating sounds. It was after one such meal that Hansen informed his wife: "LD has many of the habits you didn't like in me—coldness, silence, oppression."

Dinner at noon was served one dish at a time, starting with soup, followed by potatoes, a vegetable, a meat dish, salad, and then the inevitable apple compote. Sometimes there was fresh fish from Acapulco. As a rule, no alcohol was served, except for a glass of wine on November 7. The fare sounds unobjectionable, but the staff without exception rated the food along a scale of bland to distasteful. Newcomers quickly learned that one had better clean one's plate, lest the Old Man and Natalia's feelings be hurt or their inquiries result in a visit to the doctor. When Trotsky finished his meal and left the room, jaws unclenched, tongues loosened, and the conversation flowed freely. Supper was, by comparison, light and mercifully brisk.

The second type of meal found Trotsky in a gregarious mood, and he would start to tease individuals at the table—about their name, their country of origin, their pronunciation of some foreign term, etc.—sometimes to fantastic lengths. On occasion the kidding would take a detour, and all of a sudden it was time to determine who was responsible for some slip-up. Those on the receiving end of this bantering, as Hansen called it, had to accept it cheerfully. Van noted that Trotsky's friendly jesting had an edge that often left friends and associates feeling wounded. One of his victims was Max Eastman, who perceived this trait as another of the vanquished revolutionary's character flaws. Trotsky, he observed, knew "no laughter but of mockery."

For Hansen the endurance test that passed for mealtime at the Blue House was a special source of concern because his wife, Reba, was planning to move down from Salt Lake City and become one of the inmates, as the staff members called themselves. "You must understand that the place is a good deal like prison," he warned her, "and the psychological pressure becomes terrific on everyone, small items are blown up into elephants, and the OM is like a volcano. For any length of time whatsoever the situation would become unendurable."

A major offender when it came to inflating elephants was Natalia, who was high-strung and used to having things her own way, especially in the kitchen. It was there that she and Rae Spiegel, Trotsky's Russian typist from New York, got into frequent scrapes. The cook and maid, Rosita, a full-blooded Indian who spoke no French and thus strained to comprehend Natalia's instructions, could be heard pining for the warmer climes of her native Tampico.

Trotsky managed to avoid these skirmishes, although on one occasion he happened to get drawn in, and with dramatic results. The incident involved Van's wife, Gaby, who arrived from France at the beginning of November 1937 with their three-year-old son. As she had at the household in Barbizon, Gaby helped out in the kitchen. A few weeks after her arrival, while she was preparing lunch, she took exception to what she felt was Natalia's overbearing manner toward Rosita. Voices were raised, and when Trotsky came upon the scene he blew up completely. Harsh words had been spoken, and Van's wife and child were soon on their way back to France. Van was devastated. Rosita wept over the separation of the family. Hansen was exasperated: "It really takes a hay rack of patience to keep things going smoothly in the house."

The atmosphere was far more relaxed at the modest home of the prodigious Fernández family in Tacuba, where Trotsky, Natalia, and members of the staff were frequent guests. The family patriarch, a schoolteacher in his fifties, and three of his sons, Octavio, Carlos, and Mario, were members of the local Trotskyist group. More than comrades, the family became warm friends of Trotsky and Natalia, offering them the only family setting they experienced during their time in Mexico. A special attraction for the secretaries was the home cooking of Mama Fernández, which was strong on chilies, tamales, and *atole*. Here the guests were introduced to fresh *pulque*, an alcoholic drink made from the fermented juice of the maguey plant and with the disconcerting appearance of watery milk. Trotsky avoided *pulque* as well as the beer, and his sensitive stomach would not allow him to eat spicy foods.

After dinner came dancing. Young comrades were called up and the Indian girls next door were invited over. The radio played a marimba band, and the living room was transformed. "Damn, can they dance," Hansen marveled, bewitched by the movements of the two blossoming Fernández daughters, Graciela and Ofelia, lithe and exotic. "They whooped it as if the meat were still smoking from the kill." Inevitably, the girls pressed the paleface visitors to dance the rumba. Resistance was difficult, and when a foot-tapping Trotsky was on hand to take the side of the girls, impossible.

For Trotsky's secretaries, these festivities often had to conclude early, either because the OM and Natalia wished to retire for the night or because there was a guard shift to cover. Hansen made a point of reassuring his wife that his sole nighttime companion was a handgun. "It is heavy, powerful, accurate, sure action, with very good safety devices. It has not yet worn a hole in my pocket, but I have grown used to the weight and feel undressed when I lay it aside. At the slightest noise at night I awake with its heaviness in my hand." This might be mistaken for coded language had Hansen not specified that the gun in his hand was "a beautiful PARABELLUM automatic." This was the same Luger that his predecessor used to dismantle at night. It had gotten rusty, but a rod and some oil had cleaned it up nicely.

Guard duty at the Blue House took only a few hours out of the day, but the toll could be significant. Sleep deprivation was a persistent condition. The nightly barking and howling of the neighborhood dogs could put nerves on edge, though the greater challenge was fending off boredom. The danger came into sharper focus on February 2, 1938, when the phony deliveryman came calling with his package of fertilizer. It was in reaction to this bomb scare that Diego bought the property next to the Blue House. At the same time, the staff prepared to implement security measures that had long been put off, such as the installation of an alarm system. Also back on the agenda was the recruitment of a full-time guard. Once again Trotsky signaled his approval, only to retreat and call for a postponement. This time, however, Van and Hansen became exasperated by his vacillation and asked New York to overrule him.

The decisive event came two weeks later, on February 16, with the news of Lyova's death. When Trotsky emerged from mourning on the twenty-second, he authorized the hiring of the new guard. In New York, Hank Stone was already packing his things. By now, however, the New York office had decided that one guard was no longer enough to meet the threat. A garrison of three comrades was required, operating independently of Trotsky's secretarial staff and on a separate budget. The guards would be lodged in the small house just acquired next door and be provided with their own automobile. The money needed to pay for this and other essential security measures had yet to be identified, but there was no time to waste and Trotsky should not be given an opportunity to change his mind.

The residents of the Blue House anticipated a frontal attack. Every contingency had to be considered. Van made inquiries about procuring gas masks and a machine gun. As it turned out, however, the next assault assumed a familiar form. In Moscow, another show trial had begun.

The "Trial of the 21," the third and most grotesque of the Moscow show trials, opened on March 2, 1938. The most high-profile defendant was Nikolai Bukharin, at one time a popular figure and a favorite of Lenin. A leading Bolshevik theorist, Bukharin was for years editor of *Pravda*, and in the 1920s the leader of the Party's right wing. Among his co-defendants in the dock were some fellow Old Bolsheviks and, a strange sight, Genrikh Yagoda, the former head of the GPU. The fact that the man who had exposed the Trotskyist conspiracy in the first trial had now been unmasked as Trotsky's agent was one of the more bizarre features of this extraordinary spectacle.

The indictment had a familiar ring. The defendants were accused of forming a "Right-Trotskyist Bloc" for the purpose of overthrowing the Soviet regime and restoring capitalism. Toward this end and acting in conjunction with the intelligence services of Germany, Japan, Great Britain, and Poland, they had conspired to carry out a variety of criminal acts, including sabotage, murder, and mass poisonings of workers. Bukharin, among his many other crimes, was charged with plotting to assassinate Lenin in 1918. All the defendants eventually confessed their guilt, and all but three were executed immediately, including Bukharin, the rest within a few years.

Once again, Trotsky was placed at the center of the conspiracy, as was Lyova, now conveniently unavailable to defend himself. Trotsky's study at the Blue House was transformed into a war room. Newspaper reports about the trial testimony were carefully scrutinized for contradictions and outright absurdities, which were then made the subject of daily press releases. For Trotsky, this was the third time around, and yet he could hardly believe his eyes as he read the fantastic confessions of the accused. "It all seems like a delirious dream," he said. This time, however, he was not nearly so isolated. The Dewey Commission had recently delivered its verdict condemning the first two trials, and Dewey himself now denounced the third proceeding from New York, where for many beleaguered sympathizers of the Soviet Union the execution of Bukharin was the last straw.

Lyova's death and Bukharin's trial gave a boost to the New York Trotskyists as they sought to raise the funds necessary to improve security at the Blue House. Trotsky personally took part in the effort, declaring, "I will not be accused that I offered a too easy victory to the G.P.U." Two days after the trial ended, on March 15, Hank Stone, the first chief of the guard, arrived in Coyoacán.

Stone, whose real name was Henry Malter, was a thirty-year-old military engineer and officer in the New York National Guard. He had been a Trotskyist since 1930, when he joined the Spartacus Youth League. In 1937 he volunteered to go to Spain with the Eugene Debs Column, which was conceived as a non-Communist alternative to the Abraham Lincoln Battalion but never went to Spain. Instead, Stone had to settle for an earful of his friend Harry Milton's hair-raising tales of the civil war.

Stone's initial inspection of the Blue House revealed several problems that needed immediate attention, including an appalling shortage of the most basic supplies. There were no tools whatsoever, not even a hammer and nails, nor any tools or spare parts for the Dodge. There were no extra bulbs or fuses and only one flashlight, with dead batteries. A household that was bracing for a machine gun raid or a bomb attack lacked a decent first aid kit.

Firearms were also in short supply. Two of the five handguns were found to be *hors de combat*. These did not include the guns of Trotsky and Natalia, which were in working order, though there was precious little ammunition for any of these weapons, none of which had been cleaned since Hansen arrived the previous autumn. "Some actually looked as though they had cobwebs inside the barrel," Hank complained.

So much for the existing arsenal. The local comrades promised a supply of new guns, but Stone became impatient and decided to go shopping. Purchasing guns in Mexico, he was happy to discover, was as easy as "buying bananas." He bought three .38 Colt revolvers to supplement the new one he had carried down with him, and a hundred shells to add to his own total of fifty. A final acquisition was a .22 Colt to use for target practice, along with a thousand shells.

The budget to support the guard was set at $100 per month, but soon Stone was reporting that he would need at least $150. In fact, during the next two years the guard fund was often low on cash or completely broke, and it became routine for the chief of the guard to harass New York about a long-promised sum of money. In a pinch, he could request a loan from Natalia, who managed the separate household fund supported by Trotsky's publishing income.

In the wake of the February events, New York arranged for the party's Minneapolis organization to underwrite the guard at the Blue House. Minneapolis became a Trotskyist stronghold in 1934, during the city's great truckers' strike, a protracted and violent conflict that ended with the unionizing of the teamsters. That same year witnessed similar strikes at the Toledo car plants and the San Francisco docks, but only in Minneapolis, home of General Drivers Local 574, led by the notorious Dunne brothers, did the Trotskyists make inroads into American labor.

Minneapolis agreed to contribute financially and also in kind, offering the services of two members of the local union defense guard. Their names were Bill and Emil, and although they were identified at the Blue House simply as the Minneapolis boys, in fact both were experienced picket-line fighters in their mid-thirties. Emil was a gentle giant. Hansen describes him as "a big meaty fellow with a dagger stabbed through a bleeding heart tattooed on his forearm, weighs about 230 or 240 pounds and carries a great paunch full of guts." Bill was also built to defend his position, an impression reinforced when he smiled, disclosing the absence of six teeth in his left upper jaw.

Hank signed on for six months, Bill and Emil for three, though at the outset it seemed doubtful that the Minneapolis contingent would last more than a few days. The first indication of trouble was their stunned reaction when Hank broached the subject of nightly guard duty. Evidently their job description said nothing about a night shift.

Soon it became obvious to Hank that while he had been warned to expect a prison regime in Coyoacán, Bill and Emil had been promised a Mexican vacation. They were quite willing to assist Hank in strengthening the front door and installing the new lighting and alarm systems. What they found objectionable were Natalia's many requests for help cleaning the kitchen and maintaining the patio, as well as other chores, such as driving into town with the mail and doing the food shopping. This gave rise to bitter complaints about having to perform "women's work."

Most troubling to Hank was Bill's delinquency as a guard. The problem first came to light one evening shortly after Bill's arrival, when Trotsky accompanied a guest to the front door at about 10:30 p.m. and found no guard on duty. When Hank questioned Bill about this, "he in polite terms told me to go jump in a lake." The next night, when Hank came on duty to relieve Bill at 4:00 a.m., a Mexican comrade reported that the recalcitrant gringo had retired at 2:00 a.m. Confronted about this dereliction, Bill threatened to return to Minneapolis. "He has a very independent spirit," Hank informed New York, "very good on a picket line—but of no value here. Here one demands discipline."

Food proved to be a major source of discontent. Bill and Emil were incredulous when they realized that their Mexican vacation meant bread without butter and coffee with milk instead of cream. Hank also grumbled about the food placed in front of him, but the teamsters threatened to strike. Since Lyova's death, Trotsky and Natalia took their meals in their room, but Natalia continued to set the menu and resisted making adjustments for the sake of Bill and Emil. Hansen was amused by their plight: "They cannot endure any food except potatoes and gravy with whatever goes with potatoes and gravy." Hank remained grim-faced: "Lack of butter on the table should not create a political crisis."

A new complication materialized in the third week of April in the person of Bill's wife, Edith, who showed up unannounced and moved in with Bill. After listening to her husband's tale of woe, Edith offered to cook. Hank relented and agreed to the operation of a separate kitchen for the guards in their spartan dwelling. The new arrangement lasted all of ten days. Edith, standing over the stove with a spoon in one hand and a *Ladies' Home Journal* in the other, served meals up to two hours late. When Hank complained, Edith refused to cook anymore. Bill and Edith then began to eat all their meals in the city, which is where Bill spent most of the day.

It was at this very time, toward the end of April, that George Mink was rumored to be in the vicinity. "I believe that a group of Stalin agents headed by 'The Mink' has arrived in Mexico, plotting to kill me," Trotsky told *Time* magazine. A Philadelphia cabby who used to work with Mink begged to differ: "He hasn't got the brains of a flea! He won't kill nobody!" This was not the perception in Trotsky's circle, where Mink was credited with having served as the GPU's station chief in Spain. He was said to have murdered Andrés Nin, kidnapped Erwin Wolf, and arrested Harry Milton. A ten-year-old photograph of the jackal was located and sent down to Coyoacán, where it was used by the guards for target practice. To judge from Hank's tales of woe, it is doubtful that the alert about Mink could have had much effect on security at the Blue House. For Bill and Emil, mention of the Butcher of Barcelona must have conjured up visions of a choice cut of beef to go with their potatoes and gravy.

Trotsky had been inspired by the prospect of being guarded by genuine American proletarians; instead, as a scandalized Van remarked about the Minneapolis boys, *"Ils se conduisent ici comme dans une maison de bourgeois."* In Van's estimation, Hank and Emil were satisfactory, while Bill was a disaster. His attitude demoralized the entire household, especially Hank, who at one point refused to communicate with Bill except through Van. Most distressing was Bill's inclination to whistle, sing, and shout in the patio as he passed by the French doors to Trotsky's study. A reprimand from Trotsky had no effect. When Bill said he would recognize no authority, this evidently included the former People's Commissar of War.

By the middle of May, Van feared that Trotsky would erupt and fire everyone on the spot. Hank's demoralization was now complete. "I no longer consider myself chief of guard or anything of the sort," he wrote to Frankel, who must have squirmed to read this statement in Hank's latest dispatch, which asked that Milton be sent to Coyoacán.

Bill and Emil were scheduled to be released on June 15, but conspired to skip out four days early. Their departure brought a collective sigh of relief; and yet, as these things happen, the goodbye was sad. The night before, the Fernández family threw a party for them and loaded them down with mementos. Trotsky presented each of them with an autographed photo. Natalia gave them presents, as did Rosita, whose cooking the boys found so uninspired, as well as Armando, the boy helper. Hansen describes an emotional send-off: "The cook cried, Armando cried, and the cook's little boy, Alfonso, too—his great black eyes spilling tears on his cheeks like the overflow from a dark pool."

Dry-eyed, Hank stayed on until mid-August, working with a new American guard, Chris Moustakis, recently of Boston, who had a master's degree in history from Harvard. Having driven down to Mexico in a Plymouth Coupe in search of adventure, he became friends with Hansen, who admired his automobile and recruited him to the cause.

Thanks to Hank, the Blue House was now equipped with an elaborate alarm system, made conspicuous by the clutter of switches and alarms spread throughout the house. Visually more impressive was the effect of the floodlights positioned all along the top of the high blue wall. At night, in the inky darkness of Coyoacán, the house stood out like a fort—or a prison. As a security precaution, two cedars and a pine rising above the wall on the street side had been cut down; the great cedar in the main patio now stood alone. The police, meanwhile, had replaced their small wooden shelter outside the house with a more permanent structure made of bricks and covered with stucco, with slots for guns on all four sides.

Summer rains now drenched the patio, and Trotsky was relieved to have some peace and quiet. The deliberations about who should take Hank's place only rankled him. From Coyoacán, Sara Weber, his Russian typist off and on since the Prinkipo days, warned New York that LD was "getting fed up with the entire matter. One of these days all the little irritations and annoyances will just get the better of him and he will flatly refuse to have anyone 'guard' him."

Yet every time Trotsky may have been tempted to let down his guard, along came a reminder of the perils of complacency. In July 1938 it arrived in the shape of a headless corpse floating in the Seine.

The victim was a twenty-eight-year-old German by the name of Rudolf Klement. As a young student from Hamburg, he arrived in Prinkipo in 1933 to serve as Trotsky's secretary, then followed him to France. He was secretary designate of the inchoate Fourth International, whose founding congress was planned for later that summer. He vanished on July 13. When his beheaded remains were identified several days later, it strengthened Trotsky and Natalia's belief that Lyova had died at the hands of the GPU.

Among the Trotskyists in Paris, a dark cloud of suspicion had thickened around Mark Zborowski, the man they called Étienne. He had first come under close scrutiny after the theft of Trotsky's archives on the night of November 6 and 7, 1936. Zborowski was one of only a few comrades who knew the location of these files, and at a tense meeting to sort the matter out, only a strong endorsement from Lyova spared him from an investigation.

The timing of this theft was deliberate: in Moscow the success of the operation was reported to Stalin that same day, the anniversary of the Bolshevik Revolution. Not long afterward, Nikolai Yezhov, Yagoda's successor as People's Commissar of Internal Affairs, presented Stalin with select items from the haul: "I am sending you 103 letters taken from Trotsky's archive in Paris." Among this trove was Trotsky's correspondence with Max Eastman for the years 1929 to 1933.

Zborowski regularly supplied Moscow with articles from the *Bulletin of the Opposition* before they appeared in print, and with copies of Trotsky's letters and manuscripts, including portions of his book-length indictment of Stalin, *The Revolution Betrayed*, which turned up on Stalin's desk before its publication in Paris in the summer of 1937. That August came Zborowski's triumph, when Lyova went to the south of France and entrusted him with a small notebook containing the addresses of Trotskyists living outside the Soviet Union. "As you know, we have dreamed about getting hold of it for a whole year," an exultant Zborowski wrote to his superiors using his code name, "Tulip," "but we never managed it before, because SONNY would never let it out of his hands. I enclose herewith a photo of these addresses."

During Lyova's absence from Paris, Zborowski stood in for him in negotiations to arrange a meeting with Ignace Reiss, the first of the GPU defectors. In making his break with the Kremlin, Reiss, the illegal resident in Belgium, had turned for help to the Dutchman Henk Sneevliet, a Communist member of parliament and trade union leader who had once been a close comrade of Trotsky's. Sneevliet, working through Zborowski, invited Lyova to meet Reiss on September 6 in Reims, France—which is where the defector might have met his end had a GPU mobile squad not machine-gunned him to death a day earlier on a rural road outside Lausanne.

When he read the news, Trotsky was incensed at Sneevliet. A GPU defector who could have drawn back the curtain on the Moscow trials had been murdered in obscurity. Worldwide publicity, Trotsky argued, would have shielded Reiss from assassination. Instead, Sneevliet had acquiesced in Reiss's plan to delay any public announcement until after his impassioned letter of resignation reached the Central Committee in Moscow. Reiss was unaware that the staff member at the Soviet embassy in Paris to whom he had entrusted the mailing of his letter had betrayed him, setting off a manhunt.

Trotsky saw deviousness, as well as ineptitude, in Sneevliet's handling of the Reiss affair. Sneevliet had not only failed to inform Trotsky in a timely way about the defection; he even appeared reluctant to bring Lyova into direct contact with Reiss. Yet while Sneevliet does give the impression of being the controlling sort, he also had the feeling that Lyova's comrades in Paris could not be trusted. And in the wake of Reiss's murder, Sneevliet's misgivings came to focus on Zborowski.

In October came another defection, that of Walter Krivitsky, the chief of Soviet military intelligence in Europe. Krivitsky, who was stationed in the Netherlands, was a childhood friend of Reiss's. The two men had discussed their disillusionment with Moscow after the execution of the defendants in the first show trial in August 1936. They returned to the subject in the spring of 1937, as the terror began to ravage the ranks of the secret police and the military. Krivitsky resisted Reiss's suggestion that they simultaneously break with Moscow, arguing that in spite of everything, the USSR still represented the best hope of the international proletariat.

Reiss's murder helped Krivitsky overcome his doubts; in fact, his friendship with the dead defector left him little choice. He applied to the French government for political asylum, which brought with it police protection. Reiss's widow, meanwhile, suspected that Krivitsky had had a hand in her husband's death—or at least had failed to warn him of the danger. For his part, Krivitsky, in attempting to establish contact with her through the French Trotskyists, became convinced that they had been infiltrated by the GPU. During a tense meeting at the office of Trotsky's Paris lawyer, Gérard Rosenthal, with Lyova, Sneevliet, and Reiss's widow in attendance, Krivitsky warned: "There is a dangerous agent in your party."

Krivitsky was wary of the Trotskyists for other reasons, as Lyova learned during a series of strenuous meetings with the reluctant defector in the final weeks of 1937. Trotsky and Lyova wanted Krivitsky to make a full and public break with the Kremlin, but he was torn about what to do next and eager to justify his past. Lyova lent him a sympathetic ear, thereby drawing the ire of Trotsky, who was impatient to seize the moment. After Lyova pressed Krivitsky to endorse the Fourth International, Krivitsky broke off their relations. Although Krivitsky had come to like and respect Lyova, he found little to admire about his milieu. Trotsky, the man, was a formidable figure, politically the equivalent of a government, he said later, whereas his followers were mere children.

The rupture may have saved Krivitsky's life. That autumn, Lyova assigned Zborowski to be the defector's contact and escort. The two men ended up taking walks together, probably conversing in their native Polish, and "Tulip" undoubtedly supplied his GPU handlers with information about the traitor's movements. On one occasion they wandered into Père Lachaise cemetery, where Krivitsky noticed some dubious-looking characters off in the distance and for a moment was convinced that the shooting was about to begin. Why their fraternization did not precipitate Krivitsky's murder is a mystery, perhaps best explained by Zborowski's instinct for self-preservation.

Lyova's death in February 1938 may have been a victory for the GPU, but for Zborowski it meant the loss of his chief defender. He used his position as "Sonny's" successor to deflect suspicion away from himself. The principal target of his intrigues was Sneevliet, who, Étienne now dutifully reported to Trotsky, had been spreading the story that Reiss's murder had resulted from Lyova's negligence. Predictably, Trotsky became outraged at the "slanderer" Sneevliet for besmirching his dead son's reputation.

Zborowski worked assiduously to create the impression in Coyoacán that he was Trotsky's most devoted comrade. The two men never met, and the obsequious manner that sometimes irritated Lyova did not come across in Zborowski's letters to Trotsky. The *Bulletin* now began to appear more regularly than it had for a long time. "You are doing a great service in publishing the *Bulletin* so punctually and with such care," Trotsky commended Étienne. "This is to your credit."

Zborowski's facility in Russian made him irreplaceable in Paris, which is why his superiors were wrong to suppose that Lyova's death opened up the possibility "to get to the OLD MAN" by transplanting "Tulip" to Mexico. Zborowski was instructed to offer his services in Coyoacán, but although he claimed that his letter to Van broaching this idea had gone unanswered, no such letter exists and it is unlikely that one was ever sent. Zborowski and his family lived in a comfortable apartment building in Paris, courtesy of the GPU, and in his spare time he was able to pursue his studies in ethnology. It is hard to imagine him going to much trouble to exchange all this for an uncertain future in Mexico alongside the ultimate outlaw.

Then came the Klement murder in July 1938. About two weeks later Trotsky received a letter purporting to be from the victim, writing as a disillusioned follower. An obvious provocation, the text accused Trotsky of collaborating with the Gestapo and of behaving in a Bonapartist manner, and it declared the bankruptcy of the nascent Fourth International. Somehow, Klement's death helped confirm Sneevliet in his suspicion that Zborowski was a GPU informant, a charge that he began to make openly that autumn. So did Victor Serge, like Sneevliet once a close confederate of Trotsky's who had lately become an irritant. Krivitsky and Reiss's widow, meanwhile, voiced suspicions about Serge, detecting the hand of the GPU in his release from Soviet exile two years earlier.

Thousands of miles away in his Mexican redoubt, Trotsky tried to weigh the significance of these conflicting indictments. When Zborowski appealed to him for advice on how to clear his name, Trotsky proposed that he challenge Sneevliet and Serge to bring their charges before an authoritative commission. "The sooner, the more decisively, the firmer, the better," he wrote without any pretense of neutrality, advocating an "energetic initiative . . . to press the accusers to the wall."

Here matters stood in the final days of 1938, when a letter arrived at the Blue House that demolished Trotsky's presumptions. The three-page letter, dated December 27 and sent from New York, was typed in Russian on a Latin-script typewriter and signed "Your Friend." The writer, who claimed to be a Russian émigré to the United States by the name of Stein, said he was a relative of Genrikh Liushkov, a GPU commissar who had defected to Japan. Stein declared that he had recently returned from visiting Liushkov, who wished to warn Trotsky that a "dangerous provocateur" lurked among his followers in Paris. Liushkov, said Stein, could remember only the spy's first name, Mark. This Mark had been close to Lyova and now published the *Bulletin of the Opposition*. He was further identified as a Jew from Poland, between thirty-two and thirty-five years old, who wrote well in Russian, wore glasses, and had a wife and young child. Trotsky understood that the person in question was Mark Zborowski.

According to Liushkov, Zborowski had kept Moscow informed about Lyova's every move, read Trotsky's letters, and was responsible for the theft of his archives in Paris. Mark presented himself as a Polish Communist, but Liushkov expressed skepticism about this and claimed that a background check would reveal that Mark had once belonged to the Union for Repatriation of Russians Abroad, a Paris-based organization run by former czarist officers, in which he had operated as a GPU provocateur. Mark met regularly with personnel from the Soviet embassy in Paris, according to Liushkov, who indicated that this could easily be verified by having him followed. "What surprises me more than anything," Stein injected reproachfully, "is the credulity of your comrades."

Nor was this all. Liushkov believed that Trotsky himself was to be Mark's next target. The GPU, he said, planned to send an assassin to Mexico, either through Mark or from Spain through Spanish agents posing as Trotskyists. Stein advised Trotsky to be extremely cautious. "The main thing, Lev Davidovich, is to protect yourself. Don't trust a single individual sent to you by this provocateur, neither man or woman."

Here was a warning that demanded to be taken seriously. On New Year's Day, Trotsky sent an "extremely confidential, extremely important, and extremely urgent" communication to Jan Frankel in New York, summarizing the contents of the letter and suggesting two possible sources: either a legitimate warning from a timid friend or a GPU provocation. In fact, the Stein letter was the cunning contrivance of an improbable well-wisher. He was Alexander Orlov, until recently one of Moscow's top spymasters.

At one time the illegal GPU resident in London, Orlov helped recruit and supervise Kim Philby, Donald Maclean, and Guy Burgess, the three original members of the infamous Cambridge spy ring, which passed top secret information to Moscow into the early years of the Cold War. At the outbreak of the Spanish civil war he was sent to Madrid to be the GPU's station chief. Officially a mere political attaché, Orlov was the top Soviet official in Spain. It was Orlov who carried out the purge of the POUM and the anarchists in the name of liquidating Trotskyism. When Andrés Nin disappeared from a prison near Madrid in June 1937, he was said to have escaped, but in fact he was abducted, tortured, and murdered by a mobile squad supervised personally by Orlov.

Orlov and his staff in Spain warily monitored Moscow's ongoing purge of the secret police and understood that fellow agents abroad were being ordered home and executed. The fatal summons for Orlov arrived in Madrid on July 9, 1938. Feigning compliance, he slipped across the border to France, collected his wife and daughter in Paris, and sailed to Canada.

Orlov determined that his best hope for survival was to blackmail

Stalin. From Canada he arranged to have a letter addressed to GPU chief Yezhov delivered to the Soviet embassy in Paris. In it he listed all the secrets he could reveal if his life were endangered or in the event of his death. The damage would include the exposure of numerous undercover agents such as the Cambridge spies, the truth about the fate of Nin, and the full story of "Tulip," in which connection he named Sneevliet and Reiss. A separate item on the list read: "All about the OLD MAN and SON."

Orlov figured that Moscow would see the wisdom of leaving him alone. Shortly afterward, he moved to New York, where he made an arrangement with the U.S. immigration authorities that allowed him to reside in the country in obscurity under an assumed name. Of course, Orlov could never be certain that Moscow would allow him to live at all, which necessitated a life of caution and deception. Above all, Moscow must not be given the impression that he had failed to uphold his end of their tacit bargain.

Orlov should simply have disappeared, but he must have had a troubled conscience or some other kind of itch because he decided to take a risk and warn Trotsky. As Orlov was aware, the first GPU man to do this—Yakov Blumkin, in his face-to-face meeting with Trotsky in Turkey in 1929—was exposed and executed. Orlov had to assume that his letter to Trotsky would end up in the hands of the GPU, and so he had to concoct it in a way that concealed his identity. By posing as a relative of the defector Liushkov, he was able to convey highly secret information while eluding detection.

Orlov's "Stein" letter was intended to reveal just enough about Zborowski to have him unmasked. He asked Trotsky to acknowledge receipt of the letter by placing an ad in the *Socialist Appeal*, the Trotskyist weekly printed in New York. Trotsky published the ad, which requested that Stein appear at the office of the Socialist Workers Party and ask to speak to "Martin," the pseudonym for James Cannon. Trotsky needed confirmation that the letter was legitimate, but Orlov was no fool and it would have been foolhardy for him to accept such an invitation.

In his urgent New Year's Day letter to Frankel, Trotsky proposed the formation of a commission of French comrades to investigate the allegations about Étienne-Zborowski. If they proved to be true, he advised, the provocateur should be denounced to the French police for his role in the theft of the archives, and in a way that would cut off all possible avenues of escape. At that moment Trotsky sounded convinced, but he could not help suspecting that he was being played. Three weeks later he speculated that the mysterious correspondent was Krivitsky, who had recently arrived in the U.S. Perhaps the defector actually remained in the service of the GPU and hoped to demoralize Trotsky's camp.

In the hands of the French Trotskyists, Orlov's warning would likely have led to Zborowski's exposure and perhaps his arrest. In the event, Trotsky's instructions never reached Paris. To blame this on the GPU would sound clichéd if not for the fact that Trotsky had singled out Cannon to take the matter up with the French comrades, and Cannon's secretary was an informant for the GPU. Or perhaps a letter from New York to Paris simply went astray, in which case the real mystery is why no one bothered to follow up on a question of such vital importance. For now, Trotsky assumed that the investigation was moving forward, and he waited to learn the results.

It was Orlov who years later told the story of how Blumkin, the GPU man ensnared in Turkey, had endured his interrogation in a Luby-anka prison cell with remarkable dignity and had faced death with extraordinary courage. "When the fatal shot was about to be fired, he shouted, 'Long live Trotsky!'" Perhaps Orlov's telling this story indicates where his sympathies lay, which might account for his effort to warn Trotsky. In any case, Orlov was now powerless to prevent what he had set in motion, and as a result, Trotsky's time was running out. As GPU station chief in Spain, Orlov had been responsible for the recruitment of a twenty-three-year-old native of Barcelona by the name of Ramón Mercader.

CHAPTER 7

Fellow Travelers

n the autumn of 1938, Trotsky began to confront a different sort of threat to his security when his friendship with Diego Rivera began to unravel. On November 2, Diego arrived unexpectedly at the Blue House. It was the Day of the Dead, and the painter was infected with the holiday spirit. "Looking as mischievous as an art student who has played some prank," as Van describes the scene, Diego walked into Trotsky's study and placed on his desk a large purple sugar skull on whose forehead was spelled out in white sugar the name of Stalin. Trotsky decided to ignore this holiday offering, which may have disappointed Diego but could not have surprised him. Their conversation was brief, and as soon as the mischief-maker had gone, Trotsky asked Van to remove the offending object and destroy it.

A year earlier, Trotsky would have found a way to accommodate this exhibition of the painter's irrepressible sense of black humor. Generally speaking, the two men remained on friendly terms. Diego was still the only person allowed to show up unannounced at Trotsky's door. But their friendship was under increasing strain. That these two disparate personalities would eventually clash was almost predictable. The sequence of events that opened the rift between them can be traced to the summer of 1938, when the French Surrealist poet André Breton paid an extended visit to Mexico.

Breton was the leader of Surrealism, whose theoretical principles he set down in two manifestos in the decade following his break with Dadaism in 1922. He had long been an admirer of Trotsky's. In 1925 in the journal *La Révolution surréaliste* he published a laudatory review of Trotsky's eulogistic volume *On Lenin*. Breton joined the French Communist Party in 1927, yet he and his Surrealist gang in Paris ultimately refused to knuckle under to the Communists. In 1934 they published a tract called *Planet without a Visa* in support of Trotsky's efforts to resist expulsion from France. Two years later, Breton joined a French commission of inquiry into the Moscow trials, which ended up serving as a European branch of the Dewey Commission.

Trotsky was pleased to have a major literary figure like Breton in the anti-Stalinist camp, though he was wary of the Surrealist project, which gave off a strong whiff of mysticism. He had not given much attention to Breton's books, however, and as the celebrated poet and essayist was about to visit Mexico, it was time to bone up. Van arranged to have the essential Breton *oeuvres* sent down from New York, courtesy of art historian Meyer Schapiro.

WHEN IT CAME to painting, Trotsky confessed he was never more than a dilettante. In the field of literature, however, he could claim to be an authority. He wrote extensively about literary fiction, beginning during his first exile to Siberia at the turn of the century, when he was a regular contributor to the Irkutsk paper *Eastern Review*. The young radical stood up for literary tradition. In an appreciative essay devoted to Nikolai Gogol in 1902, on the fiftieth anniversary of the writer's death, Trotsky defended the author of *Dead Souls*—his greatest novel, published in 1842—from those who found his social criticism too timid. When all was said and done, Gogol was the "father of Russian comedy and the Russian novel," the first "truly national writer," a forerunner of Goncharov, Tolstoy, and Dostoevsky.

"The novel is our daily bread," Trotsky once remarked. He was especially devoted to the French novelists; Balzac and Zola were among his favorites. He had a strong preference for realist works, a predilection reinforced by his Marxist philosophy. Only literature that was socially conscious truly satisfied him. In two early essays about Tolstoy, he praised the novelist's prodigious talent for invoking character and atmosphere—his "miracle of reincarnation"—but scorned his narrow focus on the familiar world of aristocrats and peasants and his flights from reality into nature and religion.

In the first decade of Bolshevik power, Trotsky became Soviet Russia's most influential literary critic and its most effective advocate of freedom in the arts. The idea of proletarian culture was then in great vogue among writers and radical theorists in Moscow and Petrograd. This movement, spearheaded by a group called Proletcult, argued that prerevolutionary art and literature ought to be tossed into history's dustbin along with the former ruling classes. Lenin, whose personal taste in art was conservative and pedestrian, resisted Proletcult's radical agenda, which had influential supporters within the Party. Trotsky joined the battle on Lenin's side.

His major contribution to this debate was one of his finest works, *Literature and Revolution*, published in 1923. Trotsky's book surveyed the lively contemporary Soviet literary scene, directing pointed criticism at the three modernist movements of the day: Symbolism, Formalism, and Futurism. His principal theme was the indispensability of tradition, even in the homeland of communism. "We Marxists have always lived in tradition," he admonished, "and we have not ceased to be revolutionaries because of it." The notion that the art and literature of past epochs merely reflected the economic interests of vanquished social classes he considered vulgar. Great art, he declared, was timeless and classless.

No less misguided was the belief that the dictatorship of the proletariat should extend its reach into the domain of culture. The proletariat's rule would be brief and transitory, Trotsky advised, giving way to a classless socialist society and with it the first universal culture. In any event, the Russian worker was now a cultural pauper. His immediate challenge was not to break with literary tradition, but rather to absorb and assimilate it, starting with the classics. "What the worker will take from Shakespeare, Goethe, Pushkin, or Dostoevsky will be a more complex idea of human personality, of its passions and feelings, a deeper and profounder understanding of its psychic forces and of the role of the subconscious, etc. In the final analysis," he said, "the worker will become richer."

The central task of the Party, in the meantime, was to foster an atmosphere of tolerance in order to allow Soviet culture to flourish. The Party must be prepared to exercise what Trotsky called "watchful revolutionary censorship" against any artistic movement openly opposed to the Revolution, but otherwise it should assume no leadership role. "Art must make its own way and by its own means," he insisted. "The domain of art is not one in which the party is called upon to command."

Literature and Revolution is one of Trotsky's most sparkling works. A tough-minded critic, he could be unsparing when dealing with artists hostile to the Revolution, as in his savage arraignment of the Symbolist poet Andrei Bely, author of the 1916 novel *Petersburg*, now widely regarded as a masterpiece. Trotsky's book showcases his full virtuosity as a writer: it is replete with aphorisms, telling metaphors, and brilliant turns of phrase. It was here that he introduced the label "fellow travelers" to designate writers who, despite their vital contributions to early Soviet letters, would be able to progress only so far along the road to socialism. Fellow travelers, Trotsky explained, "do not grasp the Revolution as a whole and the communist ideal is foreign to them."

Trotsky's reputation for tolerance in the arts left him vulnerable to charges of encouraging bourgeois individualism and spreading defeatism on the cultural front, transgressions that were added to the list of his heresies as head of the Left Opposition. After he was banished from Moscow in 1928, the champions of proletarian culture had their day, surging forward to lead the cultural counterpart to the crash industrialization and collectivization campaigns of the first Five-Year Plan. This tidal wave swept away the independent literary schools and the fellow travelers. Inevitably, the tide then turned against the proletarian writers. In 1932 the Party liquidated all autonomous literary organizations and enforced membership in a Union of Soviet Writers under the direction of the Party.

Soviet writers under Stalin were employed as instruments of state education and propaganda. They were expected to present idealized depictions of Soviet life: the struggle against kulak saboteurs during collectivization, the building of a steel plant at Magnitogorsk in the Urals, the construction of a hydroelectric station in the Ukraine, the rehabilitation of an inmate in a labor camp, and so on. The new style, which was imposed on all the arts, went by the name of socialist realism. A decade after Trotsky had canvassed a richly diverse Soviet literary scene, Max Eastman titled his scathing indictment of Stalinist culture *Artists in Uniform*.

All this upheaval took place while Trotsky was in exile. Still a voracious reader of novels, he no longer devoted himself to serious writing about literature, although he occasionally fired off a salvo in the direction of Soviet culture. He was offended most of all by the enlistment of the arts in the cult of Stalin and his minions. "It is impossible to read Soviet verse and prose without physical disgust, mixed with horror," he complained, "or to look at reproductions of paintings and sculpture in which functionaries armed with pens, brushes, and scissors, under the supervision of functionaries armed with Mausers, glorify the 'great' and 'brilliant' leaders, actually devoid of the least spark of genius or great-ness." Stalinist hegemony over the arts, he asserted, would go down in history as an era of "mediocrities, laureates and toadies."

Not all the laureates and toadies were mediocrities, however, a point Trotsky made using the example of Alexis Tolstoy, a gifted writer of science fiction and historical novels and a distant relative of the great novelist. In 1937, Tolstoy used his talents to promote the leader cult with a civil war novel called *Bread*, which portrayed Stalin and Kliment Voroshilov as heroic defenders of Tsaritsyn on the Volga in 1918. This especially embittered Trotsky, because as war commissar he had removed both men from the Tsaritsyn front for insubordination, even threatening Voroshilov with arrest. Now, as Trotsky looked on helplessly, the Soviet Tolstoy turned this history on its head, elevating the insubordinates and eliminating the true hero of the tale. "Thus, a talented writer who bears the name of the greatest and most truthful Russian realist, has become a manufacturer of 'myths' to order!"

Yet Tolstoy, like so many other artists under Stalin, was merely practicing a different form of realism. With the Great Terror filling the prisons and the labor camps, *Bread* was Tolstoy's insurance against a knock on the door in the middle of the night. Besides, he had long understood on which side his bread was buttered. Voroshilov, recently promoted to Marshal of the Soviet Union, had been made head of the Red Army back in 1925, the same year that the city of Tsaritsyn was renamed Stalingrad.

★ ★

Trotsky's reputation as a Bolshevik with an enlightened attitude toward the arts won him a loyal following in literary circles outside Stalin's

Russia. His arrival in Mexico helped to crystallize the disillusionment with Soviet Communism among a group of radical writers and critics who would later become known as the "New York intellectuals." For a brief, intense moment, these apostates, among them some of the country's literary luminaries, present and future, were drawn into Trotsky's orbit.

This was the era of the Popular Front. In the United States, the Communist Party lined up behind President Franklin D. Roosevelt's New Deal, while New Deal liberals came out in support of the Soviet Union. As the gap between radicalism and liberalism narrowed, the Communists enjoyed a spike in membership and influence. Literature followed politics, as liberal writers gravitated toward the party and its front groups, magazines, and writers' congresses.

These abrupt changes in the Party line inevitably produced disillusionment on the left. Among the disaffected were William Phillips and Philip Rahv, editors of the literary journal *Partisan Review*, founded in 1934 as the organ of the New York branch of the Communist-sponsored John Reed Club. In the autumn of 1936, Phillips and Rahv suspended publication of their magazine, before relaunching it the following year as an independent literary organ of the anti-Stalinist left. The *Partisan Review* editors were not the first leftists to renounce Soviet Communism, but their recast magazine became the most important rallying point of disillusioned radicals for whom Trotsky became a lodestar.

The initial source of their discontent was literary. Radical critics like Phillips and Rahv sought to create a Marxist literary aesthetic, but were repulsed by the "vulgarizers of Marxism" who put political before literary standards. Their chief antagonists were the hard-core radicals associated with the Communist paper *New Masses*, who insisted on a sharp break with the past in promoting a new generation of socially conscious "proletarian" writers. The *Partisan Review* editors argued the need to assimilate the literary achievements of the past, including 1920s Modernism, as exemplified by Proust, Joyce, and Eliot, which was anathema to the orthodox left. Modernism was bourgeois, they agreed, but must nonetheless be preserved as part of what Phillips called the "continuum of sensibility." The Popular Front strategy's sudden shift toward rural, nativist, and patriotic themes intended to appeal to a middle-class audience was the final indignity. Thanks to the financial backing of the painter George L. K. Morris, Phillips and Rahv would be able to publish their magazine without reliance on the Communist Party.

By 1936 these literary discontents were overshadowed by the two great political controversies of the day: the Spanish civil war and the Moscow trials. Spain was supposed to be the great anti-fascist cause, yet France's flagship Popular Front government failed to come to the defense of the Spanish Republic, while reports out of Spain told of Soviet persecution of the non-Communist left. The bizarre spectacle of the Moscow trials was the subject of endless debate among liberals and radicals. For the skeptics, Trotsky's condemnation of Stalinism as a betrayal of the Revolution showed how one could reject Soviet Communism without abandoning fidelity to Marxist principles and Leninist ideals.

A galvanizing event in the consolidation of the anti-Stalinist forces was the formation in late 1936 of the American Committee for the Defense of Leon Trotsky and the furious campaign by liberal sympathizers of the Soviet Union against the creation of a neutral commission of inquiry. "There is now a line of blood drawn between the supporters of Stalin and those of Trotsky," writer James T. Farrell told his diary three weeks after Trotsky had landed in Mexico, "and that line of blood appears like an impassable river."

At this critical juncture, the *Partisan Review* circle was an embattled minority in Union Square in lower Manhattan, the epicenter of intellectual radicalism in the United States, yet it conceded nothing in intellectual firepower. Among the writers and academics in its camp were Elliot Cohen, managing editor of the *Menorah Journal*; Edmund Wilson, the leading literary critic of the day; Lionel Trilling, who would assume Wilson's mantle a decade hence; Sidney Hook, the Marxist philosopher who helped persuade Dewey, his mentor, to head the commission of inquiry; James Burnham, Hook's colleague in the New York University philosophy department; Lionel Abel, the playwright and critic; V. F. Calverton, publisher of the Marxist journal *Modern Monthly*; and James Rorty, a founding editor of *New Masses*.

Among the coming generation who fell under *Partisan Review's* spell in these years were the Trotskyist students at City College in upper Manhattan—known as the Harvard of the Proletariat—young men such as Irving Kristol, Melvin Lasky, and Irving Howe. They retreated to an alcove of the lunchroom beneath the neo-Gothic Great Hall to argue radical politics and Marxist theory, joined there by socialist classmates such as Nathan Glazer, Seymour Martin Lipset, and Daniel Bell, all three of whom went on to become professors of sociology at Harvard.

Not all of the New York intellectuals were from New York. Farrell moved there from Chicago in 1932, the year the first volume of his breakthrough Studs Lonigan trilogy was published. Its graphic portraits of the lower-middle-class Irish on Chicago's South Side were drawn from his own experience. Farrell could take credit for enlisting writer and critic Mary McCarthy in the cause of Trotsky's defense, and he was essential in prodding Phillips and Rahv along the road toward open anti-Stalinism.

The first issue of the revamped *Partisan Review*, which appeared in December 1937, included fiction by Farrell and Delmore Schwartz, poetry by Wallace Stevens and James Agee, an essay on Flaubert by Edmund Wilson, a review article on Kafka by F. W. Dupee, and trenchant book reviews by Trilling and Hook. It was the beginning of what became for the next two decades the premier literary journal in the United States.

Most of the radicals who identified with *Partisan Review* in the late 1930s would become leading Cold War liberals, some of them evolving into the original neoconservatives; a few managed to retain their faith in a more modest vision of socialism. As leading public intellectuals in the postwar era, they founded influential magazines of their own, such as *Dissent*, *The Public Interest*, *Encounter*, and *Commentary*. But during those eventful years leading up to the Second World War, all of them were radicals bound together by their anti-Stalinism; and anti-Stalinism, as the art critic and *Partisan Review* contributor Clement Greenberg once observed, "started out more or less as Trotskyism."

By far most were *Trotskysants* rather than Trotskyist—meaning they never signed on as members of the Trotskyist party. They were fellow travelers, to use a term just then entering the American political lexicon, although without the pejorative sense it would acquire in the McCarthyist 1950s, when it was used synonymously with "pinks" to distinguish communist sympathizers from the card-carrying sort. Even Farrell never joined, although he went down to Coyoacán for the Dewey hearings, helping to barricade the windows at the Blue House until he developed sinus trouble.

Trotsky's appeal to these dissenting radicals went beyond his Marxist critique of Stalinism and his tolerant ideas about culture. He was perceived as the cultivated, Western, internationalist alternative to the peasant, Asiatic, and nationalistic Stalin. He was a man of heroic deeds as well as enlightened words. "For relaxation on the military train that bore him from one front to another, he read French novels," enthused Dwight Macdonald, a new member of *Partisan Review*'s editorial board. "Trotsky's career showed that intellectuals, too, could make history."

The fact that Trotsky happened to be a Jewish intellectual strengthened the connection. Most of the writers associated with *Partisan Review*, like the New York intellectuals generally, were disproportionately Jews. Many were the children of immigrants from Eastern Europe, like Phillips, whose father changed his name from Litvinsky, or had themselves come to America as children, like Rahv, who was born Ivan Greenberg in a small Ukrainian village in 1908.

In the summer of 1937, Macdonald, acting on behalf of the editors, wrote to invite Trotsky to contribute to the new *Partisan Review*, calling it an "independent Marxist journal." As for topics, Macdonald suggested that Trotsky might wish to apply the principles of *Literature and Revolution* to Soviet letters of the previous decade; or he could present an analysis of "the relation of the Marxian dialectic to the theories of Freud." Or something on Dostoevsky or on Ignazio Silone's new novel—whatever Trotsky might like, although Macdonald took care to point out that the magazine would emphasize literature, philosophy, and culture, rather than economics or politics.

Trotsky sensed timidity behind Macdonald's invitation, which explains his impudent response. He would be "very happy to collaborate in a genuine Marxist magazine pitilessly directed against the ideological poisons of the Second and Third Internationals," he wrote, "poisons which are no less harmful in the sphere of culture, science and art than in the sphere of economics and politics." He could make no commitment, however, until the editors produced a "programmatic declaration" spelling out their political orientation. An editorial statement was sent down to Coyoacán, but Trotsky found it too vague and decided to await the appearance of the journal's first issue later in the year.

The wooing continued in January 1938, when the editors invited him to contribute to a *Partisan Review* symposium titled "What is Living and What is Dead in Marxism?" Trotsky objected to the entire proposition, beginning with the title, which he called "extremely pretentious and at the same time confused." He may have detected the spirit of Max Eastman hovering over this event. Eastman, who produced the superb translation of Trotsky's *History of the Russian Revolution*, had announced his break with Communism at the beginning of 1937 in an article in *Harper's* titled "The End of Socialism in Russia." His defection was total. He renounced not just the Soviet experiment as it had turned out, but the October Revolution itself, portraying Stalinism not as a perversion of Leninism but as its logical outcome. Trotsky was outraged, even though for years he had been concerned about Eastman's blasphemous comments to the effect that Marxist theory was religious and metaphysical rather than scientific.

Eastman's change in outlook was a disturbing new development that Trotsky, taking his cue from his American followers, soon began referring to as the "retreat of the intellectuals." No one on *Partisan Review*'s invitation list openly subscribed to Eastman's views, but nonetheless Trotsky objected to the prospective contributors, American and European radicals of various stripes, such as Bertram Wolfe, Victor Serge, and the French socialist and former Trotskyist Boris Souvarine, author of a recent biography of Stalin. "Some of them are political corpses," Trotsky objected. "How can a corpse be entrusted with deciding whether Marxism is a living force? No, I categorically refuse to participate in that kind of endeavor."

Trotsky's harshly negative reaction was influenced by his disappointment with the first two issues of the magazine. "I shall speak with you very frankly," he lectured Macdonald. "It is my general impression that the editors of *Partisan Review* are capable, educated and intelligent people but *they have nothing to say*." A world war was looming, yet the

editors seemed content to create "a peaceful 'little' magazine" and retreat into "a small cultural monastery." *Partisan Review* ought to demonstrate its partisanship, Trotsky felt, by taking the battle directly to the
liberal apologists for Stalin at *The Nation* and *The New Republic*. Instead,
"You defend yourselves from the Stalinists like well-behaved young
ladies whom street rowdies insult."

This was too much for Rahv, who now took over for Macdonald.
He fired a respectful blast back at Trotsky, accusing him of being out of
touch with the American scene. Whatever its faults, *Partisan Review* was
the "first anti-Stalinist left literary journal in the world," Rahv pointed
out. As such, it was under "tremendous pressure," constantly attacked
in the pages of *New Masses* and *The Daily Worker* as a "Trotskyite" rag,
while independents demanded reassurance that it was in fact no such
thing. No wonder, then, that *Partisan Review* was politically tentative,
that it "leaned over backward to appear sane, balanced, and (alas) respectable." We had hoped to receive your support, Rahv told Trotsky,
but instead "you have shrugged your shoulders."

Rahv promised an unambiguous declaration of principles in the
April 1938 issue, and this time he did not disappoint. "Trials of the
Mind," which appeared over his name, equated fascism and Stalinism
and portrayed Trotsky as Lenin's true successor. This was enough to
satisfy Trotsky, who wrote a discursive letter to the editors that was
published as "Art and Politics in Our Epoch" in the August-September
issue.

"Art and Politics" argued the case for artistic freedom as the anti-
dote to the "lies, hypocrisy and the spirit of conformity" that afflicted
the cultural world, most acutely in the USSR. A case in point was
Diego Rivera, Red October's greatest interpreter, whom "the Fourth
International is proud to number in its ranks." An artist of courage and
integrity like Rivera, who stood up to the Rockefellers inside the very
temple of capitalism, could never be welcomed in the Soviet Union.
"And how could the Kremlin clique tolerate in its kingdom an artist
who paints neither icons representing the 'leader' nor life-size portraits
of Voroshilov's horse? The closing of the Soviet doors to Rivera will
brand forever with an ineffaceable shame the totalitarian dictatorship."

Nor, in Trotsky's view, did the state of affairs in the capitalist countries offer much reason for optimism. Just as Marxist theory had fore-cast, the decline and decay of bourgeois society created an inhospi-table environment for artistic achievement. "The artistic schools of the last few decades—cubism, futurism, dadaism, surrealism—follow each other without reaching a complete development." Even as he wrote these words in mid-June, Trotsky was taking the measure of Breton and his Surrealist project.

André Breton paid the price for being a writer of openly anti-Stalinist views in the era of the Popular Front. Desperate for money, he solicited from the French Ministry of Foreign Affairs a commission abroad. This is what brought him to Mexico City, where he was to deliver a series of lectures on French literature and art.

Surrealism preached the virtues of poetry over the novel, a genre that Breton called tedious. Nonetheless, he remains best known for his essays and other prose works, beginning with the first *Manifesto of Surrealism*, published in 1924, and its 1930 sequel. Surrealism's defining principle was "pure psychic automatism," which inspired a technique for the spontaneous generation of pictures or texts without any form of conscious control. For Breton and the Surrealists the key to indi-vidual freedom and social liberation was the unconscious mind, acces-sible through interpretations of dreams and explorations of madness. They prized magic coincidence, chance encounters, and "convulsive beauty"—phenomena at the heart of Breton's several autobiographi-cal adventure journals, whose ruminative account is supplemented by photographs of individuals, locations, and objects encountered by the narrator. The first of these, published in 1928, was *Nadja*, which Trotsky consulted before the author's visit and which remains his most popular work.

Breton met Freud in Vienna in 1921, and the Surrealists adopted him as their patron saint. It was a distinction Freud did not welcome. He rejected Surrealism's claims for the scientific validity of its "poetic" variation on therapeutic psychoanalysis. After meeting Salvador Dalí in London in 1938 and contemplating his hauntingly beautiful painting *Metamorphosis of Narcissus*, he confided to Stefan Zweig that until he had been introduced to the fanatical Spaniard, who impressed him with his intriguing symbolism and his "undeniable technical mastery," he considered the Surrealists to be "absolute (let us say 95 percent, like alcohol) cranks."

Breton arrived in Mexico City in mid-April with his wife and muse, the Surrealist painter Jacqueline Lamba, who was the inspiration behind his 1937 adventure journal, *L'Amour fou*. Forty-two years old, leonine and noble in appearance, Breton possessed an intimidating air of authority. A fanatical idealist with a weakness for the occult, he was by all reports a charismatic speaker. He also had the reputation of being something of a tyrant. He both charmed and terrified his followers, conducting café "excommunications" of nonconformists with the inquisitorial swagger of an art commissar. Friends and detractors alike called him "the pope of Surrealism." Breton and the beautiful twenty-eight-year-old Jacqueline—blond, lithe, and birdlike—were taken in by Diego and Frida. Breton, who was moved to tears by the majesty of Rivera's murals, was bewitched by Frida's little windows into the unconscious mind. With no guidance from Surrealism, he marveled, her art had "blossomed forth . . . into pure surreality."

Breton and Jacqueline made their first visit to the Blue House at the beginning of May. Breton later described his state of excitement as he was led across the patio, his heart racing, his mind barely conscious of the bougainvillea, the cacti, the stone idols along the walkway leading toward the French doors to Trotsky's study. Inside the well-lit room filled with books stands the living legend himself. Breton is astonished to find him looking so young, his complexion as soft as a young girl's, his eyes a deep blue, his brow impressive beneath an abundance of silver-gray hair. It begins to sound as if Breton has wandered into Madame Tussaud's, but then the figure before him begins to move: "As his visage becomes animated, as his hands express with extraordinary finesse this or that remark, he radiates from his whole person something electrifying."

Somehow Breton was able to maintain his composure, and the two men and their wives had an enjoyable visit. News was exchanged, but no major topics were discussed. Trotsky was sizing up Breton. He was keen to learn the reactions in Paris to the Moscow trials, particularly of the writers André Malraux, who had gone to fight in Spain and remained loyal to Moscow, and André Gide, who had repudiated Communism after his 1936 visit to the Soviet Union. The date of Breton's first lecture, set for the Palacio de Bellas Artes, was approaching, and Trotsky asked Van to organize an unobtrusive security force from among the Mexican comrades in case the Communists tried to disrupt the event.

Their next meeting—on May 20 at the Blue House, with Natalia, Jacqueline, and Van sitting in—was more memorable. Trotsky quickly launched into a defense of Zola, a favorite target of the Surrealists. "When I read Zola," Trotsky said, "I discover new things, things that I did not know. I enter into a broader reality. The fantastic is the unknown." Breton, taken by surprise, visibly stiffened. "Yes, yes, I agree. There is poetry in Zola," he replied, swerving to avoid a head-on collision.

Trotsky then challenged Breton's claim on psychoanalysis. "Freud raises the subconscious into the conscious. Are you not trying to bury the conscious under the unconscious?" It was a charge that Breton had heard many times before and he did not hide his impatience. "No, no, obviously not," he said. "Is Freud compatible with Marx?" he came back at Trotsky. Surrealism, its adherents professed, reconciled Freud and Marx, although Breton had always made clear that no theory, not even dialectical materialism, would be allowed to interfere with Surrealism's "experiments with the inner life." Trotsky parried Breton's thrust. Freud analyzed the individual, he said, while Marx interpreted society. "One would have to enter into an analysis of society itself."

Natalia served tea and the mood lightened, as the conversation turned to art and politics. The Nazi government had recently staged its notorious antimodernist exhibition "Degenerate Art," which opened at the Haus der Kunst in Munich and then traveled to other cities in Germany and to Austria, which Hitler had annexed two months earlier. Stalinist influence over culture, meanwhile, was being spread more insidiously, through a proliferation of centers, committees, and congresses, making it difficult, as Trotsky put it, "to trace the line of demarcation between art and the GPU." To counter this threat, Trotsky proposed the creation of a federation of independent revolutionary artists and writers. Breton endorsed the idea and agreed to draft the founding manifesto.

Then came the excursions and the road trips. Trotsky and Breton, sometimes in the company of Diego, picnicked in Chapultepec Park, ascended the Pyramid of Quetzalcoatl in Xochicalco, dined along a frozen lake in the crater of Popocatépetl, traveled to the snow-capped volcano of Toluca, and toured Cuernavaca, where they took in Rivera's frescoes on Mexican history at the Cortez Palace. Breton declared Mexico to be a "land of convulsive beauty" that was destined to become "the Surrealist place par excellence." In making this pronouncement he mentioned Mexico's mountains and flora and the mixed race of its people, though there were other Mexican treasures that he wished to claim for Surrealism. He greatly admired his host's extensive collection of pre-Columbian sculptures from Chupicuaro: ornate clay figurines representing naked women with conspicuously rendered genitals. On the road Breton and Rivera would forage through small villages in search of these three-inch ladies of Chupicuaro, while Trotsky looked on with obvious disdain.

Breton later confessed to have fallen hard for Trotsky. *"Cette séduction est extrême,"* he told an audience of Trotskyists in Paris as he recalled Trotsky's magnet pull. He also confided that there had been "skirmishes" between them, by which he meant several testy exchanges about Surrealist theory, which put Breton on the defensive. Trotsky remained skeptical of Breton's concept of the unconscious as a tool for social liberation. On one occasion he questioned whether the poet's true concern was to "keep open a little window on the beyond," his hands tracing a small square in front of him.

There were other kinds of skirmishes as well, which Breton would not have cared to speak about in public. One afternoon, the two of them, together with Van, stopped to visit a church in a small town near Puebla. The interior was low and dark, its left wall and pillars covered with *retablos*, the ex-votos, or votive offerings, painted in oil on sheets of tin. Breton was instantly captivated by these folk-art treasures—to the extent that he removed several of them, perhaps a half dozen, and tucked them under his jacket. Van could see from the expression on

Trotsky's face that he was infuriated. These were, after all, personalized religious icons left by humble people, not *objets trouvés*. Should the police discover this theft, moreover, it could be used by Trotsky's enemies to discredit him. Van braced for an explosion, but it never came. Instead, "Trotsky walked out of the church without saying a word."

It was at the start of June that Trotsky began to press Breton for the promised draft of their joint manifesto. "Have you something to show me?" he would ask the celebrated maker of manifestos each time they met. Trotsky quite naturally slipped into the role of the stern schoolmaster, which inhibited Breton completely. It came to a point where Breton pulled Van aside and asked him to do the writing assignment for him. Although sympathetic, Van wisely declined.

These were the circumstances in place when two cars pulled away from the Blue House in the middle of June and headed for Guadalajara, about 350 miles northwest of Mexico City. The lead vehicle was Trotsky's Dodge, driven by Joe Hansen, with Breton riding up front and Trotsky and Natalia in back. This seating arrangement was Trotsky's idea, since he wanted to talk with Breton. Van rode in the second car, driven by Sixto, with Jacqueline and Frida in back. Diego was already in Guadalajara painting, and the plan was to meet up with him there.

About two hours into the eight-hour trip, the Dodge slowed and rolled to a halt. The trailing vehicle did likewise, some fifty yards behind. Van got out and walked ahead to find out the reason for the unscheduled stop. Hansen, walking back toward him, said, "The Old Man wants you." Breton, meanwhile, had also gotten out. As he passed Van, "without saying a word, he made a gesture of baffled astonishment." Van took Breton's place in the Dodge, which started off again. Trotsky offered no explanation for this switch, and Van understood from his ramrod straight attitude that he had better not ask. Afterward, Hansen could shed no light, since Trotsky and Breton had conversed in French. Natalia, who understood perfectly well, was vague about it.

Arriving in Guadalajara, Trotsky's party went directly to their hotel without making plans to meet up later with Breton and Diego. Once settled in, Trotsky asked Van to arrange a meeting with the muralist José Clemente Orozco, who was painting in the city. In those days, Orozco was Rivera's only equal as a muralist. Although the two painters were not personal enemies like Siqueiros and Rivera, they were tacit rivals. By establishing contact with Orozco, Trotsky apparently intended to put some distance between himself and the Rivera-Breton group. Van was able to set up a meeting for the following day.

Like Rivera, Orozco had practiced his craft in the United States in the late twenties and early thirties, with commissions in New York City, at Pomona College, and at Dartmouth College, where in 1932 he painted a mural cycle on the history of the American continent. After his return to Mexico in 1934, at the behest of the governor of the state of his native Jalisco he moved to Guadalajara, where he painted murals at the university and in Governor's Palace. At the time of Trotsky's visit, the fifty-five-year-old artist was at work in the Hospicio Cabañas, a deconsecrated church, executing a monumental mural cycle about Mexican history, *The Spanish Conquest of Mexico*, widely considered to be his crowning achievement. In part because of his pure Spanish blood, Orozco was often called "the Mexican Goya," yet he was the only one of his great contemporaries not to have studied in Europe.

Trotsky, along with Natalia, Van, and Hansen, met Orozco at the assembly hall of the university. Pale and gloomy in appearance, an impression reinforced by his thick-lensed glasses and a bulwark mustache rising above a perfunctory smile, Orozco was the tormented introvert to Rivera's affable extrovert. His mournful demeanor, like the dark violence of his art, was often attributed to the boyhood explosives accident that had reduced his left arm to a stump at the wrist, though Orozco himself scoffed at this suggestion. Orozco did not portray pre-Hispanic Mexico as a utopia, the way Rivera did. Another contrast was Orozco's depiction of the Mexican Revolution as a tragedy marked by violent struggle, demagogy, and ideals betrayed. And unlike Rivera's, Orozco's brush dealt no less harshly with the poor than with the rich. His style is marked by sharp diagonal lines, oblique angles, and dramatic contrasts of light and dark. His tone is severe, sardonically bitter, even nihilistic.

"He is a Dostoevsky!" Trotsky exclaimed after viewing Orozco's university murals. High above, in the spacious cupola, is *Creative Man*, with idealized images of a worker, a philosopher-teacher, a scientist, and a rebel. On the walls below, in three panels, is *The Rebellion of Man*. Here, in the central panel, the exploiters are not capitalists, but prophets of a false ideology, among them figures resembling Marx, Trotsky, and Siqueiros. Art historians and tour guides continue to draw such associations, although Orozco assured Trotsky that these false prophets were not modeled on actual people, living or dead.

The time he spent with Orozco was the highlight of Trotsky's visit to Guadalajara. He started back to Coyoacán without having made contact with Diego and Breton, in part because he had tired of their bohemian antics. The two artistic couples, undoubtedly more relaxed without Trotsky there to supervise, passed the time wandering around the city in search of old paintings and photographs and antiquities— like schoolkids playing hooky, as Breton remembered it.

After the return from Guadalajara, relations between Trotsky and Breton slowly warmed up again, so that a trip to Pátzcuaro was planned for the first week of July. This time, there would be no drama en route, as Van went ahead with Breton and Jacqueline. High in the mountains of Michoacán, some 230 miles west of Mexico City, they found a quiet and charming town of narrow, dusty cobblestone lanes, large squares, and one-story whitewashed adobe houses with red-tile roofs. Three miles to the north was Lake Pátzcuaro, dotted with islands and surrounded by wooded mountains and extinct volcanoes. The hotel they had chosen was a large old house of a dozen rooms and a garden lush with ferns and flowers. Trotsky and Natalia showed up two days later, and Diego and Frida arrived separately.

Once everyone had assembled, the group took a boat ride on the lake and after sunset ate *pescado blanco* on the tiny cone-shaped island of Janitzio. The plan was to make excursions to the small lakeside villages during the day, while the evenings would be set aside for discussions of art and politics. These would then be published under the title "Conversations in Pátzcuaro," with Trotsky, Breton, and Rivera as authors.

During the first session, Trotsky did most of the talking, much of it in a utopian mode. In the communist society of the future as he envisioned it, art would wither away, as Marx had said about the state, and dissolve into life. Professional painters and dancers would become extinct, as ordinary people decorated their homes beautifully and moved about harmoniously. This is reminiscent of the uplifting passage at the close of *Literature and Revolution*, where Trotsky describes a world in which man's movements become more rhythmic and his voice more musical, a world in which "The average human type will rise to the heights of an Aristotle, a Goethe, or a Marx."

Trotsky's prophesying seems to have sucked the air out of the room. In any case, by the time he finished speaking it was his bedtime, so discussion was put off until the next evening. This was fine with Breton, who was unsettled by Trotsky's egalitarian vision. "Don't you think," he said to Van afterward as they chatted in the garden, "that there always will be people who will want to paint a small square of canvas?" The next evening, Jacqueline informed the group that Breton had come down with a fever and an attack of aphasia, leaving him unable to speak. She reassured everyone that it was not the first such instance and that he was in no danger, but the announcement dampened the spirits of the vacationers. There would be no more conversations in Pátzcuaro.

Back in Coyoacán, Breton made a surprisingly rapid recovery. No longer speechless, he had also managed to overcome his writer's block, giving Trotsky a couple of pages of text written in his meticulously beautiful handwriting, using his trademark green ink: this was the long-awaited draft of the manifesto. Trotsky then went to work, adding passages of a more polemical character to balance Breton's more theoretical approach. Trotsky's contributions amounted to about half the final text. Van translated these from the Russian and the whole was then stitched together. The manifesto, titled "Pour un Art Révolutionnaire Indépendant," was dated July 25, 1938, and signed by Breton and Rivera, as Trotsky decided to bow out in favor of the two revolutionary artists.

Breton's initial draft contained a formula borrowed from Trotsky's *Literature and Revolution*: "Complete freedom in art, except against the proletarian revolution." But Trotsky, aware of how this caveat could help produce an abomination like socialist realism, expunged the qualifying phrase and called instead for an "anarchistic" freedom in the arts. This did not mean art for art's sake, however. The manifesto closed with an exhortation: "Our aims: The independence of art—for the revolution; The revolution—for the complete liberation of art!"

To translate these principles into action, the manifesto called for the creation of an International Federation of Independent Revolutionary Artists, first with local and then national branches, leading to the convening of a world congress. As it turned out, the Paris branch, under Breton, was the largest, with sixty members; and there were small branches formed in Mexico City by Rivera and in London. But all three organizations were short-lived. In Europe, the threat of war, not the declarations of avant-garde artists, was the chief preoccupation. In the United States, *Partisan Review* published the Breton-Rivera manifesto, but the editors' efforts to organize an American chapter of the federation proved to be, as they informed Trotsky, "a resounding flop."

Trotsky and Breton parted on July 30, 1938, as the sun shone brilliantly on the patio of the Blue House. Trotsky presented Breton with the original joint manuscript of the manifesto. The poet was clearly moved by this gesture. He gave Trotsky a portrait of himself by Man Ray, with the inscription: "To Leon Trotsky, in commemoration of the days spent in his light, with my absolute admiration and devotion."

There remained one piece of unfinished business between the two men, which Breton decided to attend to in a letter he wrote to Trotsky during the voyage to France. He confessed to feelings of inhibition whenever he was in Trotsky's presence; its cause was the "boundless admiration" he felt for him. It was a "Cordelia complex," wrote Breton, invoking the name of the youngest of the three daughters of King Lear. It paralyzed him whenever he came face to face with the greatest of men—a pantheon by now reduced to Trotsky and Freud. "Don't laugh at me, it is quite innate, organic, and, I have every reason to believe, ineradicable. . . . But I won't bore you any more with these personal explanations. Let them serve merely to do justice to our misunderstanding on the road to Guadalajara, which you have every right to want to have clarified."

Trotsky, who was quite capable of delivering encomiums to Marx, Engels, and Lenin, evinced a certain queasiness at this manifestation of his own personality cult, and he let Breton know about it. "I am sincerely touched by the tone, so amicable and cordial, of your letter, dear friend, and—should I say it?—a bit embarrassed. Your eulogies seem to me, in all sincerity, so exaggerated that I am becoming a little uneasy about the future of our relations. From the danger of being embarrassed by the eulogies of friends, I am—thank heaven!—well protected by the much more numerous insults of my enemies."

As he wrote these lines in the late summer of 1938, Trotsky could not have imagined that before the year was out, he would be forced to count among his enemies the man he had recently eulogized as Red October's greatest painter.

Diego Rivera's name had been appended to the manifesto for an independent revolutionary art, even though he had not written a single word of it. Rivera had consented to this arrangement, yet looking back on how the friendship with Trotsky unraveled, it appears that this was the start of the trouble. Not long afterward, Rivera began to behave like a man with something to prove, above all to Trotsky.

Frida's absence from Mexico, which seemed to disorient Diego, no doubt influenced the course of these events. In early October she left for New York to prepare for her one-person show at the Julien Levy Gallery, on Madison Avenue at 57th Street, opening on November 1. From there it was on to Paris, where Breton had arranged an exhibition of her work. The Paris show, called "Mexique," placed Frida's work among pre-Columbian sculptures, old paintings, Surrealist photographs by Manuel Álvarez Bravo, and Breton's personal collection of what Frida called "all that junk": masks, dolls, whistles, ornate frames, sugar skulls, pottery, and an assortment of *retablos*.

An underlying source of friction between Trotsky and Rivera concerned the painter's interactions with the local Trotskyists. The Mexican League numbered only about two dozen active members, which did not inhibit them from splitting into factions, as Trotskyists were prone to do. Rivera's fame, money, and force of personality enabled him to impose his will on these comrades, although he had trouble making up his mind. His near-total absorption in his painting, meanwhile, left him little time to devote to routine organizational matters. The effect on the local Trotskyists was disruptive and demoralizing.

During his first year in Mexico, Trotsky did not perceive a problem. On the contrary, he gushed to Hansen about the painter's "incomparable political intuition and insight" and he waved off Jan Frankel's warnings that Diego was a political wild card. By the summer of 1938, however, Trotsky's opinion had changed. Rivera possessed an abundance of "passion, courage, and imagination," he observed, qualities that made him "absolutely unfit" for everyday administrative work. Several times Trotsky said to Diego directly: "You are a painter. You have your work. Just help them, but do your own work."

In order to ensure Rivera's disengagement, Trotsky arranged for the founding congress of the Fourth International, meeting in Paris in September 1938, to pass a resolution, which he helped draft, declaring that the painter would no longer be an active member of the Mexican League and would instead sit on the Pan-American committee. Comrade Rivera was a figure of international stature, too valuable to the movement to be allowed to squander his energies on the minutiae of local politics. So went the reasoning behind the resolution, yet its wording was brutal, making Rivera sound like an errant comrade being punished rather than promoted. Trotsky later lamented this choice of language, although he had endorsed it.

Eastman, in his wide-ranging criticism of Trotsky's personal deficiencies as a politician, underscored his "gift for alienating people." Its source, he determined, was "failure of instinctive regard for the pride of others, a lamentable trait in one whose own pride is so touchy." Eastman, whose own touched pride informed this judgment, might have added Rivera to the list of Trotsky's casualties, but the case of this enfant terrible defies simple explanation.

One day Diego came to a meeting at the Blue House carrying an essay he had written about art and politics which he proposed to read aloud. Trotsky demurred. His limited Spanish would enable him to comprehend only half the presentation, he explained, asking that the discussion be postponed until he had a chance to read the essay. Taking this as a snub, Diego accused Trotsky of wanting to get rid of him. "The idea of my wanting to be rid of Diego," Trotsky marveled in a letter to Frida, "is so incredible, so absurd, permit me to say, so mad, that I can only shrug my shoulders helplessly."

It was at this unpropitious moment that the O'Gorman affair erupted. Juan O'Gorman was a painter and an architect, a friend of

Trotsky and Diego Rivera, 1937.

Diego and Frida who had designed their linked homes in San Angel. Commissioned to paint frescoes inside the terminal building of the Mexico City airport, he used the opportunity to editorialize by caricaturing Hitler and Mussolini and their confederates. These images confronted the Mexican government with a political dilemma.

The previous March, President Cárdenas had nationalized Mexico's petroleum reserves and expropriated the equipment of the British and American oil companies, a coup that prompted Britain to sever diplomatic relations and to boycott Mexican oil. Germany and Italy replaced Britain as Mexico's main oil purchasers. O'Gorman's provocative murals threatened to cause a diplomatic confrontation that might lead to an economic crisis. In response, General Francisco Múgica, the minister of communications, gave the order to have O'Gorman's inconvenient artwork destroyed.

Rivera loudly condemned this act of "vandalism," which he imagined to be a reprise of the Battle of Rockefeller Center. He denounced

Múgica, who happened to be Trotsky's most important ally inside the Cárdenas government, as a "reactionary bootlicker of Hitler and Mussolini." Somehow he seems to have expected Trotsky to echo this vituperation, but Mexico's most controversial exile saw the matter differently. The O'Gorman episode had nothing in common with the fate of the Radio City mural, he instructed Rivera. The obliteration of the airport frescoes, however repugnant, was carried out in the interest of national independence. "Mexico is an oppressed country and she cannot impose her oil on others by battleships and guns." Rivera accused Trotsky of putting his asylum ahead of his principles.

This is where matters stood toward the end of December 1938, when Rivera struck the match that ignited this combustible mix. Intending to compose a letter to Breton in Paris, he asked Van to come to San Angel to serve as his typist. In the course of dictating his letter, Diego began to speak critically about Trotsky's "methods"—at which point Van stopped typing. Diego assured him that he intended to show the letter to Trotsky and he asked him to continue. "With any other person I would have left," Van explains. "But the relations between Trotsky and Rivera were exceptional." He decided to accept the painter's word that he would talk to Trotsky. "We will have it out," Diego promised.

Returning to the Blue House, Van placed the letter on his desk, where it was discovered by Natalia. She brought it directly to Trotsky, who exploded in anger. Rivera's indictment of Trotsky was based on two recent episodes in their dealings with the local Trotskyists. These were trifles, yet Rivera made them the basis of his complaint to Breton that Trotsky had carried out a "friendly and tender" coup d'état against him.

Trotsky could easily demonstrate the falseness of the allegations. Using Van as his emissary, he asked Rivera to revise his letter. Rivera agreed and made an appointment with Van but canceled at the last moment; then he arranged a new time to meet and canceled again. "He was obviously going through an emotional crisis," Van comments. "The words 'friendly and tender' in his letter to Breton show that he was still attached to Trotsky."

As the new year began, Rivera continued down this destructive path, launching a number of initiatives with small anarchist and trade-unionist groups hostile to the Trotskyists. Trotsky called these intrigues "purely personal adventures" by which Rivera intended to impress him with his political mastery. Together with Natalia, Trotsky visited him at home in San Angel and passed what he felt was as a "very, very good hour" with the painter; some time later, Trotsky met with him alone. After each conversation he assumed that their differences had been resolved, only to discover otherwise.

On January 7, 1939, Rivera sent a letter of resignation to the secretariat of the Fourth International in New York. Trotsky refused to accept it, reasoning that Rivera was too important to let go without one final attempt at reconciliation. Hoping to enlist Frida in this effort, he wrote to her in Paris, telling his side of the story and pleading that her help was essential. "Now, dear Frida, you know the situation here. I cannot believe it is hopeless." But Frida saw things differently, boasting to friends in New York that Diego "told *piochitas* (Trotsky) to go to hell in a very serious manner. . . . *Diego is completely right.*"

She may have reconsidered this opinion after her return from France in March. Within a few months she and Diego divorced, only to remarry the following year in San Francisco. There is no hint that Diego's new willfulness was brought on by a discovery of his wife's affair with Trotsky. Unable to consult Frida, Trotsky had no way of knowing this, however, and he may have been sweating it out.

Mexico's presidential politics managed to aggravate the Trotsky-Rivera imbroglio. President Cárdenas, elected in 1934, could not run again and was preparing his succession. He failed to get his party's approval for a candidate of his choice, however, and was forced to select a conservative politician instead. This caused confusion on the left about which candidate to support in the next election, more than a year away. Hoping to influence the presidential succession, Rivera founded the Party of Workers and Peasants. Taking the controls of this vehicle, he executed what Trotsky called a "series of incredible zigzags" in search of "some political magic." Trotsky now had to consider that people might think—and his enemies choose to believe—that he was collaborating with Rivera and thus breaking his promise to steer clear of Mexican politics. If only for appearances' sake, he had to separate himself from the painter.

Trotsky also decided that he could no longer remain under Rivera's roof. "It is morally and politically impossible for me to accept the hospitality of a person who conducts himself not as a friend, but as a venomous adversary," he wrote privately on February 14, one year after his guardian had mortgaged his San Angel home in order to reinforce security at the Blue House. Trotsky must truly have believed that the breach was irreparable, because he knew how difficult it would be to find an affordable home to rent that provided comparable security. Nor had the sense of danger abated. The daily *El Universal* had recently reported that about 1,500 former foreign volunteers in Spain—Poles, Germans, Austrians, and others—would be given asylum in Mexico in coming weeks. Trotsky assumed that these refugees had been selected by the GPU.

While the search for a new house was under way, Trotsky proposed to pay rent for as long as he remained in the Blue House. Rivera rejected this offer, insisting that the house belonged to Frida and thus that the proposal to pay him rent was intended as an insult. Trotsky called this assertion ridiculous—"He wishes to impose his generosity on me"—and offered two hundred pesos as a modest monthly payment. Rivera accepted the money then refused it, so it was donated to the local comrades.

In early March, Trotsky's staff found a new house, located only a few blocks away, on Avenida Viena. It would need extensive cleaning and renovation before it could be occupied. In the intervening weeks, for the sake of security, the impending move was to be kept quiet. Secrecy was maintained until the second week of April, when Diego made public his break with Trotsky in an interview with the local daily *Excelsior*, a story that was picked up by *The New York Times*. Rivera's tone was restrained and regretful. Off the record, however, he was heard to say that Trotsky's interception of his letter to Breton was typical of the methods of the GPU. Diego's promiscuous application of the GPU label had been troubling Trotsky for some time. In recent months he had similarly unmasked Hidalgo, Múgica, and O'Gorman, among other friends and enemies.

"A tremendous impulsiveness, a lack of self-control, an inflammable imagination, and an extreme capriciousness—such are the features of Rivera's character," Trotsky wrote to the Pan-American committee in explanation of Rivera's repudiation of the Fourth International.

To Frankel in New York he wrote contritely: "You warned us many times about his fantastic political ideas." Trotsky rehearsed for Frankel how Diego's "fantastic mind" had concocted his "fantastic slander" and his "fantastic letter" to Breton. "We were very patient, my dear friend. We hoped that in spite of everything, we would be able to retain the fantastic man for our movement. . . . Now we must show this fantastic personality a firm hand."

Trotsky, who preferred to attribute his setbacks to the workings of larger historical forces, was not content to cite the dark side of Rivera's artistic temperament. "In spite of the individual peculiarities," he explained to Breton, *the painter's case is a part of the retreat of the intellectuals*—by which he meant a retreat from communism. "Our painter is only more gifted, more generous and more fantastic than the others, but he is, nevertheless, one of them."

Had he lived a few years longer, Trotsky would have been forced to revise this analysis, as Rivera executed his fantastic political U-turn back toward the Mexican Communist Party and toward Stalin. After all of his Trotskyist sins, it would take Rivera several attempts before he was allowed back into the Communist fold. In other words, he had to perform more than the usual amount of groveling and self-criticism required on such occasions. In one application round, he told the tale of how he had secured Trotsky's asylum in Mexico for the purpose of having him assassinated.

On May 1, 1939, Trotsky's household made the move to the new residence on Avenida Viena. Trotsky himself was transferred on May 5. At the moment of his departure, he approached his empty desk and placed on it two or three small items, gifts from Diego and Frida. One of these, a favorite pen, had been a present from Frida, who had contrived to get a sample of his signature and have it engraved along the pen's barrel. He then turned and walked out of his study, under the gaze of Frida, who stood between two curtains holding a bouquet of flowers, in the self-portrait she had dedicated with all her love to Leon Trotsky.

CHAPTER 8

The Great Dictator

I t was March 1939, and Pavel Sudoplatov was being driven to an important meeting in the Kremlin in the company of NKVD chief Lavrenti Beria, who sat beside him. Sudoplatov was head of the Administration for Special Tasks, an elite unit that specialized in sabotage, abduction, and assassination of enemies of the people on foreign soil. Sudoplatov's predecessor had been arrested the previous November, and he feared his own arrest after being denounced by a colleague as a "typical Trotskyist double-dealer." When Beria summoned him, he suspected the worst. The car entered the Kremlin through the Spassky Gate on Red Square, and drove down a dead end alongside the old Senate building. Only then did Sudoplatov realize that Beria was taking him to meet with Stalin.

The two men entered the building and walked up the staircase to the second floor, then down a long, wide, carpeted corridor past offices behind tall doors, like rooms in a museum, Sudoplatov thought. "I was apprehensive and tense with enthusiastic excitement." He could feel his heart beating as Beria opened the door and they entered an enormous reception room, from where they were led into Stalin's office.

Stalin, dressed in his trademark gray Party tunic and old baggy trousers, invited his guests to sit at a long table covered with a green baize cloth. Nearby stood his desk, its papers arranged in perfect order. On the wall behind the desk was a photograph of Lenin; on an adjacent wall were images of Marx and Engels. Stalin appeared focused, poised, calm. Sudoplatov was impressed by his self-confidence and ease. The steady gaze of the dictator's honey-colored eyes gave the impression that he was listening to every word. Beria, dressed in a modest suit with an open collar, adjusted his pince-nez and came right to the point, recommending that Sudoplatov be appointed deputy director of the NKVD's foreign department.

Stalin frowned, a reaction that might have completely unnerved an uninitiated visitor, but Sudoplatov had seen this expression before. The pipe in Stalin's hand, though stuffed with tobacco, was not lit. "Then he struck a wooden match with a gesture known to all who watched newsreels, and moved an ashtray close to him." Stalin ignored Sudoplatov's nomination and told Beria to summarize his foreign intelligence agenda. This was Sudoplatov's third meeting with the Soviet leader, and once again he took note of Stalin's gruffness, which he assumed was "an inseparable component of his personality, just like the stern look that came from the smallpox marks on his face."

As Beria spoke, Stalin rose from his chair and began to pace slowly back and forth in his soft Georgian boots. Sudoplatov's promotion, Beria proceeded to explain, would enable him to mobilize all resources necessary for the liquidation of that most treacherous enemy of the people, the renegade Trotsky. Stalin must have been thinking it was high time.

Ten years earlier, he had chosen to banish Trotsky from the Soviet Union. At the time, he was not yet powerful enough to have his vanquished enemy executed—not openly anyway, and he could not risk an assassination. Deportation, Stalin assumed, would cut off all potential avenues for Trotsky's political comeback in the USSR. He probably figured that the exile would remain isolated, without friends or funds, and that he would become tainted by his foreign associations. Within a few years, however, as Trotsky denounced him relentlessly in interviews, articles, pamphlets, and books, Stalin came to regret having let the "chatterbox" out of his grasp.

Trotsky knew this instinctively. "Stalin would now give a great deal to be able to retract the decision to deport me," he wrote privately in 1935. "How tempting it would be to stage a 'show' trial! But the danger of exposure is too great." Once again Trotsky underestimated his adversary, who then cast him in the role of mastermind of the elaborate conspiracies exposed in three spectacular show trials. Stalin's bitterness about having allowed Trotsky to get away was assuaged by the exile's usefulness as a satanic symbol of treason and heresy. Stalin could not have invented another scapegoat like Trotsky. And alarms about one and another "Trotskyist center" in the USSR would not have served Stalin nearly so well had the traitor not been alive and living abroad.

Once the show trials were over, Trotsky had outlived his usefulness. Sudoplatov records Stalin's complaint, at their March 1939 meeting, about the "treacherous infiltrations" of the Trotskyists in the international Communist movement; once the looming European war broke out, such machinations would endanger the Soviet state by hindering its subversion operations behind enemy lines. Stalin may have portrayed Trotsky as a threat to national security for the benefit of the young intelligence officer sitting before him, but in fact he was under no illusion about the dangers posed by the tiny Trotskyist movement, either to Soviet security or to his own grip on power. Paranoia, in other words, did not influence Stalin's calculations.

Envy, hatred, revenge—these provided motivation enough for Stalin to want Trotsky dead. A few years after the Revolution he was heard to say: "The greatest delight is to mark one's enemy, prepare everything, avenge oneself thoroughly, and then go to sleep." For Stalin there was no greater object of loathing than Trotsky, that "operetta commander" who had dared to ridicule him as the "outstanding mediocrity" of the Party and denounce him as the "gravedigger" of the Revolution.

When Zinoviev and Kamenev broke with Stalin and joined Trotsky in opposition in 1926, they carried dire warnings about their erstwhile ally. As Trotsky launched into a critique of Stalin's policies toward China, Great Britain, and other countries, Kamenev interrupted him: "Do you think that Stalin is now considering how to reply to your arguments? You are mistaken. He is thinking of how to destroy you." Zinoviev and Kamenev drew up a joint testament, kept safely hidden, which warned that in the event of their "accidental" deaths, Stalin should be held responsible. They advised Trotsky to do the same.

For years Stalin had to remain content with Trotsky's mere political destruction—although in the purge that followed the Kirov murder in December 1934, he was able to strike at the exile's family members living in the USSR. After learning of his son Seryozha's arrest in Moscow, Trotsky wrote about Stalin in a diary entry: "His craving for revenge on me is completely unsatisfied: there have been, so to speak, physical blows, but morally nothing has been achieved. . . . At the same time he is clever enough to realize that even today I would not change places with him: hence the psychology of a man stung."

The idea that the dictator might choose to administer the ultimate "physical blow" still seemed improbable. "Naturally, Stalin would not hesitate a moment to organize an attempt on my life, but he is afraid of the political consequences: the accusation will undoubtedly fall on him." That was before the Terror and the trials and the cascading charges against Trotsky of treason, espionage, sabotage, and assassination. By 1939, after the bloody annihilation of the Old Bolsheviks and of the Red Army command, and with Hitler's troops capturing headlines with their occupations of Austria and then Czechoslovakia, Stalin had no inhibitions about hunting down the outlaw Trotsky in distant Mexico. The fugitive fully comprehended the danger.

When Beria was done speaking, Sudoplatov heard Stalin say that the only significant political figure in the Trotskyist movement was Trotsky himself. "If Trotsky is finished the threat will be eliminated." Previous attempts to organize Trotsky's liquidation had come to naught. Now the assignment was to be handed to Sudoplatov, an experienced killer. The year before, he had carried out the assassination of the émigré Ukrainian nationalist Yevkhen Konovalets in Rotterdam. Konovalets had a sweet tooth, and Sudoplatov, having gained his confidence, contrived to present him with a booby-trapped box of chocolates. Sitting across a restaurant table from his target, Sudoplatov removed the box from his coat pocket and laid it flat on the table. Shifting the device to the horizontal position activated the timer. The two men shook hands and Sudoplatov left the restaurant. He walked into a nearby haberdasher's shop, where he purchased a raincoat and a hat. Thirty minutes later, exiting onto the street, he heard a bang that sounded like the blowout of a tire. People began running toward the restaurant. Konovalets was dead.

Stalin instructed Sudoplatov to assemble a team of shock troops to carry out what he called the "action" against Trotsky. If the operation were successful, he pledged, the Party would always remember the service rendered by the participants, would see to their welfare and that of their families. Then Stalin stiffened and issued an order: "Trotsky should be eliminated within a year."

The last time Trotsky and Stalin saw each other was in October 1927, at the Central Committee meeting that voted to expel Trotsky from that body. On the drive back to their Kremlin apartment, Natalia did her best to calm her husband, who was highly agitated. "But they cannot tear me away from history!" he declared, a statement that was equal parts defiance and self-consolation. The fact is, however, they had already begun to alter Trotsky's role in accounts of the Revolution. The Man of October was being remade into the Judas Iscariot of the Party.

Fiercely jealous of his place in history, Trotsky was determined to put up a fight. He would be well equipped to do so, thanks in part to a misunderstanding among Stalin's policemen. The order for Trotsky's expulsion from the country said nothing about his personal archives: crates and trunks stuffed full of Soviet-era documents, including copies of his correspondence with Lenin and other Bolshevik leaders and the records of the Opposition since 1923. He was allowed to take these incriminating documents with him into exile, together with his personal library. When Stalin found out, he was incredulous. In the aftermath, several people were arrested, including three GPU agents.

The passport Trotsky was handed as he boarded the steamer *Ilich*, leaving Odessa for Istanbul in February 1929, listed him as a writer. This must have pleased him. As a youth he had dreamed of becoming a writer, but he chose instead to subordinate his literary work, like everything else, to the revolution. During the Soviet years, his extended writing projects on literature and culture offered him an escape from the stresses and strains of political life. In exile he would have the opportunity to devote himself to serious writing. He would in fact be compelled to do so in order to support himself, to pay for his protection, and to fund the *Bulletin of the Opposition*. It was these considerations, rather than vanity, that persuaded Trotsky, not long after he had settled in Turkey, to accept an offer from Charles Scribner's Sons of New York to publish his autobiography. One year later, *My Life: An Attempt at an Autobiography* was selling briskly in English, Russian, German, and French editions.

Trotsky took obvious pleasure in composing the book's early chapters, which contain vivid recollections of growing up on his father's prosperous farm in the southern Ukraine, his schooling in Odessa, his turn to radicalism, and his first prisons and Siberian exile. The writing of the later sections, however, which recount his battles with Stalin and the other "epigones," took a toll on his nerves and his health. Here Trotsky was forced to answer the favorite question of journalists, comrades, and perfect strangers, one he had come to dread: "How could you lose power?" The question was naive, he thought, as if losing power was like losing a watch or a wallet. Once more he had to explain that his defeat came at the hands not of a man but of a machine. It was not Stalin who had triumphed over him, but the ascendant bureaucracy Stalin personified.

The success of *My Life* led to a publishing contract from Simon & Schuster for a book on the Russian Revolution. Trotsky spent the better part of two years on the project, drawing on his memory and imagination, his books and archives, as well as library books that were shuttled back and forth to him in Turkey by comrades in Paris and Berlin. The result was Trotsky's masterpiece, *The History of the Russian Revolution*, a hugely detailed narrative account of Russia's upheaval, from the fall of the Romanovs to the Bolshevik coup d'état. Written in Russian, it was published in English translation in three volumes in 1932 and 1933.

The *History* is best appreciated as a work of literature. The narrative pulses with drama and coruscates throughout, as Trotsky switches effortlessly back and forth between the movements of armies and of crowds and the actions of individuals. There are powerful set pieces. An encounter on a Petrograd street during the February days between a demonstration of 2,500 Petrograd workers and a detachment of Cossacks, the czar's enforcers, is especially memorable, as is the Red Guards' assault on the Winter Palace during the October insurrection. The portraits of individual actors are sharply drawn. Trotsky subjects the opponents of the Bolsheviks—be they monarchists, liberals, or socialists—to his corrosive blend of irony, sarcasm, and mockery. Not only are they invariably found guilty of being on the wrong side of history; they are typically both wicked and stupid. George Bernard Shaw once remarked that "When Trotsky cuts off his opponent's head, he holds it up to show that there are no brains in it."

Trotsky's *History*, while free of jargon, is unmistakably the work of a Marxist historian. The author claimed to be objective in his presentation of facts, but he did not pretend to be impartial. Despite the mounting suspense he is able to sustain throughout his narrative, the outcome is never in doubt. Russia must overcome its backwardness by leaping over the bourgeois stage of history directly into socialism. The Provisional Government, personified at the pivotal stage by the charismatic socialist lawyer and politician Alexander Kerensky, is doomed to defeat, as are the Mensheviks, the Socialist Revolutionaries, and the other rival parties of the Bolsheviks in the Petrograd Soviet.

The masses are the collective heroes of the drama, yet ultimately only the Bolshevik Party can lead the way and seize power in the name of the workers and peasants. It was Trotsky himself who directed the October putsch, but here he goes out of his way to remove himself from the narrative. Instead, as he did in *My Life*, he deliberately places himself in Lenin's shadow. Without Lenin, Trotsky states explicitly, the Bolsheviks would not have taken power in October, and probably not at all—a remarkable statement from someone who believed that impersonal social forces determined the course of events.

Trotsky idolized Lenin, and yet here his elevation of the Bolshevik leader was in part an act of self-aggrandizement. Trotsky's name was inseparably linked to Lenin's in the context of the Revolution. Trotsky was Red October's chief of staff, Lenin's second-in-command. Thus, in exalting Lenin, he was by implication also lifting himself onto the pedestal. This was intended as a slap at Stalin's historians, who had begun to portray the dictator as Lenin's right-hand man from the moment the Party's leader arrived in Petrograd. Stalin had been famously described as a "gray blur" in 1917. Trotsky's account leaves him in obscurity.

On the strength of the clamorous reception and respectable sales of *The History of the Russian Revolution*, the American publisher Doubleday, Doran & Company signed Trotsky to a contract for a biography of Lenin. When he began the new project, he was living in Barbizon, France, some thirty miles south of Paris, where he conducted his research using books brought to him by Lyova. During a sedate autumn and winter of 1933–34, he wrote the initial chapters covering Lenin's youth. Further progress was stalled when his asylum came under hostile scrutiny and he was forced to move, first within France and then to Norway. In retrospect, the villa on the island of Prinkipo seemed like a writer's paradise.

When Trotsky arrived in Mexico in January 1937, *Time* magazine gave its readers the impression that the exile was eager to return to work on his biography of Lenin. But Trotsky's life was in a state of upheaval and his financial situation was extremely precarious. He owed the Norwegian government hundreds of dollars in taxes and he had left behind unpaid medical and legal bills totaling hundreds more. It was only thanks to the generosity of Diego Rivera that he was comfortably situated in Coyoacán.

Trotsky had been counting on income from the sale in the United States of a small book about Stalinism he completed in Norway just before the first trial in August 1936, a work that had already appeared in France as *The Revolution Betrayed*. Instead, he learned that his literary agent in New York, Max Lieber, had failed to sell the manuscript. Nor, it appeared, had he even tried. Moreover, Lieber's elusiveness had jeopardized potentially lucrative deals for interviews and articles. Trotsky was flummoxed: his agent, he said, was behaving like a "counter-agent."

"What is the matter with Lieber?" he inquired impatiently of a comrade in New York. "Has he perhaps become connected with the Stalinists?" Indeed, he had. Lieber's literary agency served as a front for Soviet espionage activity, including that of Whittaker Chambers, shortly to become the most important American defector from Communism.

Once Lieber was dropped, an agreement was quickly reached with Doubleday to publish *The Revolution Betrayed*, which came out in March 1937. But Doubleday was insisting that Trotsky complete his biography of Lenin, for which he had been paid his full advance of $5,000 three years earlier. Yet Trotsky needed income. He figured that a book on the Moscow trials could be the best-seller that would rescue him financially. He began to cobble together from his recent short articles and other odds and ends a counterindictment he called "Stalin's Crimes." Harper & Brothers agreed in principle to bring out such a book, but when plans were made to publish the transcripts of the Dewey Commission hearings, Trotsky felt compelled to abandon his project.

In the summer of 1937, the need for money inspired Trotsky to try his hand at writing magazine articles, but he was quick to realize that his style was "not sufficiently adapted to the average man on the New York street." He floated the idea of updating *My Life* to include the years since 1929, but he himself was reluctant to take it up. He decided instead to move forward on the Lenin biography, but that effort was cut short early in September by the departure of his Russian typist, who suddenly decided to get married. In December, still adrift and without a typist, Trotsky warned the New York office that his financial position was "extremely acute."

On February 16, 1938, the day Lyova died, a breakthrough occurred in New York, where Trotsky's new agent, Alan Collins, of the well-regarded Curtis Brown literary agency, worked out an arrangement whereby Harper & Brothers would buy out Trotsky's contract with Doubleday. The new deal would require him to write two biographies: first a popular life of Stalin, followed by the monumental study of Lenin. Trotsky would receive $5,000 for the two books.

Overwhelmed by grief at the death of his son, Trotsky could hardly imagine undertaking a biography of the man he assumed had just had him killed. Yet the monetary reward was tempting. By this point, the household was on the edge of insolvency. Natalia was borrowing funds from the Mexican comrades and becoming extremely worried, and there was only so much of this that she was able to hide from her husband. From New York, Jan Frankel wrote an anxious letter to Van saying that unless Trotsky accepted the Harper proposition, they would be unable to implement the plan to increase the guard at the Blue House.

Ten days after Lyova's death, Trotsky finally relented, signaling to Harper that he found their proposal "totally acceptable." The truth is, he was not in a position to refuse. As Van advised Frankel: *"Le vieux semble disposé (à contre-coeur)."*

Having been warned by the Doubleday editors about Trotsky's inability to meet a deadline, Harper decided on a hardheaded arrangement for paying its new author. The $5,000 advance for the two books would be spread out in ten payments of $500 each, delivered at two-month intervals. The British publisher, Nicholson and Watson, would divide its payment of $2,500 for the Stalin book into four installments. The Stalin biography was to be 80,000 words in length and be completed within six months; the Lenin book would be 150,000 words, written during the subsequent eighteen months. The details of the contract were still being negotiated when the first advance check arrived in Coyoacán at the end of April, just as the work was getting under way.

At the time, it was commonly assumed that Trotsky set out to write a life of Stalin as a way to settle scores with his old foe. But in fact, in signing his book contract he raised not the proverbial sword, but the shield. To defend himself from Stalin's assassins, Trotsky would have to write his biography.

"Beginning in 1897, I have waged the fight chiefly with a pen in my hand," Trotsky wrote in his autobiography. In 1902, during his first escape from Siberia, he was nicknamed Pero, Russian for the Pen, a tribute to his journalistic achievements in exile. Over the years, he was always obsessing about his pen. And yet most of his literary output since the early 1920s, from his correspondence to his books, he produced by dictation, a practice that enabled him to draw on his skills and experience as an orator.

Listening to Trotsky's resonant voice as he gave dictation, one could imagine its power when he harangued his troops without the aid of a microphone. Dictating in Russian, he would pace the floor of his study, speaking without interruption for an hour or two, sometimes longer. His secretaries marveled at his ability to conjure up lengthy passages of beautifully crafted prose from a few pages of notes in his hand. The clicking keys of the typewriter signaled their concurrence and urged him forward. Punctuation was left entirely to the discretion of the typist, who understood that Trotsky hated to be interrupted. If asked to stop or repeat something, he would easily lose his train of thought and his patience.

Trotsky was known as a literary stylist and he worked hard at it. As he wrote to Cass Canfield, the president of Harper & Brothers, after he began work on the Stalin biography: "At least one-third of my working time is devoted to the literary form of the book. I must have a perfect translation." But a perfect translation is always elusive, and Trotsky's Russian presented special challenges. He took full advantage of the freedom afforded by Russian syntax to manipulate the word order within a sentence in order to express emphasis or nuance or for dramatic effect. He refused to concede a trade-off between precision and style, and was always trying to bend the rules of English, French, and German grammar. He complained that Max Eastman's translation of his *History of the Russian Revolution* was full of errors, despite its magnificent style.

For the Stalin biography, a scholar of Russian literature was hired to translate Trotsky's chapters in New York as each was completed. Unlike in the past, Trotsky himself would not be able to consult books borrowed from a major library. Instead, a comrade in New York would serve as his researcher, while queries could be sent to the Paris comrades, who had better access to old Russian newspapers and other obscure sources. The research and translation phases of the work were thus in good hands, but Trotsky had yet to find a replacement for his Russian typist. After fifteen years of dictation, he had lost the habit of writing by hand, except for short texts. Without the services of a Russian collaborator, he would not be able to meet his deadline.

While the search was under way, a comrade from New York with serviceable Russian was asked to fill in, but Trotsky chafed at the slow pace and the constant interruptions. She was replaced in the first week of May by Sara Weber, who had worked for Trotsky in Turkey and during the previous summer in Coyoacán. She arrived from New York in the first week of May, intending to stay for six months, through the completion of the Stalin book, but a family illness forced her to scale back this commitment. So, while the work proceeded, Trotsky continued to be preoccupied by the search for a permanent typist.

Security concerns complicated the search. In May, Frankel sent word about a candidate in his native Czechoslovakia, a young woman of eighteen who reportedly was an expert typist in Russian. The hitch was that the woman in question, whose parents had emigrated from the Soviet Union, was thought to be a Communist. Van entered Trotsky's study and gave him the news. With a theatrical sweeping motion of his left arm, Trotsky exclaimed: "Let her come! We shall win her over!"

Van and Frankel advised against it, but Trotsky was undaunted. "She is a quite young girl of eighteen," he wrote to Frankel. "I do not believe that she can be a terrible agent of the G.P.U. Even if she comes with some sympathies for the Stalinists and with some wicked intentions against us (which I consider impossible, for nobody would entrust diabolical schemes to a little girl without experience), even in that case we feel strong enough to watch her, to control her, and to re-educate her." Trotsky, who tended to see the hand of the GPU in unlikely places, here detected nothing but an expert typist with perfect Russian. A month later, he was still importuning Frankel: "A girl of eighteen cannot make conspiracies in our home: we are stronger. In two or three months she would be totally assimilated."

In August, just as Sara Weber was preparing to leave Coyoacán, Trotsky discovered the Dictaphone. He had resisted the idea a year earlier, but when it turned out that Diego owned one—an Ediphone in need of minor repairs—he felt he ought to give it a try. Like most people of a certain age, Trotsky was initially skeptical of the new technology. Hansen says he behaved "like a peasant shying away from an optician—grandpa never wore no glasses." Whenever he got stuck, he would run into the patio calling, "You see? You see? Your American machine . . . it don't work." After Hansen showed the Old Man what he had done wrong, "he would make a little hissing sound in his teeth and then settle down again to more dictating."

Once Trotsky got the hang of it, his enthusiasm for his recording machine was unbounded. He began to bring it into his bedroom so he could dictate at night. A Russian émigré living in Mexico was hired to replace Sara, but she was a novice typist, and soon Trotsky had created a backlog of recorded text. New York was asked to send down a fresh supply of wax cylinders, as well as a shaving machine so they could be erased and reused.

The work moved ahead over the summer and into the autumn,

despite the frequent intrusion of more pressing matters: the French police investigation into Lyova's death, the battles with Lyova's widow over
custody of grandson Seva and Trotsky's papers, the Klement murder in
Paris, travels with Breton and Diego, and preparations for the founding congress of the Fourth International in Paris that September. There
was the usual stack of mail to answer, but Trotsky had sworn off "guerrilla polemics" until further notice. For considerable stretches, the long
workdays were devoted exclusively to the Stalin book, with breaks for
a meal, a siesta, and sometimes an evening discussion.

Like all historical research, Trotsky's job was one of excavation. In
this case, however, progress was slowed by multiple layers of falsification about the youth and early career of Joseph Djugashvili. The man
who became Stalin was born a year before Trotsky, in 1878, to a violent,
drunken cobbler in Gori, a small town in Georgia, in the Caucasus,
the mountainous southern reaches of the Russian Empire. In 1894 he
entered the seminary in Tblisi, the Georgian capital, where he was exposed to nationalist and other subversive influences. He was expelled
from the seminary in 1899, the year he became a professional revolutionary and adopted the pseudonym Koba, taken from the hero of a
Georgian novel, an intrepid, avenging Caucasian outlaw. In 1902 he was
arrested and exiled to Siberia, the first of seven such banishments, from
which he made six escapes. The collapse of the Russian autocracy freed
him for the final time.

Unlike the better-educated bourgeois revolutionaries living in
European emigration, like Lenin and Trotsky, Stalin operated in the
underworld of revolutionary politics inside Russia. As his biographer,
Trotsky faced the challenge of establishing even the basic facts about
Stalin's movements and activities before 1917. This required a careful
sifting through the differing versions put forward in a succession of
official histories and in memoirs both friendly and hostile. When did
Stalin become a Marxist? Which of the Party congresses abroad did he
attend? When did he first meet Lenin? What was his involvement in
the bank robberies carried out in the Caucasus to raise funds for the
Party? Was Stalin in fact an agent of the czarist secret police, as nagging
rumor had it?

In piecing this story together, Trotsky had to accommodate the awkward truth of Lenin's promotion of Stalin to the top ranks of the Party. In 1913, a year after Stalin was brought onto the Central Committee and the year he adopted the pseudonym Stalin, meaning Man of Steel, Lenin enthused about his "wonderful Georgian" in a letter to writer Maxim Gorky. Stalin had just visited Lenin in Cracow, where the two men worked on the protégé's article about Marxism and the nationalities question in Russia. This now became his area of expertise, and after the Revolution he was named People's Commissar of Nationalities. He may have been a gray blur on the revolutionary stage in 1917, but when the Party's new elite decision-making body, the Politburo, was created the following year, Stalin took his place on it, alongside Lenin, Trotsky, Kamenev, and Nikolai Krestinsky.

These were Trotsky's preoccupations in the middle of 1938, as he hustled to meet his publisher's deadline. On July 1, he sent off the first chapter of a projected dozen. A second chapter was finished in mid-August and a third in mid-September, at which time he wrote to his agent: "I must inform you that the whole book will be far more than 80,000 words, I believe 120,000." The reason, he explained, was the need for completeness. In truth, however, Trotsky was finding it impossible to write a popular biography of Stalin. Too often, the author felt compelled to pause the narrative in order to discredit Stalin's flatterers. As a result, the writing was tedious and repetitive, as though written for the Society of Old Bolsheviks instead of the Book of the Month Club.

Alan Collins, his agent, voiced concern that Trotsky had changed the conception of the book and that at his current pace he would not be able to meet his November 1 deadline, in which case Harper would withhold further payments. Collins was not placated by Trotsky's announcement, made at the end of September when he delivered a fourth chapter, that he hoped to finish the book by February 1.

After several comrades in New York echoed Collins's apprehension, Trotsky responded with a vigorous defense. Stalin was not a figure from the Middle Ages, after all, but a present-day tyrant. He was, moreover, a man whose life and career had been systematically falsified and distorted. "I knew the situation well enough before I went into this work, but at every page I am two or three times surprised, astonished, bewildered by this international conveyor of historical, theoretical, and literary frame-ups." Because of this deplorable state of affairs, said Trotsky, his handling of historical sources had to be fully transparent, even at the cost of disrupting the narrative. Otherwise, Stalin's liberal sympathizers would accuse him of being partial and subjective. "*My* book on *Stalin* must be unattackable, or better not to be written at all."

Despite his agent's unease, in mid-October Trotsky received his third payment from Harper, together with the third installment from his British publisher. A month later he sent off a fifth chapter, informing his agent that he was now working simultaneously on the sixth and seventh, which took the coverage of Stalin's life through the Russian Revolution. The writing would proceed more quickly now, Trotsky assured Collins, because as the story entered the Soviet period, he would be able to draw on his own experiences and memories of his subject.

As it turned out, however, crossing the threshold of 1917 had the opposite effect on Trotsky. For when he began to narrate the story of his contest with Stalin, his health gave out—just as it had fifteen years earlier, at the crucial juncture.

"Health is revolutionary capital and must not be wasted," Trotsky was always admonishing his staff. The need for vigilance about one's health was a Bolshevik principle that originated with Lenin. His obsession with matters of health and fitness—both his own and that of his comrades—inspired the convention whereby the physical well-being of Party officials was the business of the Politburo.

Like Lenin, Trotsky believed in a strict regimen and physical exercise. He was passionate about hunting and fishing, although life as an endangered exile restricted his opportunities. In Turkey, he occasionally shot quail, but fishing became his regular form of exercise. He liked to set out well before dawn, dragging along guards and secretaries, to cast lines or nets into the Sea of Marmara, which teemed with fish. Fishing for Trotsky was strenuous work. He threw his entire being into it.

Returning with his catch from these exhausting labors, he began the workday refreshed and energized.

After Trotsky left Turkey in 1933, he had fewer occasions to hunt and to fish. This was especially the case in Mexico, where his movements were restricted by concerns about his safety. During his first year at the Blue House, he often paced the patio. Then came the security renovations in the spring of 1938, which led to the landscaping of the back patio. Trotsky decided to lend a hand, and he quickly realized that this was the kind of demanding physical exercise that he had been missing. He became a gardening addict, the master of the hoe, the trowel, and the shovel. He planted several species of fern, and it was not long before nasturtiums grew luxuriantly throughout the back patio.

In late summer, the cactus was Trotsky's new obsession. Cactus expeditions became a routine activity, with the entire staff pressed into service. When the Dodge and the Ford came to rest at the designated location, Trotsky and his helpers emerged with picks and buckets and got to work. Trotsky was especially fond of a species called the *viejo*— Old Man Cactus—a phallus-shaped plant covered with long snowy white strands of hair. Some of the species singled out for removal were fiercely armed with heavy spines. The largest weighed nearly two hundred pounds, and the workers' sweat poured freely under the blazing sun as the cacti were uprooted and then loaded into the automobiles. Looking on, Natalia made jokes about this punishing form of "penal labor." Trotsky said it was the next best thing to hunting.

During these and similar outings, Trotsky liked to regale his young friends with hunting stories from his Soviet days. Many were simply humorous episodes, like the time Lenin had to drag an unwilling Zinoviev out of a haystack by his boots. It did not take much to get LD reminiscing, as Hansen could testify. Over New Year's 1938, he and Trotsky were hiking through a field near Taxco. "We flushed a covey of mourning doves and that started the Old Man telling me about hunting trips in the Caucasus mountains." In the telling, Trotsky made it sound like "the best hunting ground in the world for variety and size of birds such as quail, sage hens, and pheasants."

Inspired by these recollections, the following day they bought a supply of 12-gauge shells and drove to a lake twenty miles outside of town.

Trotsky on a cactus hunt, winter 1939–40.

Game was scarce, yet Trotsky, who had lost none of his quickness and accuracy with a shotgun, shot four mourning doves. The next evening Hansen escorted him to the same spot. "We tried to get some ducks but they were wary and flew out into the lake. A couple of snipe he shot fell into the water near shore where it was swampy." Hansen waded out to retrieve these trophies and was promptly relieved of his shoes. As it was getting dark, they had to return to the car. Hansen would have to drive back to Taxco in his bare feet.

This gave Trotsky an opening to poke fun at his vulnerable secretary. It was the first time in his life, he said, that he had a chauffeur without shoes. How did Hansen expect to get past the Taxco authorities bare-footed? How could he prove that he was the owner of the automobile when he did not even own a pair of shoes? The Old Man was making the most it, and yet the sight of Hansen's bare feet triggered inescapable memories of a different kind of hunting story, one that changed the course of Soviet history.

ON A SUNDAY in October 1923, Trotsky was in the marsh country north of Moscow, in a region called Zabolotye—literally Beyond the

Swamps. Here the Dubna River would spill its banks and flood the surrounding countryside for miles, creating lakes, swamps, and marshes densely bordered by tall reeds. "In the spring," Trotsky remembered, "the place is visited by geese, storks, ducks of all kinds, curlew, snipe, and all the rest of the swamp brotherhood." Trotsky's boatman, Ivan Vasilevich Zaitsev, was the duck lord of this territory, like his father, his grandfather, and his great-grandfather before him. "He has no interest in moorcocks, woodcocks, or curlews. 'Not my guild,' he will say cursorily. But he knows the duck through and through, its feathers, its voice, its soul."

On that frosty Sunday morning, at the conclusion of the hunt, Zaitsev was delivering Trotsky to his automobile, which was waiting for him on a rise of land. "From the canoe to the automobile I had to walk about a hundred steps, not more. But the moment I stepped onto the bog in my felt boots my feet were in cold water. By the time I leaped up to the automobile, my feet were quite cold." Once inside the car, Trotsky took off his boots and tried to warm his feet by the heat of the engine. He came down with the flu and a high fever, later diagnosed as a paratyphoid infection.

In Moscow, the doctors ordered Trotsky to bed, and that is where he spent much of the rest of the autumn and winter of 1923–24. Lenin was mortally ill, and the succession struggle had broken out into the open, with Zinoviev, Kamenev, and Stalin closing ranks against Trotsky. Owing to Trotsky's confinement, Politburo meetings took place at his Kremlin apartment. From an adjacent room, Natalia could hear her husband's heated arguments and the impassive responses of his rivals. "After each of these meetings, L.D.'s temperature mounted; he came out of his study soaked through, and undressed and went to bed. His linen and clothes had to be dried as if he had been drenched in a rainstorm."

Trotsky's illness continued to plague him for years. He would experience low-grade fevers, sometimes exceeding 100 degrees, that lasted for days, even weeks, at a time. They were accompanied by fatigue, headaches, numbness, and pains in his limbs, but it was the fevers that debilitated him. As he recalled in his autobiography, "my high temperature paralyzed me at the most critical moments and acted as my opponents' most steadfast ally." The decisive moment occurred in January 1924, in the days following the death of Lenin.

Feverish and exhausted and needing to escape Moscow, on January 18 Trotsky headed south for the Black Sea resort of Sukhumi. The rail journey was slowed by drifting snows, but Trotsky's load seemed to lighten the farther he traveled from the capital. On Tuesday, January 22, as his train stood in the station of Tblisi, Georgia, a grim-faced aide walked into the working end of Trotsky's railroad car and handed him a telegram. The message, signed by Stalin, informed him that Lenin had died the previous evening. Trotsky immediately cabled back: "I deem it necessary to return to Moscow. When is the funeral?" The reply came about an hour later: "The funeral will take place on Saturday. You will not be able to return on time. The Politburo thinks that because of the state of your health you must proceed to Sukhumi. *Stalin.*"

Lenin had been seriously ill for the better part of two years, and a stroke he suffered the previous March had left him severely disabled. Still, his doctors had held out hope for his recovery, and Trotsky, already deeply depressed, took the news hard. A delegation of local officials came to his car and urged him to write a tribute to their fallen comrade. "But I had only one urgent desire," he recalled in *My Life,* "and that was to be alone. I could not stretch my hand to lift the pen." Nonetheless, while the train was held up for a half-hour, Trotsky wrote his eulogy. The text was sent by wire to Moscow and published two days later in *Pravda* and *Izvestiia* under the title "Lenin is no more." Having performed his duty, Trotsky resumed his journey toward the Black Sea.

In Moscow, Lenin's body lay in state for four days not far from the Kremlin inside the ornate Hall of Columns, in an eighteenth-century neoclassical building that was once the Club of the Nobility and would later serve as the venue for the Moscow trials. More than half a million people entered the vast hall, which was draped in black and red banners and ribbons, and filed past Lenin's bier. Outside, bonfires burned day and night to warm the unbroken stream of mourners, who stood in line for hours in the extraordinary cold in order to pay their last respects to their beloved "Ilich," Lenin's patronymic and affectionate nickname.

All the major Bolshevik leaders were observed beside Lenin's open coffin—all, that is, except for Trotsky. Reporting from Moscow for *The*

New York Times, Walter Duranty described a series of false rumors that Trotsky was about to return from the Caucasus and at last take his rightful place among the mourners. "More than once crowds assembled to greet him at the station, and official photographers were sent to wait chilly hours before the Hall of Columns to film his entry." Trotsky's absence generated not only puzzlement but also resentment among those who took it as a sign of disrespect.

With Party and government officials journeying to Moscow from across the country—some from farther away than Tblisi—the funeral was postponed by one day until Sunday. No one informed Trotsky of the postponement, and he later concluded that Stalin had lied to him about the day of the funeral in order to keep him away from Moscow. Trotsky maintained that he could not possibly have arrived in time for a Saturday funeral: The distance from Tblisi combined with the severe weather conditions made it impossible. "I had no choice," he explained.

But Trotsky was no ordinary traveler. He was the head of the Soviet military. Extraordinary measures could have been taken in order to speed his journey to the capital. Max Eastman, who had befriended Trotsky and was living in Soviet Russia at that time, came to the conclusion that the embattled warlord had no desire to return to Moscow. His mysterious fevers were psychosomatic. What sickened him were the intrigues, the name-calling, and the backstage politics at which he proved to be completely inept. In Eastman's view, Trotsky did have a choice. "In ten minutes he could have had a locomotive on the other end of the train and been on his way north to attend the funeral and make a funeral oration that might have been crucial, and would certainly have been historic."

The funeral was held on Sunday, January 27, on an arctic Red Square, where the temperature hovered at thirty-five degrees below zero. Starting at ten o'clock in the morning, thousands of mourners, bundled up against the cold, some carrying banners, flags, and portraits of Lenin, walked past the coffin and the makeshift mausoleum. The smoke from the bonfires merged with the frozen breath of hundreds of thousands of people to produce an icy fog that hung over the square, in Duranty's phrase, "like a smoke sacrifice."

Confusion as to Trotsky's whereabouts also hovered over Red Square.

"To the last many believed he would come," Duranty reported. "A dozen times came a cry from the throng around the mausoleum, 'There's Trotsky,' or 'Trotsky's here,' as anyone in a military greatcoat faintly resembling Trotsky passed before us." But Trotsky was a thousand miles away, in Sukhumi, among the mimosa and the camellias in the bright, warm January sun. At the rest house where he and Natalia were staying, the veranda looked out upon enormous palm trees and to the sea beyond.

At 3:55 p.m., as the Moscow sky darkened, Lenin's coffin was lifted and carried into position by eight pallbearers: Stalin, Zinoviev, Kamenev, Bukharin, Molotov, Tomsky, Rudzutak, and Dzerzhinsky. At four o'clock, the coffin was lowered into the vault. At that moment, an explosion of sound erupted, as factory sirens, steamship whistles, train whistles—everything that could make noise—blared for three minutes, punctuated by salvos of rifle and cannon fire. The effect was deafening.

At four o'clock precisely, every radio broadcast and every telegraph line in the country transmitted the same message: "Stand up, comrades, Ilich is being lowered into his grave!" Lying on the veranda covered with blankets, Trotsky heard the thunderous booms of an artillery cannonade coming from somewhere on the shore below and wondered about the reason. "It is the moment of Lenin's burial," he was told.

Delayed by the snow, the Moscow newspapers began to arrive in Sukhumi carrying memorial speeches, obituaries, and articles about the funeral. The mail brought disconsolate letters from comrades in Moscow, none more anguished than that of seventeen-year-old Lyova. Bedridden with the flu and a high fever, he had left his sickbed to visit the Hall of Columns and see Lenin for the last time. There he waited and waited for his father to appear. His letter conveyed his agonized incomprehension. Trotsky tried to blunt the force of his son's reproach by ascribing it to youthful despair: "I should have come at any price!"

This was the historical terrain Trotsky was about to revisit when he fell ill in December 1938. His cryptogenic fever and related symptoms had plagued him during much of his time in France and Norway, but since arriving in Mexico he had managed to remain generally healthy. Under the stresses and strains of the first few months, he experienced episodes of nervous agitation, sweats, and the usual persistent insomnia. He lost ten pounds, in part because his colitis was aggravated by new foods and unfriendly bacteria. That September he complained about headaches and chest pains, as he did in the tumultuous days of February 1938, but on both occasions a medical examination detected no signs of a heart problem.

Now, in the winter of 1938–39, the fifty-nine-year-old exile began to experience symptoms of his old illness—lethargy, dizziness, headaches—although not the mysterious fever. He did not leave the house for more than two months that winter. His only exercise came from tending to his newly acquired rabbits and chickens, housed in coops and hutches in the back patio. When Frankel offered to send down an American doctor, Trotsky responded stoically: "Nothing is new other than an aggravation of the chronic things. The general name of my illness is 'the sixties' and I do not believe that in New York you have a specialist for this malady."

Writing and dictating helped Trotsky maintain his inner balance, so the loss, yet again, of his Russian typist was an additional setback. In December she became ill, then briefly returned to work, until a car accident on January 1 landed her in the hospital. Trotsky pleaded his hard luck case to his agent, who was able to convince Harper to pay out the fourth and final installment of the Stalin advance, which arrived in the middle of February. Still the typist failed to return, and Trotsky found it impossible to write by hand. In mid-March he again appealed to New York for help: "I am almost desperate."

Trotsky's sense of despair was magnified by his feud with Rivera and the subsequent preparations to move out of the Blue House. The new residence, several blocks away at Avenida Viena 19, was in dilapidated condition. Painters and masons would have to be hired. The walls surrounding the expansive garden would need to be augmented and lights and alarms installed. A room would have to be made ready for grandson Seva, expected to arrive from Paris in the summer. These renovations would be expensive, and no longer was Diego on the scene to help out with a loan.

In the first week of May, just as Trotsky was preparing to move into the new house, he found a Russian typist, barely qualified but acceptable. Four weeks later, he managed to send off another section of the manuscript, at which point he advised his agent that the coverage of the story after 1917 would be less detailed and more "synthetic" than the earlier chapters. He hoped to finish the book within a few months, he said, "if nothing extraordinary happens," a reference to the ominous war clouds gathering over Europe.

By now Trotsky was thoroughly disgusted with his subject and with the product he was turning out, and he wanted simply to be done with it. As he knew, the style of the work was ponderous and forbidding, marred by crude formulations such as "Stalinism is counterrevolutionary banditry." The narrator's tone was that of an aggressive prosecuting attorney, introducing hearsay and innuendo to demonstrate cruelty in Stalin or suggest duplicity, darkly hinting, based on no hard evidence, that the cunning Georgian was at one time an agent provocateur.

The Man of the Apparatus remains a gray figure, with dull face, yellow eyes, and guttural voice. His outstanding qualities are insatiable ambition, exceptional tenacity, and "never-slumbering envy." This hardly seems an adequate description of the man Lenin sized up as one of the two most capable Bolshevik leaders, together with Trotsky. That was in 1922, the same year that Stalin, with Lenin's endorsement, was named general secretary of the Party. It was not long afterward, in Trotsky's account, that Stalin stepped out from behind a curtain onto history's stage "in the full panoply of power." He was already a dictator, although not even he realized it as yet. The outcome of the struggle to succeed Lenin was decided by the time of his death. Trotsky, in other words, never had a chance.

Only one person could have stopped the Stalinist juggernaut, Trotsky suggested, and that was Lenin himself. In the final year of his life, Lenin became alarmed by the power Stalin had accumulated and sought to remove him as general secretary. He failed to do so only because he ran out of time. And his time was cut short, Trotsky now realized, by Stalin himself. In order to ensure his victory, Stalin had hastened Lenin's death.

How had Trotsky arrived at this shocking conclusion? He remembered that in the final year of his illness, Lenin had asked Stalin to obtain potassium cyanide for him to use should his suffering became unbearable. Stalin informed a Politburo meeting of this surprising request, conveyed to him by Lenin's wife, Nadezhda Krupskaya. No vote was taken, but the consensus in the room was that Lenin's appeal must be rejected. At the time, Trotsky saw nothing sinister in this episode.

His thinking began to change under the influence of the Moscow trials, with their bizarre accusations of poisonings by Kremlin doctors, under the direction of secret police chief Yagoda. At the time of the second trial in January 1937, his son Seryozha was arrested for allegedly attempting a mass poisoning of workers. A year later came Lyova's mysterious death, the result, Trotsky assumed, of poisoning by the GPU. In February 1939, the death of Lenin's widow, Krupskaya, further fueled his suspicions. By now, Trotsky's view of the more distant past had come into sharper focus. In 1922, Lenin had warned about Stalin, "This cook will prepare nothing but peppery dishes." In fact, Trotsky now concluded, "They proved to be not only peppery but poisoned, and not only figuratively but literally so."

Trotsky had made this determination by the summer of 1939. At the time, he had been trying, without success, to earn money by selling articles to American magazines. His fortunes changed dramatically with the signing of the Nazi–Soviet Pact on August 23 and the outbreak of war in Europe with the invasion of Poland a week later. Suddenly, he was in great demand as a commentator on world affairs. *Life* magazine, in its October 2 issue, which appeared as the Soviet and German armies were dividing Poland between them, published his appraisal of Stalin as a statesman.

Perhaps emboldened by the Soviet dictator's sudden ignominy as Hitler's new ally, in a sequel article Trotsky presented the evidence for his belief that Stalin had poisoned Lenin. "I realize more than anyone else the monstrosity of such suspicion. But that cannot be helped, when it follows from the circumstances, the facts and Stalin's very character." Trotsky drew a direct connection between Lenin's untimely demise and his own absence from the funeral. Stalin, Trotsky hypothesized, feared that he would recall Lenin's request for poison the year before and suspect a link to Lenin's death. Arriving in Moscow for the funeral,

Trotsky might interrogate Lenin's doctors, perhaps even demand a new autopsy. Stalin could not take such a risk, so he maneuvered to keep him away. Unfortunately for Trotsky, this kind of speculation was more suited to *True Crime* than *Life*, whose editors demanded more facts and less conjecture. In response, Trotsky accused *Life* of caving in to "the Stalinist machine."

TROTSKY'S MORBID SUSPICIOUSNESS had been exacerbated by the text of an anonymous letter he received at the beginning of May 1939, shortly after he moved into the new house. The letter, typed in Russian and mailed from San Francisco, was sent by someone who claimed to travel back and forth to the Soviet Union. The correspondent offered no further clues—genuine or fictitious—about his identity. His only motivation, he claimed, was to warn Trotsky about a possible assassination attempt.

The writer said he had heard a radio news report that Trotsky was leaving Diego Rivera's house for a larger dwelling, in connection with the expected arrival of his thirteen-year-old grandson. The radio report indicated that the boy was the son of the deceased Lyova, but the writer said he knew that Lyova had no children. He surmised, therefore, that Trotsky was expecting the arrival of a grandson from the USSR. "This has aroused in me great alarm."

The anonymous friend presented a series of terrifying scenarios for Trotsky to contemplate. The GPU might arrange for a provocateur to escort Trotsky's grandson to Mexico: the impostor would claim that he had succeeded in deceiving the Soviet authorities by passing off the grandson as his own son. Or the GPU might contrive to send Trotsky a different boy posing as his grandson, with the assignment to kill him. Even if the real grandson were delivered, the writer warned, the GPU had had the time and the means to indoctrinate in him the necessity to commit a "heroic" act—in other words, an act of terrorism.

This was followed by a list of emphatic instructions: Trotsky must not allow anyone escorting his grandson to enter his home, whatever trust he may have felt toward that person in the past. When Trotsky's grandson arrived, he must be searched to ensure he was not carrying poison. The boy must not be allowed access to weapons. Nor should entry be permitted to the boy's friends, who could supply him with a weapon or poison, or who might directly carry out a terrorist act.

The writer closed by apologizing in the event that he had misheard the radio broadcast, or if his letter had the effect of spoiling Trotsky's familial feelings toward his grandson. He asked that his communication be kept secret, because he planned to return to the USSR. "I wish you health and success in the struggle."

These alarms could easily have been dismissed as a prank or a provocation, but Trotsky, who a year earlier had brushed aside the danger posed by an eighteen-year-old girl Stalinist, decided that they deserved to be taken seriously. His hunch was that the writer was the same person who had sent him the anonymous warning about his Paris comrade Mark Zborowski several months earlier, a letter he had strongly suspected to be the work of a provocateur. As it happened, Trotsky's guess was correct: the author of the second anonymous letter, like the first, was GPU defector Alexander Orlov.

Orlov was now living in Los Angeles, still under a false name. In the spring of 1939 he was in San Francisco with his wife and daughter visiting the Golden Gate International Exposition. Orlov may have sincerely believed that young Seva was to be brought from the USSR. More likely he was feigning ignorance as a way to help obscure his identity, in which case his purpose once again was to arouse Trotsky's suspicion against the spy Zborowski, who might be asked to escort Seva to Mexico. Whatever the case, the warning was meant in deadly earnest. The notion of a teenage boy being reprogrammed into a parricidal killer might seem far-fetched, but Orlov could tell hunting stories that would send shivers up your spine, including one about a booby-trapped box of chocolates.

At the time Orlov mailed his letter, he tried to reach Trotsky by telephone. He called in the evening asking for Natalia or for a secretary who spoke Russian. The call came on a different telephone system from the one installed in the new house. In order to take the call, Natalia would have had to go outside. It was after dark. Trotsky figured it was probably a hoax, and he had to assume it was a plot. Two days later the anonymous letter arrived from San Francisco in two copies, one each for Trotsky and Natalia.

Trotsky sent a copy to Frankel, saying it appeared to be legitimate. What possible motive could the GPU have in sending him such a letter? He supposed that the author was Walter Krivitsky, another Soviet defector living in hiding in the United States. If both anonymous letters came from the same source, Trotsky told Frankel, then the first letter merited more serious consideration. He wondered why he had not been informed of the results of the investigation of Zborowski he had ordered.

As Trotsky puzzled over the identity of his anonymous well-wisher, in room 735 of the Lubyanka, the headquarters of the NKVD in Moscow, spymaster Sudoplatov was putting together a team of operatives to carry out the "action" Stalin had assigned to him. To head the task force Sudoplatov recruited Leonid Eitingon, Orlov's deputy and successor as chief of Soviet intelligence in Spain. Eitingon was a logical choice because the operatives would be recruited from the agency's Spanish network.

The details of the operation were finalized on July 9, 1939. The plan to assassinate Trotsky was code-named Operation Utka, Russian for Duck. It envisioned an assortment of possible methods: "poisoning of food, of water, explosion in home, explosion of automobile using TNT, a direct strike—suffocation, dagger, blow to the head, gunshot. Possibly an armed assault by a group." Which is to say, whatever it took to achieve the stated goal: "the liquidation of Duck." Sudoplatov and Eitingon identified the Spanish comrades who were to carry out this very special task. They requested a budget of $31,000 over six months. In the first days of August, Stalin authorized the operation.

CHAPTER 9

To the Finland Station

On August 8, 1939, Trotsky's grandson Seva arrived in Coyoacán from France. He was escorted by Alfred and Marguerite Rosmer, old friends of Trotsky and Natalia. Alfred, Trotsky's contemporary and one of the founders of the French Communist Party, was a supporter of the Left Opposition until he broke with Trotsky in 1930 and withdrew from politics. The two men, whose friendship survived their political break, had not seen each other since a visit to Prinkipo in 1929. Trotsky and Natalia felt relieved to be reunited with Seva, now thirteen years old and perhaps their sole surviving family member. And they were rejuvenated by the appearance of their old friends. There was much reminiscing about Paris at the turn of the century and much discussion about Europe in the looming shadow of war.

Two days later, the family and their guests headed to Taxco for an extended stay, taking advantage of an arrangement Trotsky had with the pioneering American historian of Latin America, Hubert Herring. Herring put his Taxco home at Trotsky's disposal in exchange for his participation in Herring's occasional Mexico seminars. The Taxco idyll was interrupted on August 21 by the shocking news that Hitler's foreign minister, Joachim von Ribbentrop, was headed to Moscow to conclude a nonaggression pact between the Third Reich and the Soviet Union. Two days later, in a festive late-night ceremony in the Kremlin, Ribbentrop and his Soviet counterpart, Vyacheslav Molotov, put their signatures on the Nazi-Soviet pact, as a beaming Stalin looked on jubilantly. The world was stupefied. The Nazis and the Communists, supposedly ideological opposites, had declared their mutual friendship. The treaty cleared the way for Germany to invade Poland, whose security had been guaranteed by Britain and France. War in Europe, long anticipated, was now imminent.

Joe Hansen was at Trotskyist headquarters in New York City, having been replaced as Trotsky's American secretary guard–driver by Irish O'Brien, Hansen's close friend from Salt Lake City. O'Brien assumed that the news of the pact was the signal to break camp in Taxco and return to Coyoacán to monitor the crisis. To his surprise, Trotsky insisted that the pact was of secondary importance. He would not budge.

Back at the house on Avenida Viena, O'Brien's wife, Fanny, was deluged with requests from news organizations all over the world for Trotsky's analysis of Stalin's treaty with Hitler. Unable to get through by telephone, she took a bus to Taxco in order to alert Trotsky to the urgency of his return. Nonetheless, and to O'Brien's bafflement, "the OM refused to be disturbed." Only when O'Brien showed him an anxious letter from Hansen saying that the American comrades were waiting for his assessment and his guidance did Trotsky snap to and give the order to start packing.

O'Brien, who was less inclined than Hansen to hero-worship the Old Man, was chagrined at his nonchalance. By delaying his return from Taxco he had squandered an opportunity to make a sizable sum of money from interviews and articles for the major newspapers, news services, and magazines. By the time the vacationers arrived back in Coyoacán on August 30, the offers were drying up. Two days later, Germany invaded Poland. On September 3, Britain and France declared war on Germany.

There was a hint of smugness in Trotsky's show of imperturbability at this historic moment. For years he had been predicting a rapprochement between Stalin and Hitler, a prospect that began to appear more likely after Germany's absorption of Austria in the Anschluss of March 1938, and especially after the Munich Agreement had sanctioned Germany's annexation of Czechoslovakia's Sudetenland. Munich failed to appease Hitler, of course, and when German troops marched into Prague in March 1939, Trotsky felt certain that a Nazi-Soviet accord was in the works.

Trotsky tragically underestimated Stalin, but from early on he was keenly sensitive to the danger posed by Hitler. His writings of the early 1930s that sounded the alarm about the Nazi menace were among the most perceptive and prescient he ever produced. He denounced the Comintern's policy of labeling the German Social Democrats "social fascists," which he predicted would facilitate the rise of the National Socialists. He took *Mein Kampf* seriously, warning that if Hitler came to power, the Red Army should immediately be mobilized.

As the Nazis consolidated dictatorial control in 1933, Trotsky changed his mind about remaining inside the Comintern as the Left Opposition. The only way forward, he decided, was to build a Fourth International, to replace the Communist Third International, which had superseded the Socialist Second International, the successor to Karl Marx's original. The goal was to unite the Trotskyists—the self-styled "Bolshevik-Leninists"—into an organization that would become the true standard-bearer of proletarian internationalism.

Although for the next several years Trotsky and his comrades referred to themselves as members of the Fourth International, formally the organization came into existence only in the summer of 1938. The moment was hardly propitious. Worldwide there were only a few thousand Trotskyists, spread out among numerous marginal organizations in many countries and often riven by factionalism. The French Trotskyists, the most important "section" of the embryonic Fourth International in the early thirties, had been crippled by a factional split. By the time Trotsky arrived in Mexico in 1937, the United States was home to what was easily the largest of the Trotskyist groups, although its total membership probably never exceeded two thousand.

Nor did the movement's growth prospects appear at all promising in that summer of 1938. In the Soviet Union, the Trotskyists had been either liquidated or banished to the camps. In Germany, Austria, and Italy, fascism reigned. In Spain, where Franco's Falangist armies were pressing their offensive against the Republican Loyalists, the Trotskyists had been purged or forced to flee the country. In Asia, the Trotskyists were without a significant foothold, least of all in China, which had been fighting for its independence since the full-scale Japanese invasion the year before. It is no wonder, given this depressing state of affairs, that many Trotskyists were skeptical that this was an appropriate moment to launch a new International.

Trotsky himself, however, was the voice of supreme optimism. Western capitalism was in the throes of an economic depression from which it could not recover. Just as Marx had predicted, capitalism's internal contradictions were ripening, most portentously in the United States, where President Roosevelt's New Deal could only postpone the inevitable. Just as the First World War had carried the Bolshevik Party to power in 1917 on a wave of revolution, the next world war would precipitate a revolutionary tidal wave that would sweep the Bolshevik-Leninists to victory. Trotsky's view of the matter was summed up by the title he gave to his new organization's programmatic statement: "The Death Agony of Capitalism and the Tasks of the Fourth International."

The scene of the founding congress belied such optimism. It took place at the home of Alfred Rosmer in Périgny, a village outside Paris, on September 3, 1938. Twenty-one delegates were in attendance, representing Trotskyist sections in eleven countries. Max Shachtman, of the American group, presided. The delegates elected Lyova and two of Trotsky's one-time secretaries, Rudolf Klement and Erwin Wolf—all three presumed murdered by the GPU within the past year—as honorary presidents. As a security precaution, it was arranged for the conference to complete its business in a single day. Votes were taken on various reports and resolutions, most of them written by Trotsky, with little time for genuine discussion. Only the Polish delegates openly questioned the wisdom of establishing a new International at a time when the political outlook was so grim.

At the end of the day, a press release announced the historic initiative, although in order to keep the GPU off the trail of the dispersing delegates, the congress was said to have taken place in Lausanne. This deception accomplished nothing, though, because the Russian section was represented at the congress by the Soviet spy Mark Zborowski, who provided Moscow with a complete report, which included his own canny contribution to the proceedings. Upon the election of the International's executive committee, Zborowski protested that the Russian section had not been given a seat. In response, the congress designated Trotsky as a secret and honorary member of the executive. Since

Trotsky could not directly participate in the executive's work, his place was filled by the GPU provocateur.

THE FOURTH INTERNATIONAL was not the only historic congress involving the Trotskyists who gathered in Paris that summer. Among them was Sylvia Ageloff, a Brooklyn native in her late twenties. A short, frumpy dishwater blonde, with a pointed nose and a broad, lipless smile that suggested a smirk, Sylvia was the oldest of three Ageloff sisters, daughters of a Russian émigré father, all of them active in the Trotskyist movement. She was accompanied to Europe by a friend named Ruby Weil, who invited herself along as Sylvia's traveling companion. The Ageloffs were aware of the rumor that Ruby had joined the Communist Party. What they did not know was that she was working for the GPU.

In Paris, Ruby looked up a friend of her sister, a Belgian in his mid-thirties by the name of Jacques Mornard, and introduced him to Sylvia. This encounter led to others, as Jacques took the ladies sightseeing in his Citroën and entertained them lavishly. In his excellent English, Jacques told them that he was studying journalism at the Sorbonne, that his generous supply of spending money came from his aristocratic parents, and that his father was a high-ranking Belgian diplomat. He proved to be the perfect dilettante, with a smattering of knowledge about art, music, and literature—only politics did not interest him.

Before long, Ruby decided to return to New York, leaving the field to Sylvia. Jacques was tall, lean, and muscular, with swarthy good looks. He took Sylvia to his favorite restaurants, always insisting on ordering the finest wines. The Belgian playboy intoxicated the homely Brooklyn social worker, and he seduced her. In other words, he did exactly what was expected of a penetration agent.

Jacques Mornard's real name was Ramón Mercader. He was born in Barcelona in 1914, the son of a Catalonian father and a Cuban-born mother, Caridad. She acquired a taste for radical politics not long after she left her husband and moved to Paris with her four children in 1925. The children were shifted back and forth between mother and father, France and Spain. At age fourteen, Ramón entered a hotel management school in Lyon; he later returned to Barcelona and became assistant chef at the Ritz, the city's premier hotel.

After the Spanish revolution in 1931, when the monarch fled the country and the Republic began its precarious existence, Ramón enlisted in the Spanish army, where he remained for two years and attained the rank of corporal. In 1934, he took part in the Catalonian uprising against rule by Madrid, serving with the Communist forces. After this rebellion was suppressed, Ramón was active in an underground cell of Communist youth in Barcelona. Arrested in June 1935, he was released when the Popular Front government was elected in Madrid at the beginning of 1936.

That summer, Franco and his generals launched their military assault on the Republic from Spanish Morocco. Ramón's flamboyant mother, now a fervent Communist, distinguished herself by leading an impromptu attack on Francoist machine gun positions in a central plaza in Barcelona, a ferocious onslaught with homemade grenades and rifle fire that wiped out the Francoist units at the cost of many lives. Caridad and her sons Ramón and Pablo were among the first to enlist in the Republican people's militia. Ramón served as a political commissar with the 27th Division on the Aragon front, with the rank of lieutenant.

The Spanish civil war became the NKVD's training ground for political terrorism. It organized six schools for saboteurs, the largest with upwards of 600 students. Leonid Eitingon, the NKVD's deputy resident in Spain, was responsible for training new recruits for commando and sabotage operations and for organizing detachments to carry out sabotage and terrorist acts deep inside enemy territory. Eitingon and Caridad Mercader became lovers, which made Ramón, who was recruited by the NKVD in February 1937, one of his special students. After serving for several months with a commando unit, Ramón was brought back from the front with a wounded arm.

Late in 1937 Eitingon sent Ramón to Paris. His forged Belgian papers identified him as Jacques Mornard. His NKVD code name was "Raymond." Sylvia's visit to Paris was a windfall for Ramón's handlers. Although she had come as a tourist, after making contact with American Trotskyist friends there, she was asked to serve as a translator at the founding congress of the Fourth International. Jacques was absent from Paris at the time, which was a relief to Sylvia, because she was worried that her Trotskyism might alienate her lover and had, for the time being, chosen to conceal it from him.

The fact that Sylvia was keeping such a secret—or thought she was—made her less inclined to question some things about Jacques that did not add up: stories about his family in Belgium, his life in Paris, and his sudden absences. Sylvia had a master's degree in psychology from Columbia, and when she expressed an interest in finding a job in Paris, Jacques arranged for her to ghost-write weekly synopses of books on psychology for a French newspaper syndicate. She was handsomely paid for her work, although she never saw the published results and Jacques refused to put her in direct contact with the syndicate. Sylvia sensed that her "job" was merely a tactful way for her lover to support her in Paris.

For the NKVD, it was money well spent. Sylvia herself was an insignificant figure in the Trotskyist movement, but Trotsky was especially fond of Sylvia's sister Ruth. Ruth had been in Mexico at the time of the Dewey hearings and proved to be of enormous help as a translator, typist, and researcher. She did not live at the Blue House, but she visited almost daily and was considered a reliable and devoted comrade. Ramón's NKVD controllers understood that Ruth's sister would be welcomed into Trotsky's home. Now they had to maneuver Sylvia—and Ramón—to Coyoacán. And the road to Coyoacán led through New York City.

A few weeks after the conclave in Paris, on Friday, October 28, 1938, at 8:00 p.m., the American Trotskyists gathered in the main auditorium of the Center Hotel, the future Hotel Diplomat, just off Times Square, on West 43rd Street. They came together to celebrate the founding of the Fourth International and the tenth anniversary of the American Trotskyist movement. New York had supplanted Paris as the center of Trotskyism, and on this night Times Square was its epicenter. An overflow crowd of up to 1,400 Trotskyists, sympathizers, and the curious packed the hall, including both galleries, paying twenty-five cents in order to witness the celebration and, more importantly, to hear Trotsky speak.

The hall was festively decorated with banners and streamers honoring the Fourth International and its American section, the Socialist Workers Party. Above the speakers' platform hung a six-by-four-foot charcoal drawing, draped in red, of Lenin and Trotsky. The Trotsky youth, some fifty strong, most in their early twenties, performed ceremonial duties, outfitted in uniforms of blue denim and red ties, with red armbands that read "Young People's Socialist League, 4th International."

The mass meeting began with the singing of the "Internationale." The program included speeches from the party's leaders, each accompanied by mounting anticipation of the performance of the evening's headliner, Comrade Trotsky, who naturally was saved for last. When the moment finally arrived, it was after ten o'clock. The audience fell silent as thirty male comrades came forward and positioned themselves below the front of the stage. They stood with arms folded and faces hardened in an attitude of defiance. The organizers were taking no chances, remembering that Trotsky's attempt to address an audience at the New York Hippodrome the year before had been sabotaged.

The lights were extinguished and a spotlight beam illuminated a photographic portrait of Trotsky placed at the center of the stage. "I hope that this time my voice will reach you and that I will be permitted in this way to participate in your double celebration," Trotsky began in his heavily accented English. The Bolshevik-Leninists, he continued, were genuine Marxists, governed not by wishful thinking but by an objective evaluation of the march of events. Trotsky's analysis of those events, which lasted close to fifteen minutes, came across clearly, despite some hiss and the occasional crackle from the gramophone record.

There was certainly no mistaking Trotsky's revolutionary optimism. The Communist International, he reminded his audience, had become a "stinking cadaver." The Fourth International had replaced it as the world party of socialist revolution. Its victory in the coming revolution was assured. "During the next ten years the program of the Fourth International will become the guide of millions and these revolutionary millions will know how to storm earth and heaven. Long live the Socialist Workers Party of the United States! Long live the Fourth International!" The hall erupted in tumultuous applause.

Trotsky's uplifting message notwithstanding, the Socialist Workers Party, the nucleus of the newborn Fourth International, was divided against itself. The party, all of ten months old, had been founded after a near-decade of Trotskyist splits and mergers. It was led by three able men of widely different backgrounds and talents: James Cannon, Max Shachtman, and James Burnham.

Cannon, the party's leader, was born in 1890 in rural Kansas, the son of Irish immigrants with strong socialist convictions. In his youth he was an itinerant organizer for the trade unionist Industrial Workers of the World and a member of the Socialist Party of America. He belonged to the Socialist left wing, which in 1919 broke away to form the first Communist party in the United States. Cannon sat on the presidium of the Comintern in Moscow in 1922 and 1923, and he attended its sixth congress in 1928. Shortly afterward, he along with Shachtman and a third comrade were expelled from the party for their Trotskyist sympathies, and together they formed the Communist League of America, the original American Trotskyist group.

By the late 1930s, Cannon, with his stocky build, thick gray hair, and florid complexion, fit the stereotype of the jovial, hard-drinking Irishman. He operated out of the party's headquarters near Union Square, but his political base was the Teamsters organization in Minneapolis. A forceful public speaker, he spent eight months in 1936 and 1937 agitating among the seamen and cannery workers on the California coast, and he maintained ties to the unionized automobile workers in Ohio and Michigan.

Max Shachtman was born in 1904 in Warsaw, Poland, then part of the Russian Empire, and was brought to New York as a small child. He was the party's leading journalist and its most brilliant orator. Shachtman, like Cannon, had a keen wit, and he was able to exploit his Yiddish accent in the service of a laugh line, especially when addressing his constituents in the Bronx.

Unlike Cannon and Shachtman, James Burnham was not a professional revolutionary. The son of an executive of the Burlington Railroad, Burnham, a relative newcomer to the movement, was born in Chicago in 1905. He was educated at Princeton and Oxford, and taught philosophy at New York University, where he came to national attention as co-editor of the journal *Symposium* and as coauthor of a well regarded textbook, *Introduction to Philosophical Analysis*. Burnham was the party's leading theorist, and he and Shachtman edited its monthly journal, *New International*. His Manhattan address, 34 Sutton Place, testified to his privileged circumstances, as did his occasional appearance at political meetings in a tuxedo, donned earlier in the evening for some high society social gathering.

Burnham and Cannon coexisted uneasily. Cannon was wary of Burnham's social and academic status, while Burnham objected to Cannon's authoritarian management style, his anti-intellectualism, and the crude invective he hurled at his opponents: "scoundrels," "bloodhounds," "sons-of-bitches," "shysters," "miserable," "contemptible," "sniveling," "stinking," and so on. Burnham also criticized Cannon for blindly following Trotsky's lead. "The tendency in your letters to lump together all our opponents as 'Stalinist agents,' " he complained to Cannon in June 1937, "(analogous to, and perhaps copied from, T's recent habit of calling everyone who disagrees with him a 'G.P.U. agent') seems to me unprofitable."

The real trouble between the two men started when Burnham began to challenge Trotsky's position on what was known in the movement as the "Russian question." Trotsky had long maintained that despite the repressiveness of the Soviet bureaucracy, even the purges and the Terror, the fundamental achievement of the October Revolution—the abolition of the private ownership of the means of production—remained intact. The USSR, Trotsky said, was a "degenerated workers' state," deeply flawed but still salvageable for socialism and therefore deserving of "unconditional defense" should it come under military attack. Any attempt by Bolshevik-Leninists to deny the proletarian nature of the Soviet Union, Trotsky had warned, would be regarded as "treason."

In 1937, Burnham, together with another comrade, Joseph Carter, began to argue that the Soviet bureaucracy was no mere caste, as Trotsky insisted, but a new exploiting class, and that therefore the USSR could not be characterized as a workers' state, not even a degenerated one.

Burnham and Carter described the Soviet system as "bureaucratic collectivism." An increasing number of comrades thought this analysis made sense, to the point where, toward the end of 1937, Cannon alerted Trotsky that the party was experiencing "a little epidemic of revisionism." From this and other reports reaching him, Trotsky learned that the opposition was centered in New York, and was especially strong among the youth.

At the founding congress of the Socialist Workers Party, which convened in Chicago on December 31, 1937, Burnham and Carter's statement on the Russian question received only four out of seventy-five votes. Cannon hoped this would end the matter, but the controversy became more acute under the pressure of events, including the third Moscow trial in March 1938 and continued Soviet treachery against the non-Communist left in Spain. What was the difference, a growing number of comrades openly began to ask, between Hitler's Germany and Stalin's USSR?

Here the dissenters could support their arguments with quotations from Trotsky's recent book *The Revolution Betrayed*, where he described Stalinism and fascism as "totalitarian" twins bearing a "deadly similarity." The essential difference, in Trotsky's view, was that the Soviet government had nationalized the means of production. But for an increasing number of comrades, this was a distinction without a difference. A factional fight was brewing by the summer of 1939, even before the announcement of the Nazi-Soviet pact.

Trotsky failed to appreciate the enormous shock the pact produced on his followers. For many of them, the aftershocks were no less disorienting. Trotsky had been predicting that the Kremlin would reach a purely defensive agreement with Nazi Germany, as a way to keep the war off Soviet territory for as long as possible. The Soviet invasion of Poland, which began on September 17, only two and a half weeks after the German assault from the west, demonstrated that the pact was no mere nonaggression treaty, but an aggressive military alliance. This confounded the Trotskyists and, it seems clear, staggered Trotsky himself.

The Germans had launched their blitzkrieg with a massive attack from the air that destroyed the Polish air force and communication lines. As bombs rained from the sky, German armored columns plunged deep into the Polish interior, up to thirty miles ahead of the infantry, scattering civilians, spreading terror, and leaving the Poles no chance to mount a coordinated defense. In three weeks, western Poland was entirely overrun. Only Warsaw managed to hold out for another week under the Luftwaffe's relentless aerial bombardment.

By comparison, the Soviet invasion from the east was more like an occupation. The Poles there had been ordered not to fight because it was believed—or at least hoped—that the Red Army was entering to join the fight against the Germans. Instead, the Soviets arrested and deported hundreds of thousands of Poles to remote regions of the USSR. Tens of thousands more Poles were executed. The most infamous episode came to be known as the Katyn Forest Massacre, in which more than 21,000 Polish reserve officers who had been mobilized at the outbreak of the war—the large majority of them teachers, doctors, lawyers, and other members of Poland's intelligentsia—were shot to death and buried in mass graves. The Soviets would later attempt to place responsibility for this atrocity on the German armies that invaded the Soviet Union in June 1941 as part of Operation Barbarossa.

As the German and Soviet armies erased Poland from the map, Trotsky dictated a long article called "The USSR in War," which he completed on September 25. Much of it was devoted to a theoretical discussion about whether Stalinism, fascism, and even the New Deal constituted a new political paradigm, so-called bureaucratic collectivism. Trotsky turned back this theoretical challenge, but in doing so he said something entirely unexpected. Socialism, he announced, was about to face its ultimate test. If the Second World War did not spark a proletarian revolution in the West, or if the proletariat were to take power but then surrender it to a privileged bureaucracy as in the USSR, this would confirm the emergence of a new form of totalitarianism. In that case, Trotsky acknowledged, "nothing else would remain except only to recognize that the socialist program based on the internal contradictions of capitalist society ended as a Utopia."

Less than a year earlier, Trotsky had presented a vision of the Bolshevik-Leninists preparing to storm heaven and earth. Now he appeared to be harboring doubts about the entire socialist project, and this admission took his followers by surprise. It also undermined their confidence in his analysis of the Soviet occupation of eastern Poland, which was another surprising part of his article. In Trotsky's view, the Red Army, far from behaving like a mirror image of the Wehrmacht, was serving as a vehicle for progress in Poland by expropriating the large landowners and nationalizing the means of production. In other words, despite the reactionary nature of the Stalinist bureaucracy, the Soviet Union was objectively spreading the features of socialism abroad. Most of Trotsky's American comrades found this judgment difficult to square with what common sense told them about the Soviet subjugation of Poland.

"The USSR in War," instead of uniting the Socialist Workers Party, served to sharpen its discord. Shachtman had now joined forces with Burnham. Together they declared that the Soviet Union could in no sense whatsoever be classified as a workers' state, that the Soviet invasion of Poland was an act of imperialism, and that the party should disavow its pledge to defend the USSR unconditionally. A serious factional fight had broken out. Trotsky now put everything aside in order to devote his energies to preventing the party from splitting in two. Anyone familiar with his past record as a conciliator in factional politics could have anticipated that disaster lay ahead.

At the zenith of Trotsky's glory, after he had masterminded the Bolshevik insurrection in October 1917 and then led the Red Army to victory in the civil war, Anatoly Lunacharsky, the People's Commissar of Education, wrote a profile of him. Among the inevitable comparisons with Lenin, one came out decidedly in Trotsky's favor. Lenin, although irreplaceable as the chief executive of the Soviet government, "could never have coped with the titanic mission which Trotsky took upon his own shoulders, with those lightning moves from place to place, those astounding speeches, those fanfares of on-the-spot orders, that role of being the unceasing electrifier of a weakening army, now at one spot, now at another. There is not a man on earth who could have replaced Trotsky in that respect."

And yet, Lunacharsky testified, "Trotsky was extremely bad at organizing not only the Party but even a small group of it." The same charismatic personality that swept people off their feet was "clumsy and ill-suited" to working within a political organization. He could electrify crowds, but not persuade individuals. "He had practically no whole-hearted supporters at all."

Lunacharsky based this judgment on Trotsky's career since 1902, after his first escape from Siberia and arrival in London. It was Lenin who arranged for him to be brought to western Europe and who introduced him into the émigré circle of the Russian Social Democratic Labor Party. In March 1903, at Lenin's suggestion, Trotsky was co-opted onto the editorial board of the party's organ and power center, *Iskra*. Lenin sized up the twenty-three-year-old Trotsky as "a man of exceptional abilities, staunch, energetic, who will go further."

The honeymoon ended abruptly four months later, at the second congress of the Russian Social Democrats. The delegates assembled in Brussels, but then transferred to London to escape the attentions of the Russian secret police. The congress was attended by forty-three delegates representing twenty-six Marxist organizations. Trotsky held the mandate of the Siberian social democrats. Several issues divided the delegates, most importantly the definition of party membership. Lenin advocated a strictly centralized party, with all members participating in revolutionary activity—participating, as opposed to merely cooperating, which was the less restrictive formula proposed by Julius Martov and Pavel Axelrod, close colleagues of Lenin who now closed ranks against him.

This fundamental disagreement was compounded by others. When Lenin proposed to reduce the *Iskra* editorial board from six members to three, the move was seen by his opponents as a further attempt to consolidate his control over the party. After a vote taken toward the end of the conference was won by Lenin's supporters, they became known as *Bolsheviki*, Russian for Majoritarians, while Martov, Axelrod, and the others were labeled *Mensheviki*, the Minoritarians. On the crucial votes, Trotsky was with the Mensheviks.

Trotsky's reaction to the split was self-contradictory. He declared himself in favor of party unity yet launched extremely bitter polemical strikes at Lenin, whom he accused of behaving imperiously and of advocating a dangerous centralism. The most violent of these attacks took the form of a lengthy pamphlet called *Our Political Tasks*, published in Geneva in 1904. Here Trotsky called Lenin "malicious," "hideous," "dissolute," "demagogical," and "morally repulsive," among other epithets. He compared Lenin to Robespierre and, more trenchantly, a "slovenly lawyer."

A strong proponent of social democracy as a mass movement, Trotsky was genuinely repulsed by Lenin's centralism, which placed professional revolutionaries in the vanguard and seemed to assume that workers were a hindrance to the revolution. Trotsky was thus a proponent of Menshevism against Bolshevism, yet in September 1904 he announced his break with the Mensheviks. The Russian Revolution of 1905, which catapulted him to fame as a leader of the short-lived St. Petersburg Soviet, validated his status as a revolutionary free agent.

As Czar Nicholas II called in the army and the police to crush the revolution, Trotsky was arrested, tried, and sentenced to a second term of exile in Siberia, from which he again escaped, landing in Vienna in 1907. There, for the next seven years, he made his living from journalism, much of it devoted to bringing about a reconciliation of the Bolsheviks and Mensheviks, who remained factions of the same party. His ineptitude as a conciliator served to isolate him further. Although he had closer personal ties to the Mensheviks, he managed to alienate them, even as he continued to earn the animosity of the Bolsheviks. After Lenin consummated the schism in 1912 by declaring the Bolsheviks to be a separate party, Trotsky bitterly denounced him: "The entire structure of Leninism is at present based on lies and falsification, and carries within it the poisonous seeds of its own destruction."

Then came the world war and a string of catastrophic defeats for the Russian army led by Czar Nicholas, resulting in the collapse of the Russian autocracy in the February Revolution of 1917. Trotsky arrived in Petrograd in May, not long after Lenin. At first Trotsky turned down Lenin's offer to join the Bolsheviks, but changed his mind in July, a few weeks before he was elected chairman of the Petrograd Soviet. It was

Trotsky's idea to cloak the Bolshevik coup against the Provisional Government in democratic legitimacy by timing it to coincide with the opening of a national Congress of Soviets about to gather in Petrograd. On the night of October 25, the congress was informed about the seizure by the Red Guards of the Winter Palace. Some delegates walked out of the hall. The Mensheviks accused the Bolsheviks of carrying out a putsch, and protested that some kind of political compromise ought to be agreed.

Trotsky showed no sympathy for his vanquished former comrades, only mockery and disdain. Taking the platform, he delivered History's cruel verdict. The Bolshevik triumph, he declared, was a mass insurrection, not a conspiracy. "Our rising has been victorious. Now they tell us: Renounce your victory, yield, make a compromise. No, here no compromise is possible. To those who have left and to those who tell us to do this we must say: You are bankrupt. You have played out your role. Go where you belong from now on: into the dustbin of history!"

From that moment on, Trotsky held tightly to the myth of Red October as a workers' revolution. Try as he might, however, he could not obscure his long history of anti-Bolshevism, which his enemies in the Party preferred to characterize as his "Menshevism." This goes a long way toward explaining Trotsky's passivity in the struggle to succeed Lenin: as an outsider, he made a fetish of Bolshevik unity. He was, in any case, poorly equipped to lead a Party faction. He could not overcome his isolation. He had never acquired the habits necessary for working within a political organization, let alone for maneuvering in the corridors of power.

Max Eastman, who was electrified by the description of Trotsky in John Reed's *Ten Days That Shook the World* and who then had the opportunity to watch Trotsky in action as head of the Left Opposition in Moscow, sized him up much the way Lunacharsky had done back in 1919. "In the time of revolutionary storm, he was the very concept of a hero," Eastman observed. "But in calmer times he could not bring two strong men to his side as friends and hold them there." In Eastman's view, this more than anything else explained Trotsky's loss to Stalin in the factional fight, and then the hopelessness of his efforts to organize an international opposition to Stalinism. "He could no more build a party than a hen could build a house."

Cast out of the Communist Party and then the Soviet Union, Trotsky saw no irony in the fact that he ended up "sharing the bitter fate he had meted out to the Martovs and the Axelrods," as one historian has put it. On the contrary, Trotsky's account of the October events in *The History of the Russian Revolution* dramatized his banishment of the Mensheviks as a moment of triumph, showing no trace of remorse.

Eastman carried with him the section of Trotsky's *History* that contained this passage—in the form of the publisher's proofs of his English translation—when he and his wife visited Prinkipo in the summer of 1932. The Eastmans were paying a social call, but the visit would also give Trotsky an opportunity to verify the accuracy of Eastman's version of the book. At forty-nine years of age, Eastman was strikingly handsome: tall, trim, and tan, with a shock of white hair and dark, pensive eyes. He had not seen Trotsky for several years, and once again he remarked on the pale blue color of his eyes, which a long line of journalists mysteriously kept insisting were black. On the second day of Eastman's visit, they were incandescent with anger.

Trotsky had been disturbed by Eastman's unorthodox views of Marxist theory, notably his debunking of the concept of dialectical materialism. Eastman's visit to Prinkipo was an opportunity for Trotsky to set the amateur philosopher straight. When Eastman stood his ground, their argument threatened to spiral out of control, as neither man allowed the other to finish a sentence. "Trotsky's throat was throbbing and his face was red; he was in a rage," Eastman wrote in his diary. Natalia became worried as the altercation spilled over from the tea table into the study: "she came in after us and stood there above and beside me like a statue, silent and austere. I understood what she meant and said, after a long, hot speech from him: 'Well, let's lay aside this subject and go to work on the book.' 'As much as you like!' he jerked out, and snapped up the manuscript."

Eastman was a contrarian by nature. Born in upstate New York to two unconventional, liberal-minded Congregationalist ministers, he was educated at Williams College, and then studied philosophy at Columbia University under John Dewey, completing the requirements for a doctorate. He chose not to accept the degree, evidently because doing so might compromise his self-image as a revolutionary poet. He settled in Greenwich Village and became an influential figure in American radical politics and culture. In 1913 he published *Enjoyment of Poetry* and became an editor of *The Masses*, the pioneering magazine of socialist politics, art, and literature.

In 1917 *The Masses* was forced to close as a result of the tightening wartime censorship, and the following year Eastman and his fellow editors were twice tried and twice acquitted for violation of the Sedition Act, in connection with the magazine's outspoken opposition to U.S. participation in the world war. He and his sister and fellow suffragist, Crystal, then founded a successor, *The Liberator*, which published John Reed's initial reports from Petrograd on the Bolshevik Revolution. The first of these conveyed an invitation from Lenin and Trotsky: "Comrades! Greetings from the first proletariat republic of the world. We call you to arms for the international Socialist revolution."

In 1922 Eastman went to Soviet Russia to see the experiment for himself. There he met the leading Bolsheviks, including Trotsky, who agreed to help Eastman write his biography. In 1924, Trotsky and the Oppositionists provided Eastman with the text of Lenin's still-secret political testament, which Eastman then published in the West, creating a sensation. In the aftermath, Trotsky felt compelled to disavow Eastman in order to placate Stalin, but thereafter the two men were closely identified in the West, even before Eastman became his translator.

That explains why Trotsky was so disturbed by Eastman's public repudiation of the dialectic, a principle of change conceived by Hegel in the early nineteenth century. Hegel believed that history unfolds in a logical process of inner conflict, in which change occurs because antagonistic forces collide and their antagonism is resolved in new, higher forms. Marx applied Hegel's concept to human society, where this inner conflict takes the form of class struggle.

In Marx's theory, the material base determines the relations of production—technology, inventions, systems of property—and these in turn determine the philosophies, governments, laws, cultural tastes, and moral values that dominate a society. As material conditions evolve, tensions build up until a point is reached where quantitative changes have qualitative consequences. That, said Marx, was how society advances. The great breakthroughs took the form of revolutions, which he called "history's locomotives." Marx labeled his philosophy "historical materialism"; the term "dialectical materialism" was introduced after Marx's death by Engels to denote an extension of dialectics beyond society to the world of nature.

Even before he went to Soviet Russia, Eastman was puzzled by the connection between Marx's social theory of class struggle and the concept of the dialectic, which Marx claimed made his philosophy scientific and proved the inevitability of socialism. If that were the case, Eastman thought, then all one had to do was sit back and wait for socialism to arrive. Yet Eastman knew that Marxism was about changing the world, not just understanding it. Otherwise, why did Marx and Engels, in the *Communist Manifesto* of 1848, call upon the workers of the world to unite and throw off their chains? And why did Lenin, in the seminal pamphlet *What Is to Be Done?* published in 1902, call for a vanguard party of professional revolutionaries? Surely Lenin did not believe in socialism's inevitability?

In Moscow, Eastman sat down in the library of the Marx-Engels Institute and applied his as yet rudimentary Russian to a study of the influence of the Hegelian dialectic on Marxism and on Bolshevism. To his dismay, he discovered that the Bolshevik leaders, Lenin included, were indeed true believers in the "science" of dialectical materialism as a universal law of motion, despite the fact that it was based on no empirical observation and, as Eastman saw it, belonged to the realm of metaphysics or religion rather than science.

Leaving the Soviet Union for the West, Eastman became obsessed with the idea of exposing Hegelian dialectics as a pseudo-scientific fraud. He made his case in a 1927 book called *Marx and Lenin: The Science of Revolution*, a work that predictably came under attack from the left. Eastman's most formidable and relentless critic was Sidney Hook, a protégé of John Dewey, ally of the Communist Party, and Marxist professor of philosophy at New York University. Hook and Eastman were Dewey's "bright boys" and two tenacious combatants, and their dispute over Marxist theory continued for several years.

Trotsky monitored the Eastman-Hook debate from Turkey. As he saw it, both men were afflicted with that peculiarly American disease, a pragmatist conception of empirical science. It was no mere coincidence, he thought, that both men were students of Dewey, one of pragmatism's founding philosophers. In Trotsky's eyes, Hook's attempt to reduce Marxism from a science to a social philosophy was bad enough. Much worse, though, was Eastman's outright dismissal of Hegel's dialectic as an example of pre-Darwinian speculation and wishful thinking.

This is what provoked Trotsky's fury when Eastman visited Prinkipo. As Eastman describes the action, "he became almost hysterical when I parried with ease the crude clichés he employed to defend the notion of dialectic evolution. The idea of meeting my mind, of 'talking it over' as with an equal, could not occur to him. He was lost." For Trotsky there could be no meeting of the minds about Marxism. Those who sought to revise Marxist theory, he said, wished "to trim Marx's beard." Eastman was trying to decapitate him.

Not long after their face-off in Prinkipo, Trotsky published a letter in the American Trotskyist paper *The Militant* calling attention to Eastman's "petty-bourgeois revisionism." Eastman's translation of the *History* was brilliant, of course, and Trotsky had thanked him for it. "But as soon as Eastman attempts to translate Marxian dialectics into the language of vulgar empiricism, his work provokes in me a feeling which is the direct opposite of thankfulness. For the purpose of avoiding all doubts and misunderstandings I consider it my duty to bring this to the knowledge of everybody."

Four years later, Trotsky was still performing this duty. When Shachtman and George Novack met his ship in Tampico in January 1937 and escorted him on the train to Coyoacán, they found him fixated on the subject of Eastman's heresy. While Trotsky was pleased to learn that Dewey was one of his defenders, he expressed grave concerns about the dangers of Dewey's pragmatism as manifested in Eastman's revisionism. His vehemence took Shachtman and Novack by surprise. Here was a man who had just landed in a new world after being refused asylum by all the countries of Europe. Hunted by the Soviet secret police, only hours earlier he had been reluctant to disembark his ship out of fear for his life. Yet here he was, carrying on as though his real nemesis was not the tyrant Stalin but the infidel Eastman. "There is nothing more important than this," he exhorted his two comrades. "Pragmatism, empiricism is the greatest curse of American thought. You must inoculate the younger comrades against its infection."

Nor, meanwhile, had Eastman forgotten the Prinkipo face-off and Trotsky's subsequent public rebuke. At the time Trotsky was vilifying him on the train from Tampico, Eastman had recently completed a translation of *The Revolution Betrayed*, a book he discussed in an article in *Harper's* in March 1938. He endorsed its damning description of Stalin's USSR, but he could not accept its argument that the Revolution's betrayal was the result ultimately of Russia's backwardness and isolation, an interpretation Trotsky supported by invoking dialectical materialism. Where had Trotsky been all these years? wondered Eastman. Had he learned nothing about the human condition from modern psychology, biology, or sociology? "It is not a question, as Trotsky thinks, of being 'frightened by defeat' or 'holding one's positions.' It is a question of moving forward or being stuck in the mud. No mind not bold enough to reconsider the socialist hypothesis in the light of the Russian experiment can be called intelligent."

Eastman had now added insult to heresy. Furious, Trotsky turned to Burnham and urged that Eastman be dealt with "mercilessly." It appears that Trotsky was unaware that Burnham himself had for years been displaying symptoms of the American disease. Burnham was prepared to defend the October Revolution and Marxism, but not dialectical materialism, because it falsely guaranteed the inevitability of socialism. Trotsky woke up to this fact at a moment when a number of American intellectuals began to distance themselves from Marxism, among them Hook, who now decided that the dialectic belonged to mythology after all, and Dewey, who saw in Marxism's Hegelian origins a strain of theology. Edmund Wilson, at work on a monumental history of socialist and communist thought from the French Revolution to the Russian Revolution, weighed in with an essay in *Partisan Review* called "The Myth of the Dialectic."

For Trotsky, this was a disturbing trend, but for him the "greatest blow" came from Shachtman and Burnham in their rebuttal to the revisionists, a major article published in the January 1939 issue of *New International* called "The Intellectuals in Retreat." Instead of attacking Eastman, Hook, and the others head on, the authors limited their criticisms to the realm of politics, and allowed that dialectical materialism was not essential to Marxist theory or practice. Trotsky was scandalized. This was, he scolded Shachtman, "the best of gifts to the Eastmans of all kinds."

Burnham's slide continued in spring 1939 with an article in *Partisan Review* that likened the dialectic to a human appendix: what Marxism needed was the intellectual equivalent of an appendectomy. Hansen wrote to inform Trotsky about this latest act of desecration, and also to report that there was much confusion throughout the party's ranks concerning dialectical materialism, especially among the youth. Few comrades even professed to understand its meaning. Hansen passed along the remark of a comrade to the effect that "Trotsky does not write on the dialectic or on philosophy because he is *incompetent* to do so." This must mean that Trotsky does not actually use the dialectic and that it really was a "metaphysical trapping," just as Eastman said.

Hansen's baiting letter helped convince Trotsky, once the factional fight in the Socialist Workers Party got under way, that the struggle against the opposition must be waged as a defense of Marxism's core principles against the insidious American infection.

Events in Europe in the autumn of 1939 served to deepen the factional divide among the American Trotskyists. As the Red Army completed its occupation of eastern Poland, the Soviet government demanded from Latvia, Lithuania, and Estonia the right to establish military and naval bases and garrison troops on their soil. By mid-October, all three Baltic states had acquiesced. In New York, the Trotskyist Minority, led by Shachtman and Burnham, insisted that the party condemn these Soviet moves as acts of aggression, and they proposed a referendum on the unconditional defense of the USSR. Cannon rejected the idea and Trotsky backed him up. At a meeting of the party's Political Committee on November 7, the vote was eight to four in favor of the Majority.

Then came the Soviet invasion of Finland, which followed the Helsinki government's refusal to grant the territorial concessions demanded by Moscow. The Soviets first staged a series of border incidents accompanied by a loud propaganda campaign. The Red Army attacked on November 30. The Finns put up a stiff defense and battled the initial five-pronged Soviet assault to a stalemate. The Soviets then regrouped and concentrated their offensive against the Mannerheim Line, in southern Finland facing Leningrad: eighty-five miles of defensive fortification—snowbound trenches, pillboxes, and reinforced concrete structures—spanning the Karelian Isthmus, from the Gulf of Finland to Lake Ladoga. The Finnish defenders, donning white camouflage suits and employing mobile units on skis, had the initial advantage over the underprepared and underequipped troops of the Red Army, but everyone understood that time was not on their side.

In the United States, the cause of "little Finland" drew enormous public sympathy. Trotsky's followers once again looked to Coyoacán. Surely the Finnish events would force their leader to revise his thinking about the USSR. Three weeks after the invasion, Trotsky's analysis arrived in New York in the form of a long and scathing polemical article titled "A Petty-Bourgeois Opposition in the Socialist Workers Party." Its entire first half was devoted to dialectical materialism and related theoretical questions, framed as an attack on Burnham and the revisionists. Taking Hansen's cue, Trotsky included a primer on theory called "The ABC of Materialist Dialectics."

Even those few party members who claimed to understand Trotsky's discussion of the dialectic were not sure how it related to current debates. There was also considerable skepticism about Trotsky's decision to emphasize the "petty-bourgeois" nature of the Minority: that there were too many clerical workers and too few factory workers in its ranks. Industrial workers, Trotsky observed, had a natural "inclination toward dialectical thinking." That explained why the Majority held the correct position on the basic theoretical questions: "Cannon represents the proletarian party in the process of formation."

Trotsky at his desk in his study, winter 1939–1940.

Trotsky's focus on Marxist theory and sociology elicited much head-scratching among comrades in both factions. But it was his analysis of the Finnish events that raised eyebrows. According to Trotsky, the Red Army in Finland was engaged in the expropriation of the large landowners and the introduction of workers' control in industry, a preliminary step toward the expropriation of Finland's capitalists. A Finnish civil war was now beginning, Trotsky contended, with the Red Army on the side of the workers and peasants, and the Finnish army on the side of the exploiters. In view of all this, Trotsky said, the Bolshevik-

Leninists must continue to lend the USSR their "moral and material support."

This interpretation of the Soviet invasion of Finland struck most comrades as utterly fantastic—even many in the Majority camp, although they could not say it openly. The notion of a civil war breaking out in Finland contradicted the known facts. So did the idea that the Red Army was imposing workers' control over Finnish industry: everyone knew that the Russian workers themselves did not have that. And if the Finnish masses were rising up in support of the Red Army, why was the Soviet Union losing the war? A member of the Minority said privately what many other comrades were thinking: the Old Man had gone "completely haywire."

Trotsky's oddly timed preoccupation with the dialectic seemed designed to divert attention away from the inconvenient facts about the events in Finland. From New York, Sherman Stanley, whose allegiance to the Minority had cost him an assignment as secretary-guard in Coyoacán, complained directly to Trotsky. He was appalled by Trotsky's argument that because the Soviet Union had nationalized the means of production, its "rape of Finland" was worthy of support. "Is not this the most monstrous and shameful *non-sequitur* in the history of our movement? Is not this worthy of the brazen 'dialectical' twisting so familiar in the history of Stalinism?"

Stanley and the Minority comrades were indignant that Trotsky had characterized them as "petty-bourgeois," a time-honored Bolshevik term of abuse that Stalin had used to censure Trotsky during their struggle for power. To criticize party comrades this way, effectively labeling them class enemies, seemed unforgivable. Manny Garrett, of Brooklyn, called Trotsky's article "disloyal, inaccurate, dishonest, and false to the core. L.D. has laid the gauntlet. We are ready to reply in kind."

When Trotsky learned of these reactions he wrote to Cannon that the Minority comrades were now behaving like "enraged petty-bourgeois." And since Cannon routinely circulated the texts of Trotsky's letters at party headquarters, this comment was inflammatory, as was Trotsky's remark about the need to unmask "Stalinist agents working in our midst" to provoke a split.

In fact, Trotsky was doing well enough on his own. Both factions interpreted his article as laying the basis for a split, an assumption that seemed confirmed by a statement he made to Shachtman: "I believe that you are on the wrong side of the barricades, my dear friend." Yet farther along in the same letter, Trotsky indicated that he had not yet given up on Shachtman: "If I had the opportunity I would immediately take an airplane to New York City in order to discuss with you for 48 or 72 hours uninterruptedly. I regret very much that you don't feel in this situation the need to come here to discuss the questions with me. Or do you?"

Instead, Shachtman carried on the debate in New York with Hansen, who was generally regarded as Trotsky's proxy and whom the Minority reviled as an enforcer for the "Cannon clique." Hansen's recent articles and speeches had earned him a reputation for heavy-handed sarcasm, pompous irony, and a vulgarized Marxism. Cannon decided it was time to dispatch Hansen to the Bronx, Shachtman's stronghold, where the Minority outnumbered—and outshouted—the Majority two to one. Cannon called this mostly Jewish rabble "declassed kibitzers" and "petty-bourgeois smart alecks." After his first appearance, Hansen described the scene to Trotsky as a "madhouse." "Where's the civil war in Finland?" was the favorite Bronx jeer. Howls of laughter greeted his every mention of the dialectic, as if he had told a hilarious joke. And every reference to Minneapolis ignited an outbreak of heckling. In the Bronx, Hansen explained, the Minneapolis comrades were considered "provincials, blockheads, stupid yokels who don't know anything but trade-union work and whose hands fly up like semaphores on a railroad whenever Cannon passes by."

In Coyoacán, Trotsky gritted his teeth. He had lived in the Bronx during his brief sojourn in America in 1917, residing in a small row house on Vyse Avenue, in a working-class neighborhood that was home to Irish, Italian, and Jewish immigrants. This experience now gave his mind's eye a vivid picture of the problem. "The oppositionists, I am informed, greet with bursts of laughter the very mention of 'dialectics.' In vain," Trotsky thundered, sounding eerily like a visitor from the Red Planet sent to warn the earthlings that resistance was futile. "This unworthy method will not help. The dialectic of the historic process has more than once cruelly punished those who tried to jeer at it." He urged that "the Jewish petty-bourgeois elements of the New York local be shifted from their habitual conservative milieu and dissolved into the real labor movement."

As for the "petty-bourgeois disdain" directed at the Minneapolis comrades, Trotsky offered a little history lesson. "At the Second Congress of the Russian Social-Democrats in 1903," he recollected, "where the split took place between the Bolsheviks and the Mensheviks, there were only three workers among several scores of delegates. All three of them turned up with the majority. The Mensheviks jeered at Lenin for investing this fact with great symptomatic significance. The Mensheviks themselves explained the position the three workers took by their lack of 'maturity.' But as is well known it was Lenin who proved correct." The behavior of the Minority, Trotsky cautioned, bore a strong resemblance to the struggle of the Mensheviks against Bolshevik centralism.

Through the winter of 1939–40, Trotsky's polemical battle against the Minority continued, thousands upon thousands of words, with major thrusts directed at Shachtman's misreading of Bolshevik history and Burnham's "brutal challenge" to Marxist theory. The effect they produced was entirely the opposite of what Trotsky intended. Even Cannon had to marvel at how "each contribution by the OM brought a worse reaction than the one previous." In any case, minds had already been made up. As Burnham said, "The Finnish events were absolutely decisive."

Cannon was now eager to toss the "petty-bourgeois windbags" into the dustbin of history. Trotsky, however, still held out hope for unity. When the Minority announced that it would convene its own national conference in Cleveland at the end of February 1940, he advised Cannon that the proper response was "a vigorous intervention in favor of unity by the majority." "Back to the Party!" he exhorted his wayward comrades. Cannon, meanwhile, was denouncing those same comrades as "enemies and traitors" who had to be "fought without mercy and without compromise on every front" and subject to "the most ruthless punishment in the form of a war of political extermination."

A special convention of the Socialist Workers Party was held in mid-April, with eighty-nine delegates and sixty alternates present, representing a total membership of 1,095. The Majority won every vote by the same total, 55 to 34. Shachtman announced that, backed by a large preponderance of the youth, the Minority had the support of at least half the membership and that it intended to form a separate party. Cannon could breathe a sigh of relief: the schism was accomplished. Stanley's postmortem captured the general feeling of bitter regret on the Minority side: "The war broke out and we did nothing. The OM did nothing. One of the most important events of our epoch took place, and we were asleep. And we stayed asleep."

THE SPLIT FOUND Jan Frankel on the other side of the barricades. He moved from New York to Los Angeles to work for the Minority. A comrade there described him to Trotsky as sick, jobless, next to penniless, and extremely disheartened by the war, which had crushed his native Czechoslovakia. Frankel blamed the split entirely on Trotsky. "The present fight in the American Party has been carried out in the traditional manner of the old *Iskra* days when Lenin and the OM engaged in their bitter polemics," he said. These methods were completely inappropriate to radical politics in the United States in the 1930s. "The proof is in the split."

Trotsky was shaken by the loss of yet another close comrade, and regretted that he could not sit down with Frankel and talk it over. Only one member of the Minority had so far made the pilgrimage to Coyoacán. She was Sylvia Ageloff, the Brooklyn social worker. She had come to Mexico City as a tourist, more or less. After she arrived, she wrote a note to Trotsky passing along greetings from her sister Ruth.

Sylvia was invited to come to the house on Avenida Viena the next day, January 26, 1940, to join a discussion about the factional struggle with another visitor, Farrell Dobbs, the Teamsters organizer whom Cannon had recently brought from Minneapolis to manage the New York office. Sylvia was a minority of one that day, among Trotsky, Dobbs, the guards, and the staff. Chief of the guard Harold Robins said that her remarks "beautifully indicate the attitude of the petty bourgeois Menshevism of the minority."

The swaggering account of this meeting that Robins sent to New York, against the backdrop of Trotsky's repeated warnings about Stalinist agents stirring the factionalist pot, raised anxieties there about Trotsky's safety. Among those who felt a sudden sense of alarm was John Wright, Trotsky's research assistant on the Stalin biography and a stalwart member of the Majority. Wright warned the staff in Coyoacán that "the factional struggle provides a *perfect* cover for the penetration of GPU provocateurs and assassins to Trotsky." He urged that "utmost caution" be exercised. As an "absolutely iron bound" rule, even the most loyal visitor must be subjected to a personal search. "We are all very anxious on this point."

Sylvia was not a threat, of course, certainly not as a debater, a fact that had prompted an outburst of bravado from Robins on Trotsky's behalf: "The old man is dying to have a fighting minority supporter and he would consider it a pleasure I am sure if you would send him one with a bit of guts in him." It turns out that the GPU was already calculating along the same lines. As Sylvia left Trotsky's home late in the afternoon of January 26, waiting by the police guard house to drive her back to the city was the Canadian businessman Frank Jacson, otherwise known to her as Jacques Mornard. The penetration agent was now just outside the gates.

CHAPTER 10

Lucky Strike

On February 27, 1940, Trotsky sat down to write his last will and testament. "My high (and still rising) blood pressure is deceiving those near me about my actual condition," he began. "I am active and able to work—but the end is apparently near." Trotsky, who had recently turned sixty, was convinced he had advanced arteriosclerosis, to the point where, as he wrote in an addendum on March 3, "the end must come suddenly, most likely—again, this is my own hypothesis—through a brain hemorrhage. This is the best possible outcome that I can hope for." In the event that his illness threatened to become protracted and make him an invalid, he said, he would exercise his right to determine his own time of death.

Trotsky became preoccupied with his health after a recent examination by his doctor, a German refugee named Alfred Zollinger. That, in any case, is how Trotsky explained to Natalia his decision to write a will. Should his Stalin biography earn him any income after his death, he wanted to ensure that the money would go to her, so that she could support herself and see to Seva's education. Alarmed at the sudden blackness of her husband's mood, Natalia spoke to Dr. Zollinger, who denied he had given his patient any cause for concern. Zollinger examined Trotsky again and tried to reassure him that his health was fine. His outlook brightened, although Natalia wondered if this was merely for her benefit.

In vowing to avoid a prolonged state of illness, Trotsky was mindful of the fate of Lenin, who experienced a series of strokes that left him incapacitated in the year before his death, at age fifty-four. This fixation on Lenin had become second nature to Trotsky. After the two men put aside their differences in 1917, Trotsky was unable to resist the force of Lenin's personal charm. "This magnetism is colossal," the Bolshevik portraitist Lunacharsky testified in describing the remarkable spell Lenin cast over untold individuals, from intellectuals, such as the writer Maxim Gorky, to the humblest peasant visitor admitted to his Kremlin office. "People who come into his orbit not only accept him as a political leader but in some strange fashion fall in love with him."

Trotsky fell hard, perhaps owing to their long years of mutual animosity. Of course, Trotsky's later veneration of Lenin happened to serve his purposes, and yet there is little doubt that it was genuine. "He was my master," Trotsky said of the Party's original Old Man. With an eye on how history would regard his relationship with Lenin—and knowing that Stalin's historians were busy falsifying the record—Trotsky presented a series of carefully composed vignettes to commemorate their historic collaboration.

Trotsky's version of the story begins in London in October 1902. After escaping from Siberia and making his way across Europe, Trotsky arrives outside Lenin's spartan two-story apartment building on Holford Square, in a lower-middle-class neighborhood near King's Cross. Although it is near dawn, the impatient young caller knocks vigorously on the outside door—three times, as he has been instructed. Lenin's wife, Krupskaya, hurries downstairs to greet him. "The Pen has arrived!" she announces from the doorway, then goes out to pay the cabbie while Trotsky shows himself inside. Lenin sits up in bed and listens raptly to the animated visitor's report on the Russian revolutionary underground and the circumstances of his flight. As Trotsky describes Lenin, "the kindly expression of his face was tinged with justifiable amazement."

Fifteen years later, the two men join forces in the decisive hours of Red October. Lenin, who has been hiding in Finland in order to avoid arrest, returns to action on the night of the Bolshevik coup d'état. The scene is the imposing, neoclassical Smolny Institute, where the Congress of Soviets has convened and the Bolsheviks plan to declare victory in the name of the proletariat. Late in the evening, as they wait for the session of the congress to begin, Lenin and Trotsky try to get some rest in a room adjacent to the hall, making use of a blanket and pillows that have been put down on the floor for them. "We were lying side by side; body and soul were relaxing like overtaut strings," Trotsky remembers. They were too excited to sleep, so they conversed in hushed tones, Lenin with "a rare sincerity in his voice."

The following year, the Revolution is in peril, as the White armies rise up and advance on Moscow, which has replaced Petrograd as the capital of Soviet Russia. In August, Simbirsk on the Volga falls to the Whites and, farther north, Kazan, the ancient Tatar capital, is under siege. Trotsky, who is about to leave for the front to command the Red forces, visits Lenin and finds him dispirited. "It's a bowl of mush we have, and not a dictatorship," Lenin laments. Trotsky assures him that his political commissars will enforce iron discipline on the Red Army, and then departs to prove his point.

After the Reds recapture Kazan and Simbirsk in September 1918, Trotsky returns from the front and pays a call on Lenin, who is convalescing from gunshot wounds to the neck and shoulder after an assassination attempt by a disillusioned radical. "Lenin was in a fine humor and looked well physically," Trotsky writes of their reunion. "It seemed to me that he was looking at me with somehow different eyes. He had a way of *falling in love* with people when they showed him a certain side of themselves. There was a touch of this being 'in love' in his excited attention. He listened eagerly to my stories about the front, and kept sighing with satisfaction, almost blissfully." Trotsky's achievement is not limited to the battlefield. "The game is won," Lenin declares. "If we have succeeded in establishing order in the army, it means we will establish it everywhere else. And the revolution—with order—will be unconquerable."

The twin Bolshevik stars, their names inseparably linked in the public mind, still have their quarrels after 1917, some of them stormy. "Lenin and I had several sharp clashes," Trotsky explains, "because when I disagreed with him on serious questions, I always fought an all-out battle." Of course such episodes would later be used against Trotsky by his rivals in the succession struggle. "But the instances when Lenin and I understood each other at a glance were a hundred times more numerous."

In the period of mourning after Lenin's death in January 1924, Trotsky takes great comfort in a private letter from Lenin's widow. Krupskaya writes to say that a few weeks before the end, Vladimir Ilich dwelled appreciatively over a passage Trotsky had written comparing him and Marx as world-historical figures. "And here is another thing I want to tell you. The attitude of V.I. toward you at the time when you came to us in London from Siberia did not change until his death. I wish you, Lev Davidovich, strength and health, and I embrace you warmly." To Trotsky, this simple, heartfelt letter stood as a refutation of all the combined slanders hurled against him by the epigones.

In Lenin's testament, which he dictated in December 1922 and which came to light after his death, he singled out Trotsky and Stalin as "the two most eminent leaders of the present Central Committee," and voiced concern that their rivalry would cause a split in the Party. He called Trotsky the "most able" of the Bolshevik leaders, but qualified this endorsement by remarking on his "excessive self-confidence" and his "disposition to be too much attracted by the purely administrative aspect of affairs"—a euphemistic allusion to Trotsky's well-known authoritarian manner.

As for Stalin, Lenin warned that the general secretary had concentrated enormous power in his hands and expressed concern that he would know how to use it properly. Subsequent events, including Stalin's insolent behavior toward Krupskaya, prompted Lenin to add a compelling postscript, dated January 4, 1923: "Stalin is too rude and this defect, although quite tolerable in our midst and in dealings among us Communists, becomes intolerable in a General Secretary. That is why I suggest that the comrades think about a way of removing Stalin from that post and appointing another man in his stead."

This would seem to have tilted the scales in Trotsky's favor, but Stalin and his allies were able to restrict the circulation of the document. When the text was leaked by the Left Opposition to Max Eastman and he published it in the West in 1925, Trotsky was pressured by Stalin into signing a statement repudiating Eastman and denying the very existence of Lenin's testament. Trotsky took this step in order to avoid a premature clash, and he reversed it a year later when the factional struggle broke out in earnest, but it was an act of political expediency that continued to dog him. Indeed, at the very time he sat down to write his own testament, the Minority opposition in the United States had resurrected the story of Trotsky's shabby treatment of Eastman, forcing him yet again to explain himself.

Just at that moment, in February 1940, a living reminder of those fateful events, in the person of Max Eastman, accompanied by his wife, Eliena, arrived in Coyoacán. Trotsky had by now written him off for the Marxist movement, so the onetime comrade could be welcomed into the house for a conversation about the good old days. Eastman says they spent two "light-hearted" hours together, although he did probe Trotsky, delicately, on a certain sensitive topic, with unsurprising results. "His faith in the disguised religion, or 'optimistic philosophy' as he called it, of dialectical materialism was absolute," Eastman confirmed. Yet this time, there was no angry explosion to drive home the point. In general, Eastman found that Trotsky had grown "more mellow," despite his increasing isolation and the waning prospects for the success of his revolutionary project.

Eastman's visit, Dr. Zollinger's examination, Lenin's example—these provided the inspiration for Trotsky to take up his pen and write his own testament. After naming Natalia as his heir, he touched upon matters of politics and ideology. He expressed the conviction that a future revolutionary generation would rehabilitate him and his fallen comrades by repudiating the "stupid and vile slander of Stalin and his agents." He thanked his collaborators over the years, too numerous to mention individually, although he made an exception for his closest comrade, Natalia. "For almost forty years of our life together she has remained an inexhaustible source of love, generosity, and tenderness. She experienced great sufferings, especially in the last period of our lives. But I take comfort in the fact that she has also known times of happiness."

He then reaffirmed his ideological faith: "For forty-three years of my conscious life I have been a revolutionary; and for forty-two I have fought under the banner of Marxism. . . . I will die a proletarian revolutionary, a Marxist, a dialectical materialist and, consequently, an irreconcilable atheist. My faith in the communist future of mankind is no less ardent, indeed it is even stronger now than it was in the days of my youth."

As he wrote these lines, seated at his desk in his study, he looked over to his left, out through the French windows and into the patio, where he saw Natalia approaching. The scene inspired him to close on a lyrical note: "Natasha has just come up to the window from the courtyard and opened it wider so that the air might enter more freely into my room. I can see the bright green strip of grass beneath the wall, and the clear blue sky above the wall, and sunlight is everywhere. Life is beautiful. Let the future generations cleanse it of evil, oppression, and violence, and enjoy it to the full."

Four weeks later, toward the end of March 1940, Trotsky appeared to discover his Fountain of Youth in Veracruz harbor. On a three-day visit there, he took advantage of an opportunity to go deep-sea fishing, the first such outing since he arrived in Mexico. The experience seemed to revive his spirit—at least that is the clear impression conveyed by the photographs and motion pictures made of this excursion.

Trotsky, dressed in a dark jacket and a soft white cap, is seen walking toward the boat, as he explains to one of his guards how he used to go fishing on the Sea of Marmara. Trotsky puts on his gear, then consults on the pier with the majordomo before they embark and set off. The chief of Trotsky's Mexican police guard, Jesús Rodriguez Casas, is seen steering the boat. Trotsky, manipulating reel and rod, appears vibrant. His famous white goatee juts forward; his round tortoiseshell glasses are speckled with sea spray. Not since the Prinkipo days has he seemed so invigorated.

Trotsky's vacation in Veracruz was preserved for the ages by an American comrade named Al Young, who came to Coyoacán for what was supposed to be a brief visit but stretched into five months. Young was born Alexander Buchman to an affluent family in Cleveland, Ohio, in 1911. He earned a degree in aeronautical engineering from the Case School of Applied Science in Cleveland in 1933. After graduation, he escaped the unemployment rolls by moving to Asia, spending most of

Trotsky fishing in Veracruz harbor, March 1940.

the next six years in Shanghai, where he worked for various foreign news agencies. A camera enthusiast, he extensively photographed and filmed daily life in Shanghai, including the Japanese invasion and occupation of the city in 1937.

As Young was leaving China in 1939, two Trotskyists he had gotten to know there arranged for him to visit Trotsky in Mexico in order to show him the nearly three hours of film he took in Shanghai. Young arrived in Coyoacán in November 1939 with his Leica and an 8mm Bell & Howell. As a member of Trotsky's entourage, he took several hundred black-and-white and color photographs and some fifty-five minutes of moving images.

Young's cameras recorded scenes from several of Trotsky's cactus-hunting picnics in the winter of 1939–40. Trotsky, wearing heavy work gloves and wielding a pickax, is seen digging out one and then another cactus, assisted by Sergeant Casas and a bodyguard. The prize specimens, one of them chest-high, are wrapped in thick coats of newspaper so as to protect their needles. Extra soil is collected in sacks for the replanting back home. The visual record makes plain why the guards were leery of these physically exhausting expeditions. Any hint of flagging energy might provoke another round of badgering from the Old Man, who at one point turns and accuses the cameraman of loafing.

The job was not done until the hunting party returned home and the last cactus had been replanted in the patio at Avenida Viena 19, Trotsky's residence since May 1939. Trotsky's home was only three blocks from the Blue House, yet it was relatively isolated, at the end of a dirt road lined with adobe hovels. Only one wall was attached to a neighboring property. Running parallel to Avenida Viena, on the opposite side of the house, was the Churubusco River, more like a creek and mainly dry.

The house was a run-down villa that had been built as a summer residence at the end of the previous century. Although the grounds were originally enclosed inside brick-and-stone walls, in the weeks leading up to Trotsky's occupation of the house they were raised to a minimum height of nearly fourteen feet. The iron gate that had once served as the central entrance on Avenida Viena was bricked in.

The house itself was shaped like a letter T. The top of the T, running perpendicular to Avenida Viena, constituted the east wall of the roughly rectangular enclosure. It contained the library, the dining room, the kitchen, a water closet, and a bedroom, all with interconnecting doors. Three tall windows, protected by iron grillwork, looked out onto a barbed wire fence, a strip of no-man's-land, and a corn field beyond. The stem of the T, which jutted into the patio, housed three interconnected rooms: Trotsky's study, Trotsky and Natalia's bedroom, and, at the base of the T, Seva's bedroom.

The entire structure was built on one story, except in the northeast corner, where a two-storied tower overlooked the river. This tower, built as an observation post during the revolutionary decade that began in 1910, was dwarfed by an enormous eucalyptus tree standing inside the north wall. Beneath it and abutting the wall was a row of adjoined brick outbuildings constructed to house the guards. While the size of the property was about the same as that of the Blue House, the guards were less comfortably billeted, and they had to be quieter because there was only one patio.

Outside on Avenida Viena, at the southeast corner of the property, the police built a brick *casita* with a loophole. In all, ten policemen, working in two shifts of five men, were assigned by the Mexican government to guard Trotsky. Visitors entered the property through the southwest corner, where heavy doors, bolted and guarded from the inside, led into the garage, from which another door gave into the patio, where grass plots abounding with cacti and agaves were crisscrossed with stone walkways.

There, over by the west wall, Young's camera found Trotsky feeding his rabbits and chickens, part of his daily routine and his principal form of exercise. This zoological hobby began at the Blue House, but the new residence—which Trotsky at first rented and later arranged to purchase, thanks to gifts totaling $2,000 from American sympathizers—gave him the freedom to take it to obsessive lengths. The stock of chickens was augmented with additional Leghorns, Plymouths, and Rhode Island Reds, among other varieties, while the rabbits proliferated on their own, so that by the autumn of 1939 there were fifty in all.

In January 1940, the burgeoning bunny population was transferred to new three-decker cages that Trotsky had designed. He was so eager to use them that he decided he could not wait until they were fully painted. Young's Bell & Howell captured the moment, as Trotsky, with the help of the young Mexican handyman, Melquiades, gently delivered the docile creatures to their new quarters. They all seem to know the Old Man, who likes to let the chief buck take a hard bite on his glove. When Eastman visited the house the following month, it seemed to him "so amusingly strange to be introduced to a flock of rabbits by the War Commissar and Commander-in-Chief of the Red Army."

By then the rabbits numbered well over a hundred. Trotsky liked to quiz the guards about this at the dinner table, where rabbit and chicken were occasional items on the menu. The guards were enlisted in the care and feeding of the animals, while Trotsky took great pride in the careful preparation of their diet. Young's films show Seva grinding corn

Trotsky feeding one of his rabbits, winter 1939–40.

for the chickens, as Trotsky and Farrell Dobbs, who arrived in mid-January on his first visit from the United States, look on. Trotsky is seen giving water to the chickens, then feeding the chickens and then the rabbits. At last, he turns to the cameraman and says, silently: "Well, that's all there is. I can't act for you anymore. Is it OK?"

Young had intended to stay only long enough to show Trotsky his China films, but after inspecting the alarm system at the house, he offered to draw on his engineering expertise to improve it. This system had been rigged up by Van, who had departed for the United States in October 1939 after marrying a visiting comrade from New York. His intention was to return in six months, at most a year, though by the time he left Mexico, he was ready to get out from under Trotsky's shadow, and the feeling was mutual. Easily the longest-serving secretary-guard, Van was the only member of the household with the authority to insist that he always be present when Trotsky met with visitors, no matter who they were. "You treat me as though I were an object," Trotsky liked to complain, dissimulating his impatience with a jab of sarcastic humor.

Young found Van's "quasi-electronic security system" to be "messy and complicated," such that its repair was beyond the technical ability of anyone in the household. The guards joked that it looked like a Hollywood version of Sing Sing. Tripwires running along both sides of the walls were connected to a set of electric light bulbs mounted on a board inside the guardhouse, a narrow wooden structure standing inside the patio against the garage. Each bulb designated a particular area of intrusion. Young rewired the control panel and provided simple diagrams to facilitate troubleshooting.

Still, there was only so much that Young could do on a severely limited budget. Julius Klyman, a reporter from the *St. Louis Post-Dispatch* who visited the house in March, intentionally exaggerated the security regime at Trotsky's "Mexican stronghold," out of sympathy for the exile's vulnerable position. Klyman, the rare American journalist to win Trotsky's trust, came upon him in the patio watering the grass. "See," Trotsky said upon greeting him, "they will not let me into your country. So I have become a farmer. Like your President. Where is he a farmer?" President Roosevelt's "farm" was in Hyde Park, Klyman deadpanned, pleased to find Trotsky in a relaxed mood. "He is a calmer, more composed man than when the writer first met him in January, 1937, three days after his arrival from Norway," Klyman reported. "Then, while completely assured, he seemed tense, a tightly wound spring waiting for its release. Now, after three years in Mexico, he seems at ease—perhaps as much at ease as a man of his dynamic intellectuality ever can be. He is continuously in mental motion."

Klyman took his readers inside Trotsky's study, a room about fifteen feet square in the stem of the T. Trotsky's "great work table," made of several planks joined and placed on heavy legs, ran away from the tall French windows, which let in ample light. The walls were lined with books, and a side desk was piled high with newspapers and magazines from many nations. There were several rush-bottom chairs bought in the marketplace in Coyoacán and a cot for his afternoon nap. On the wall behind the desk was a large map of Mexico; a smaller map of Europe had been put up after the war broke out. "The room is pleasant, informal but serious."

Klyman performed a service for Trotsky by describing his home as impenetrable, but an attentive reader would have noticed a chink in the armor. It is revealed in a passage where Klyman draws a twin contrast between the circumstances of his two visits. In January 1937, he recalled, Trotsky's English was rather weak. "A man of extreme conversational preciseness, he at that time frequently found it necessary to turn to an ever present secretary to find the exact word he wanted." Now, in March 1940, not only was Trotsky's spoken English vastly improved, but Klyman had Trotsky all to himself. "This time we held our conversations alone."

AFTER AL YOUNG had revamped the alarm system, he was asked to move into the house and take up guard duty. The idea appealed to him because he wanted more time to take photographs and films of Trotsky. Young happened to be a good driver and had an instinct for diplomacy, so he was assigned to chauffeur Natalia on her food-shopping trips.

At the time Young came on board, in January 1940, the overlapping guard and secretariat were overstretched, and remained so even after the number of guards was increased from four to five the following month. Van had been replaced as European secretary by Otto Schüssler, who had served as Trotsky's German secretary in Turkey and now lived at the house with his wife Trude, both refugees from Hitler. The English-language secretary was now Charley Cornell, a young schoolteacher from Fresno, California, who also served as Trotsky's chauffeur. The chief of the guard was Harold Robins, although Robins had just barely escaped being fired by Trotsky.

Robins was born Harold Rappaport in New York City in 1908, the son of Russian immigrants. Except for a two-year hiatus, he had been with the American Trotskyists since 1928. He was that rare breed, a worker-intellectual who had read the Marxist classics. He was arrested for his involvement in a riot sparked by the 1934 Waldorf-Astoria Hotel workers' strike and served nine months in Sing Sing. In 1937 he helped organize the wave of sitdown strikes that paralyzed the automobile plants in Detroit and Flint, Michigan.

Robins was tall and lanky, with dark brown hair combed back from a widow's peak and a perpetual stubble on his long sunken cheeks that gave him the look of a desperado. Joe Hansen, from the headquarters of the Socialist Workers Party in New York, recommended him as "quiet, very calm, one of the coolest militants in a difficult situation the party has." Yet all this was merely secondary. "The main consideration," Hansen made clear, "is that he is the best available driver." Robins had driven a cab in New York City from 1929 to 1934, then a truck in upstate New York. Behind the wheel he was "cool, skillful, careful but quite capable of emergency driving at top speeds."

Robins would have been an ideal candidate, except that he had a wife and child, and Hansen indicated that Mrs. Robins would follow her husband to Mexico in six months. This raised a red flag for Trotsky, who agreed that driving expertise was "the most important condition," but did not want to run the risk of having yet another wife join the household, even at some indefinite future time. Nonetheless, Trotsky was prevailed upon, and Robins was sent down in September 1939. Three months into the job, he disclosed that his wife was making arrangements to move to Coyoacán and that their child might soon be joining them.

This was too much for Trotsky. On January 2 he unburdened himself in a letter to his former secretary Sara Weber, instructing her to communicate its contents to Cannon and his associates. He wrote it by hand, in Russian, in order to bypass his secretarial staff and make himself clearly and forcefully understood. The comrades had to realize, Trotsky wrote bitterly, that the female companions of the guards had made life miserable for him and especially for Natalia Ivanovna. "The wives quarrel and sulk. NI has to think about sheets, personal tastes, warm compresses etc. It's worse than hard labor. We have had four or five 'wives' at a time." All of them start out saying they will live apart, but instead they eat, sleep, and take sunbaths at the house and then move in after a couple of days, "so that the result is a never-ending vacation. We have decided to prevent a repeat of this no matter what!" There were already two couples living in the house: the Rosmers and the Schüsslers. "That's enough!"

Trotsky demanded that the party immediately recall Robins, as well as his wife, who was already on her way to Mexico. "Comrades who come here must know they are coming *to a prison, not to a resort,*"

Trotsky admonished. "Work also becomes difficult for me on account of the talk, noise, running about. We want now to reduce the population and the expenses to a minimum and rest up a little in our home. We are not getting any younger." Whatever Trotsky's blast had to do with it, Mrs. Robins visited only briefly, and the storm passed.

In the first days of March, Young announced that he intended to return to the United States, and New York was asked to send down a new man. Funds for the guard remained extremely tight, and replacing a guard meant additional expenses. The increase to five guards put a further strain on the finances, as did the purchase of more and better-quality guns and ammunition, as recommended by Farrell Dobbs during his visit that winter. At remote locations in the countryside, Dobbs gave the guards training in small-arms fire and with the lone Thompson submachine gun, which had a tendency to jam and had to be sent out for repair.

Shortly after Young arrived in Los Angeles from Mexico in late April 1940, he wrote to Dobbs in New York to brief him on the state of affairs he left behind at Avenida Viena. In his letter, Young felt duty-bound to report a strain in relations between the guards and the household over the quality of the food. None of the staff thought very highly of the meals served at the Trotsky ranch, but some of the guards were tactless in complaining about it at the dinner table in the Old Man's presence. Most of this American slang sailed right over Trotsky's head, though he caught its drift. Young called this "the height of folly." If any ill will was intended, it must have been meant for Natalia, who set the menu and whose attitude toward the guards fluctuated wildly. Robins complained, "at one time we are the cream of the earth, at another the crumb."

Whatever lay behind this uncouth behavior, its effect was poisonous. "I do know that the OM was really fed up," Young told Dobbs. "A week previous and up to my departure he hardly spoke with anyone because of this." Among the culprits was Young's replacement, Bob Shields, a twenty-five-year-old New Yorker and graduate of Duke University. Dobbs himself had recommended him as a dedicated comrade and a hard worker, notwithstanding the fact that he was the son of a wealthy businessman. Dobbs made no mention of the new man's driving ability. In this case, the main consideration was his willingness to pay his own way to Mexico and to cover his board and personal expenses in Coyoacán.

Shields was his party name. His legal name was Robert Sheldon Harte, and his family called him Sheldon. To the NKVD, which recruited him in New York, he was known by the code name "Amur," after the prodigious river in the Russian Far East. He traveled to Mexico City by airplane. When he took up his duties in Coyoacán on April 7, 1940, the NKVD had a mole inside Trotsky's stronghold.

By the time Harte arrived in Coyoacán, the NKVD had two networks in place in Mexico City. The first group was called "Mother," the code name of Caridad Mercader. Its chief asset was her son Ramón, now posing as a Canadian businessman named Frank Jacson. This change of identity, which took effect in Paris the year before, was an unforeseen development that complicated his mission. Sylvia Ageloff, his lover and the NKVD's gull, had returned to New York from Paris in February 1939. Ramón was to follow her there, but a problem with his identity papers as the Belgian Jacques Mornard resulted in the United States rejecting his application for a visa. The NKVD then provided him with a passport in the name of the fictitious Frank Jacson, a Canadian citizen who was Yugoslav by birth.

Ramón received his U.S. visa and left France on September 1 as war broke out in Europe, arriving a week later in New York. There he explained his change of identity to Sylvia as a necessary step to avoid being drafted into the Belgian army. She apparently never questioned him about the unorthodox spelling of his new last name, which he pronounced in the French style, with the accent on the second syllable. It proved to be a felicitous choice: This would allow her, they agreed, to continue to call him Jacques—spelled simply "Jac"—without compromising his new identity. He explained to Sylvia that he was now a businessman, working for an international entrepreneur named Peter Lubeck, who traded in oil and sugar. This was all a fiction, of course.

Ramón's real boss was Leonid Eitingon, the operational commander of Operation Duck and the lover of Caridad Mercader.

On October 1, Ramón said goodbye to Sylvia and left for Mexico City on a business trip. Eitingon followed in mid-November, around the same time as Caridad. As the Christmas holidays approached, Sylvia contrived to get sick leave from her job as a New York City social worker on the strength of a doctor's note saying she suffered from a sinus condition and required a warm climate in order to convalesce. She flew down to Mexico City on New Year's Day 1940, and moved into Ramón's apartment there. This unfolded in accordance with the NKVD's optimal scenario, as did Sylvia's next move, which was to contact Trotsky and, using the connection of her sister, receive an invitation to his house.

During her second visit to Avenida Viena, early in February, Sylvia encountered Alfred Rosmer in the patio as she was leaving. The two had met in Paris in the summer of 1938 at the founding congress of the Fourth International. Sylvia invited Rosmer and his wife, Marguerite, to visit the apartment, where they were introduced to Jacson.

Three weeks later, Sylvia and Jacson were invited on a picnic to Mount Toluca, about fifty miles west of Mexico City, joining the Rosmers, Otto and Trude Schüssler, and Seva, with Al Young along as chauffeur. Jacson was considered good social company, although he was regarded as a superficial person and a political lightweight. The Rosmers asked him why, if he was a Canadian, he spoke Parisian French, even the current slang terms, and he explained that he had been educated in Paris.

The Rosmer connection proved to be crucial to Ramón after Sylvia had to return to her job in New York City at the end of March. There, after hearing nothing from her lover for what seemed like an eternity but was in fact less than two weeks, she asked Marguerite Rosmer to check up on him. Marguerite met Jacson at a coffee shop in the city, and afterward she was able to notify Sylvia that her Jac was fine, just busy.

On May 1, Alfred was admitted to the French hospital in Mexico City to undergo a minor operation, remaining there for ten days. Jacson offered his services as chauffeur, driving Marguerite back and forth to the hospital, and he himself checked in on Alfred. He never inquired about Trotsky or asked to enter the house, giving the impression that he understood why this would be impossible. His most effective weapons, for now, were his Buick Sedan and his patience.

AS PART OF its operation to assassinate Trotsky, the NKVD established a second and much larger network in Mexico City, this one called "Horse," its code name for the muralist David Alfaro Siqueiros. Siqueiros, along with Diego Rivera and José Clemente Orozco, was one of Mexico's Big Three muralists. Born in 1896 in Chihuahua, in northern Mexico, the son of a well-known lawyer, Siqueiros attended the San Carlos Academy of art in Mexico City, until his education was cut short in 1913 by Mexico's revolutionary upheaval. He joined the forces of Gen. Alvaro Obregón, a foe of Pancho Villa, becoming the general's messenger and later rising to the rank of first lieutenant.

When it was over, the Mexican government gave Siqueiros the opportunity to resume his studies abroad, and he went to Paris late in 1919. There he got to know Rivera and fell under the influence of Cubism, counting Braque and Léger among his friends. He took in the art of Italy in the company of Rivera, then moved to Barcelona, where in 1921 he published an influential manifesto on the need for Mexican art to rediscover its native roots.

Siqueiros returned to Mexico in 1922, joining his fellow muralists in spearheading the country's cultural renaissance. At the formation, that same year, of the Syndicate of Technical Workers, Painters, and Sculptors, Siqueiros served as its general secretary and its strident mouthpiece. He was also co-editor of its publication, *El Machete*, and designed the paper's famous masthead, a woodcut of a hand gripping a machete, with the paper's name inscribed in bold letters along the blade.

Siqueiros was a swashbuckler by temperament, high-strung and bombastic, with a theatrical appearance to match. He had an unruly shock of curly black hair, gray-green eyes, and a Cupid's bow mouth. His elongated face, of pallid complexion, was accentuated by a prominent nose whose nostrils were set in a provocative flare. It is easy to see why his friends called him Caballo, Spanish for the Horse. Siqueiros was a member of the Communist Party, and in the second half of the 1920s, when he was better known for his manifestos than his murals, he put aside his art to devote himself to the labor movement, becoming a union organizer among the silver miners and peasants in the state of Jalisco, often based in its capital, Guadalajara. This union work brought him to Moscow in March 1928, as a delegate to the International Congress of Red Trade Unions.

Not long afterward, Siqueiros managed to run afoul of both the Communist Party and the law. In the spring of 1930 he was expelled from the party for a breach of discipline. On May Day, he was seized in a police sweep after an attempt on the life of the Mexican president. After six months in prison, Siqueiros was sent to live under house arrest in Taxco. There, for the next year and a half, he painted more than a hundred oils, including portraits of poet Hart Crane and composer George Gershwin, among other visitors to this popular vacation spot.

When his Taxco sentence had been served, in the spring of 1932 Siqueiros was forced to go abroad. He moved to Los Angeles to teach and to paint, and stirred up controversy by creating two politically charged outdoor murals, both of which were promptly whitewashed. The more infamous of these, at the Plaza Arts Center, was called *Tropical America* and depicted a Latino figure bound to a cross surmounted by an American eagle.

After six months, Siqueiros left Los Angeles for South America, where his union activities got him expelled from Argentina. He then returned to the United States, this time to New York, where he set up an experimental studio that pioneered the use of synthetic paints and spray guns, and encouraged other unorthodox practices, such as dripping and flinging paint onto the canvas. One of the workshop participants was Jackson Pollock, who pioneered the application of these materials and techniques in the vanguard of Abstract Expressionism. It was during this New York sojourn, in May 1934, that Siqueiros published a savage attack on Rivera, in *New Masses*, accusing him of pandering to commercial tastes and attributing the inferiority of his art to his political support for Trotsky.

In January 1937, Siqueiros sailed to Spain and enlisted in the International Brigade. For a time he served in the 5th Regiment, whose political commissar was Carlos Contreras, the nom de guerre of the Italian Communist Vittorio Vidali, whom Siqueiros had known in Mexico a decade earlier and who was fast acquiring a reputation as a Stalinist executioner. Siqueiros later commanded a brigade and then a division of the Republican Army, attaining the rank of lieutenant colonel. When he returned to Mexico in January 1939, he became chairman of the Mexican section of the Society of Veterans of the Spanish Republic, and spent a great deal of his considerable energy lobbying President Cárdenas to throw open Mexico's doors to Spain's civil-war refugees.

In August 1939, Siqueiros was commissioned to paint a mural for the new headquarters of the Mexican Electricians' Union. He assembled a team of Mexican and Spanish artists to help design and carry out the project, which left him time to prepare for an exhibit of his new oil paintings scheduled to open in January at the Pierre Matisse Gallery in New York. Among his collaborators on the mural were Luís Arenal and Antonio Pujol, both of whom had worked with him at his New York studio. Arenal, who first met Siqueiros in Los Angeles, was an artist for *New Masses* when he lived in New York. Siqueiros was married to his sister Angelique. Pujol had accompanied Siqueiros to Spain and served with him in the International Brigade.

Siqueiros and his team chose as the site of their mural the landing of the building's main staircase, in part for the technical challenge of creating a unified composition on four surfaces—three walls and a ceiling—at right angles to each other. As originally conceived, the mural's theme was antifascist, but in the aftermath of the Nazi-Soviet pact this had to be reworked into a more generic anticapitalism, as suggested by its innocuous title, *Portrait of the Bourgeoisie*. This "portrait" is the stuff of nightmares. The photomontage-like imagery, dark and violent, offers an apocalyptical vision of fascism armed with the modern machinery of warfare. The swastikas are missing, but the symbolism is nonetheless unmistakable. *Portrait of the Bourgeoisie* is a quintessential work of the late 1930s and, in the words of art historian Desmond Rochfort, "one of the great moments in twentieth-century mural art."

A decade earlier, in her classic study of the Mexican renaissance, *Idols behind Altars*, Anita Brenner remarked of Siqueiros that he did not distinguish between his artistic and his political endeavors, passing from one to another "without noting a difference between a brush and a gun." In May 1940, he had not yet completed work on *Portrait of the*

Bourgeoisie when he was called away to lead a different kind of undertaking, this one commissioned by the NKVD.

THE MAN HOLDING the reins of the Siqueiros network was Iosif Grigulevich, an ethnic Jew born in 1913 in Lithuania, then part of the Russian Empire. He had learned his Spanish living in Argentina, where he was a Comintern activist in the mid-1930s, and then in Spain, where he arrived in 1936 after the outbreak of the civil war. His facility in several languages—including Lithuanian, French, German, Polish, and Russian—led to his rapid advancement. Initially assigned to the 5th Regiment as adjutant to commissar Carlos Contreras during the defense of Madrid, he later came to the attention of Alexander Orlov and was recruited to the NKVD.

Grigulevich participated in the bloody suppression of the anarchists and the POUM in Barcelona in May 1937. When the POUM leadership was rounded up in mid-June on trumped-up charges of spying for Franco, he took part in the operation to arrest Andrés Nin, who was taken to a Republican jail in a suburb of Madrid. After Nin refused to confess to his "crimes" under a brutal interrogation conducted by Orlov and Contreras that left him severely beaten, Stalin, who may have actually believed the Soviet propaganda line that Nin was Trotsky's agent, ordered the POUM chief's liquidation.

On the night of June 22–23, a group of men dressed in Republican Army uniforms burst into the heavily guarded prison and kidnapped Nin. Grigulevich assisted Orlov in the operation and served as his translator. Orlov and Grigulevich were part of a mobile group that included three Spanish NKVD agents who tortured and murdered Nin; his body was buried in an unmarked grave along a country road.

For their protection, Grigulevich and others involved in this "wet" operation were then withdrawn from Spain and brought to Moscow, where they underwent an NKVD training course. In the spring of 1938, Grigulevich and a colleague were sent to Mexico City to conduct surveillance of Trotsky, and if possible to penetrate his circle. They rented an apartment a few blocks from the Blue House and set up an observation point from which they could watch the comings and goings.

In the first months of 1939, Grigulevich, using the code name

"Felipe," recruited Siqueiros, an acquaintance in Spain, as well as Siqueiros's wife and her brother Leopoldo Arenal, brother of the artist Luís. Leopoldo Arenal was a fanatical anti-Trotskyist. He came up with a plan to deliver Trotsky a booby-trapped potted cactus: the bomb concealed in its soil would be triggered to explode during the transplanting. This proposal was passed on to the NKVD resident in New York, who rejected it for fear that the bomb might not reach its intended target.

After Operation Duck was launched in the summer of 1939 and as its principal agents were being maneuvered into position in Mexico, "Felipe" was summoned to Moscow. This might have been the end of the road for Grigulevich, whose name was closely associated with that of the defector Orlov. But in Moscow he impressed his superiors with his detailed knowledge of the Mexican terrain and of Trotsky's situation on Avenida Viena. He brought with him a plan to storm the villa, and urged that Siqueiros be named to lead the fighting group. He was taken to meet with Beria, who approved the idea and ordered him to return to Mexico to see to its execution. Grigulevich arrived back in Mexico City in February 1940 and sat down with Eitingon to coordinate the operational details.

As these preparations were under way, Trotsky had once again become the object of a vicious slander campaign in Mexico's left-wing press. He had unwittingly provided the pretext for this latest and most ferocious onslaught, by agreeing to appear before the House Un-American Activities Committee of the U.S. Congress, better known at the time as the Dies Committee. Trotsky was asked to testify about the history and methods of Stalinism, and his decision to accept this invitation caused great consternation among his American followers.

As an anti-Communist, Martin Dies, a Democrat from Texas, was the Joseph McCarthy of his day, a self-promoting, red-baiting opportunist who was bent on tying the American Communist Party to the Kremlin in order to expose the party's leaders to prosecution. Trotsky justified his decision to testify by saying he would use the reactionary Dies Committee as a tribune, much as he had used the liberal Dewey Commission two years earlier. There was more to it, however, because Trotsky saw Dies as his laissez-passer into the United States, where he might be able to turn a six-month visa into permanent residence.

Stories in the American and Mexican press claimed that Trotsky was to testify about Mexican and Latin American Communism and on the sensitive subject of Mexico's oil industry. This gave an opening to the Mexican Communists and their sympathizers to portray Trotsky not only as a meddler in Mexican politics, but as a tool of the oil companies and Wall Street. Until recently, Trotsky had been caricatured as an agent of the Gestapo, a Judas branded with a swastika, just as he was in the Moscow papers. In the wake of the Nazi-Soviet pact, however, he had to be recast as an agent of Yankee imperialism. Even though Congressman Dies eventually withdrew his invitation, the episode facilitated the transformation of Trotsky from a tool of the Gestapo into a tool of the FBI.

In the winter and spring of 1940, the tone of the anti-Trotsky campaign in Mexico turned violent. Meetings of the Communist Party and its front organizations were punctuated by shouts of "Death to Trotsky!" This slogan was adopted by the party during its congress in March, when it conducted a sweeping purge of its top leadership, which was accused of Trotskyism. Trotsky understood that such a purge could only have been ordered by Moscow. He guessed that the man acting as the Comintern's supervisor on the scene was Carlos Contreras, the GPU enforcer in Spain who now surfaced in Mexico City as a member of the Communist Party's honorary presidium.

On May Day, the party organized a march through the city of some 20,000 uniformed men and women shouting slogans such as "Throw out the most ominous and dangerous traitor Trotsky." Reliable reports from Trotsky's Mexican friends told of a concentration in the city of Stalinist killers from Spain. Trotsky called a meeting of the guard to warn of the danger that an armed attack was in preparation.

Among the guards listening to Trotsky was Robert Sheldon Harte, the quiet, bookish, intense young man with the kinky red-brown hair, acne-scarred face, and cleft chin. At Duke, Harte had published politically conscious short fiction, and even now he harbored literary aspirations. Still, he was unable to see the plot line of the real-life thriller in which he had become entangled. He failed to fully grasp the connection between Trotsky's warning of mortal danger and Harte's own clandestine meetings with a Soviet agent named Felipe. The NKVD's real objective, Harte understood from Felipe, was the destruction of Trotsky's archives—together with the manuscript of the slanderous biography of Stalin that Trotsky was preparing, based in part on forged documents supplied by Hitler.

The rain fell heavily at times during the night of May 23–24, and the dirt roads of Coyoacán turned muddy. At 10:00 p.m. Siqueiros and a half-dozen confederates, including fellow artists Luís Arenal and Antonio Pujol, gathered at a house on Calle de República de Cuba. Toward midnight, several men arrived with police uniforms and weapons, including a Thompson submachine gun, four revolvers, and two Thermos bombs, along with rubber gloves to prevent fingerprints. Siqueiros told his comrades to try on the police uniforms, while Pujol put on the lone military uniform, that of a lieutenant. The men laughed and joked as though it were a costume party.

Siqueiros then went out. He returned toward 2:00 a.m. dressed in the uniform of an army major. Dark glasses and a fake mustache completed the disguise. He modeled his uniform for his comrades, provoking great hilarity. "How does it suit me?" asked the *pintor a pistola*. "Very well," they replied with laughter.

An hour later, these uniformed and well-armed men crammed themselves into Siqueiros's car and drove toward Trotsky's home. Siqueiros assured them that all would go well because one of the guards had been bought. "And if this guy betrays us and we get machine-gunned?" one of them asked. Siqueiros smiled and replied, "There's no danger of that!" En route, he handed each man an envelope containing 250 pesos, about $50. They parked one street over from Avenida Viena and waited, as Siqueiros kept looking at his watch.

Toward 4:00 a.m. "Major" Siqueiros ordered his men to get out of the car. They surprised and overpowered the five policemen in the *casita*, three of whom were asleep, and tied them up. They then made their way toward the southwest corner of the property, where three other groups of men, all armed and dressed in police uniforms, converged from different directions on the entrance to the garage. Hearing Felipe's voice, Harte delivered his end of the bargain by sliding away the heavy bolt that joined the doors, as twenty raiders poured into the garage and then out into the patio.

One man stationed himself alongside the eucalyptus tree, in the vicinity of the guards' quarters. Others took up positions outside the door to Seva's room and the French windows to Trotsky's bedroom. A third contingent entered the house through the library, at the top right of the T, and made their way into the dining room, where, with a mighty heave, they forced open the locked door to Trotsky's study and continued toward the bedroom.

A burst of automatic fire tore through the bedroom door. A raider armed with a submachine gun then entered Seva's room and opened fire through the closed door connecting to Trotsky's bedroom, while a third assailant fired through the wooden shutters on the French windows, creating a crossfire from three directions. Trotsky had taken a sedative to help him sleep and was slow to realize the danger, but Natalia grabbed him from his bed and the two fell into the corner of the room beneath the window, as ricocheting bullets flew in all directions above them.

In their quarters, the guards were awakened by the gunfire and began to react. Robins opened the door to his quarters and in an instant caught sight of a man in a police uniform alongside the eucalyptus tree who turned and fired a submachine gun in his direction, spattering lead around the entrance and forcing him back inside. He heard this man— almost certainly Leopoldo Arenal—say in accented English, "Keep your heads out of the way and you won't get hurt."

Jake Cooper, who had arrived from Minneapolis only three days earlier, also heard this warning. He opened his door slightly and was met by a hail of bullets. He heard Robins yelling, "Keep your heads down!" Charley Cornell, in the room between those of Cooper and Robins, heeded this advice. Up in the tower, Otto opened the blinds of his bedroom window overlooking the patio, and as he did so, gunfire sprayed the bricks around the window, sending him to the floor. The guards could hear machine guns firing on the other side of the house—

even inside the house—and feared the worst. "Bob, where are you?!" Robins kept yelling.

The crossfire into Trotsky's bedroom lasted for several minutes. When the guns fell silent, one of the raiders entered Seva's room and threw down a Thermos bomb, the force of the explosion blowing open the door to Trotsky's bedroom and igniting a small fire. Standing on the threshold and peering into the room illuminated only by the faint glow of the flames at his feet, the intruder emptied his handgun into Trotsky's and Natalia's beds. Then he turned and ran out.

The gunfire became intermittent and more distant, as the raiders covered their retreat. They had been in control of the grounds for about fifteen minutes. Robins raced up to the roof, where he was fired on by the assailants from the street as they fled. He called down to the police in the *casita*, who appeared in the doorway, hands tied behind their backs. Charley entered the garage and found Bob's serape lying on the floor, neatly folded. The garage doors were wide open and both cars were gone. The alarm system had been turned off.

All the members of the household assembled in the patio. Seva's foot was bleeding. When the attack began he dove under his bed. A bullet fired into the bed passed through the mattress and struck him in the big toe. After the raiders withdrew, he ran out into the patio. Natalia had minor burns from smothering the fire with blankets. Trotsky received only a couple of light scratches on his face from flying glass. Everyone marveled at the family's good fortune. The Rosmers, the cook, and the housekeeper were all unharmed, as were four of the five guards. The only cause for distress was the disappearance of Bob Harte.

Within a half-hour, the chief of the Mexican secret service, Colonel Leandro Sánchez Salazar, and a team of investigators arrived on the scene. Introducing himself to Trotsky, Salazar was struck by the incongruity between the exile's famously Mephistophelian features and his bathrobe and pajamas. Salazar's men counted seventy-three bullet holes in the doors, windows, and walls of Trotsky's bedroom. Altogether well over 300 shots had been fired.

A check of the yard revealed two homemade bombs that were broken but unexploded and a third one that remained intact. On the riverbank the police found a wooden extension ladder, a manila rope ladder, a crowbar, and a portable electric saw with a very long extension cord. This evidence seemed to indicate that the raiders were not counting on the complicity of Harte, who must have been tricked into opening the door. The Ford was found two blocks away, abandoned in the mud.

Counting the bullet holes and considering the shocking ineptitude of the attackers, who could easily have killed one another in their own murderous crossfire, Salazar grew suspicious. He wondered why the guards had not fired their weapons. He questioned the calm, even conspiratorial, demeanor of the members of the household under the circumstances. Salazar asked Trotsky if he knew the identities of the assailants. Walking the colonel over to the rabbit cages, Trotsky drew him near and told him what he knew to be a dead certainty: the perpetrator of the assault was Joseph Stalin, acting through the agency of the GPU. This statement, delivered with a dramatic flare, struck Salazar as fanciful. His suspicion mounted that Trotsky himself had staged the raid.

In the days following the assault on Trotsky's home, the Mexican police guard was increased to twenty-five men on duty at all times. Every fifteen minutes through the night they signaled by whistle from each corner of the property. Inside the walls, the guards traded speculations about the fate of Bob. Was he a victim or an accomplice? They put the odds at fifty-fifty.

The testimony of the Mexican police guards was ambiguous. They saw Bob being led between two of the raiders, each one holding him by an arm as he muttered, "No, no, please don't." He was protesting but not resisting, and they could not say for sure whether he was taken against his will. The Dodge, which was discovered the following afternoon about ten miles away in the center of the city, had a tricky ignition switch, such that only Harte himself could have started it for the raiders.

Complicit or not, Harte must have opened the outside door when he heard a familiar voice. Suspicion fell on Sergeant Casas, who said he was home asleep at the time of the raid. On the day after the attack, he along with the five policemen on duty that night were arrested and held for questioning. Casas had told Trotsky's cook that the raid was a self-assault, *auto-asalto*, an expression she did not understand but which she repeated to the police, prompting the arrests. Baffled, Trotsky released a statement saying that in light of his remark, Casas was compromised and may even have been part of the conspiracy.

Jesse Sheldon Harte, Bob's father, arrived in Mexico City the day after the raid and offered a reward of 10,000 pesos, more than $2,000, for the location of his son. He met with the police investigators, and paid a call on Trotsky. He was surprised to learn that young Sheldon, who told him he had gone to Mexico on business, was one of Trotsky's bodyguards.

On May 27, after Harte senior returned to New York, the local papers published the sensational story that a photograph of Stalin, warmly inscribed, had been discovered in the missing guard's room in New York City. The source of this story was Jesse Harte. In a confidential interview conducted by a Mexican police official in the American embassy, he testified that a photograph of the Soviet dictator had been found on display in his son's room. Someone then leaked this information to the press, with the clinching detail of Stalin's inscription inserted somewhere along the way. Trotsky sent a telegram to Jesse Harte asking him to confirm the story. Harte, who was mortified to discover that this unsavory fact about his errant son was making headlines, cabled a reply that was meant to bury the story for good: "DEFINITELY DETERMINED STALINS PICTURE NOT IN SHELDONS ROOM."

Meanwhile, Colonel Salazar's investigation took a new turn. On May 28, Trotsky's household servants—the cook, Carmen, and the maid, Belem—were taken in for questioning, as they were again on the following day, when they were held for nearly twelve hours and given the third degree. With this encouragement, the cook remembered that on the eve of the attack there had been a secret meeting at Trotsky's house, from half-past three to six o'clock, and that two of the guards, Charley and Otto, had seemed very anxious the entire day. Both women signed statements declaring their belief that the raid was an *auto-asalto*.

On May 30, Salazar arrested Charley and Otto, the two guards who spoke intelligible Spanish. They were held incommunicado for two days while their interrogators pressured them to confess that Trotsky had ordered them to carry out a self-assault. During this time, the police came to arrest Robins, but decided against it when Trotsky objected, and perhaps also because Robins made it clear that he would not go willingly. Meanwhile, Trotsky had addressed an urgent letter to President Cárdenas, protesting that he was being deprived of the means to defend himself. Cárdenas intervened and Salazar released the guards.

Trotsky was greatly surprised when Salazar told him of his cook's testimony. "We are always holding conferences," he told the colonel. "Even at table at mealtimes we discuss questions of international politics. My study, in addition, is always open to any of my collaborators." As it happened, however, on May 23 Trotsky did not follow his usual routine, as he was busy all day preparing an article for the comrades in New York and worked unusually late, until eleven at night. In that case, Salazar told Trotsky, the cook lied and ought to be fired. Trotsky at first resisted this advice, but then agreed that it was the only thing to do. The maid quit a few days later.

By now, the Communist press in Mexico was portraying the raid as a put-up job, staged by Trotsky in order to malign his enemies. Trotsky countered that it was absurd to believe he would risk his Mexican asylum through such a reckless act. He turned the tables, accusing the editorial boards of the daily *El Popular* and the monthly *Futuro*, both organs of the Confederation of Mexican Workers, of taking part in the "moral preparation of the terrorist act," with the organization's president, Vicente Lombardo Toledano, orchestrating the campaign from behind the scenes. "Permit me also to assume that David Alfaro Siqueiros, who took part in the civil war in Spain as an active Stalinist, may also know who are the most important and active GPU members, Spanish, Mexican, and of other nationalities, who are arriving at different times in Mexico, especially via Paris."

On the question of Harte, Trotsky remained on the defensive. Salazar believed he was a conspirator. In his quarters the police found a key to Room 37 at the Hotel Europa, where he had spent the night of May 21 with a prostitute. She was interviewed and told Salazar that Harte was carrying a large amount of money on him that night. Salazar also learned from one of the guards that Harte had a sizable sum in American Express traveler's checks. Salazar suspected that this was payoff money.

Trotsky countered that Harte, whose family occupied a spacious apartment on Fifth Avenue overlooking Central Park, could not have been bought. Of course, Trotsky allowed, it was possible that the GPU had wormed its way into his guard, but he insisted that the facts of the raid did not support this conclusion. Assuming that Harte was beholden to the GPU, why organize twenty to thirty raiders with machine guns and bombs when a single agent could quietly enter his bedroom and knife him to death? And if Harte himself was not up to it, then why not just let in one or two attackers to do the job? Why all the commotion?

Nonetheless, Trotsky could not ignore Harte's peculiar behavior on the day before the attack. At about five o'clock that afternoon he had entered the study saying that he needed to check the alarm system. Trotsky expressed annoyance at the needless interruption and asked to be left alone. Harte was also suffering from intestinal problems that day, and Natalia gave him a hot water bottle and some medicine. This might have been nothing more than the usual Mexican flu, but the thought must have crossed Trotsky's mind subsequently that its cause was a nervous stomach.

The testimony of Trotsky's Russian secretary, Fanny Yanovitch, was especially unsettling. She usually worked only three to four hours a day, but on May 23 she stayed late so that Trotsky could complete his article. Harte, who was supposed to drive her home, seemed rattled by this change in the routine. From six in the evening he became increasingly nervous, several times asking her when she would finish and warning her to stay away from the alarmed wires at the window. On the drive home he pestered her with questions about the contents of Trotsky's biography of Stalin, which he could not decipher because it was in Russian. When this evidence was presented to Trotsky, he shrugged his shoulders and said, "Pure coincidence."

ON JUNE 17, Colonel Salazar broke the case. An overheard conversation at a bar led to the arrest and confession of Néstor Sánchez Hernández, a twenty-three-year-old former captain in the International Brigade in Spain and the author of a vicious attack on Trotsky published just days before the assault. Hernández identified Siqueiros as the leader of the operation, which he recounted in detail. His confession confirmed Salazar's suspicions about the complicity of Robert Sheldon Harte.

Harte had indeed opened the door for the raiders, Hernández testified. During the getaway, a man called Felipe, who spoke Spanish with what sounded like a French accent, ordered Hernández to accompany him in the Dodge, where Harte sat behind the wheel. The brothers Arenal joined Hernández in the backseat. Harte was greatly agitated. He must have assumed, like the escaping raiders, that Trotsky and Natalia were dead. He drove fast and erratically, and Felipe had to yell at him to calm down, instructing him in Spanish even though the American repeatedly asked him to speak English. "I had the feeling that I was taking part in a film adventure," said Hernández, whose best guess was that Felipe was a French Jew. It was obvious that Felipe and Harte had already known each other before the attack.

The Hernández confession led to the arrests of some two dozen people, all of them members or close sympathizers of the Mexican Communist Party. Among them were two women who occupied separate apartments in a building on Calle Abasolo, a few yards from Trotsky's house. Their assignment was to observe the comings and goings at the house and to become intimate with the guards, which they succeeded in doing. This information led to the rearrest of Casas and his crew of police.

The search for Siqueiros led to a farmhouse in the village of Santa Rosa, along the Desierto de los Leones road, on the evening of June 24. It was an adobe structure of three rooms, one of which overlooked the village. In the middle of this room stood an easel holding a blank canvas, alongside which were two brushes and two open pots of paint. There were several .22 caliber gun cartridges scattered on the floor, which was littered with cigarette butts. A policeman found an empty packet of Lucky Strike, which aroused suspicion because it was a luxury brand affordable only to Americans and wealthy Mexicans unlikely to inhabit such a humble dwelling.

Descending to the basement, the detectives entered a small kitchen, whose dirt floor had recently been upturned. A neighboring peasant was persuaded to use his pickax to dig up the soil. Two feet down, he uncovered the stomach of a human corpse, and within moments the investigators were overwhelmed by the stench of rotting flesh. A forensics team was brought in and the corpse was exhumed. It had been covered with quicklime, which had caused it to turn bronze. There were two bullet wounds in the head. Additional evidence, in the form of a bloodstained folding cot and quilt, indicated that the victim had been killed in his sleep.

Shortly after midnight, Colonel Salazar arrived at Trotsky's house. He brought with him a chunk of hair taken from the corpse, as well as a section of its underwear. All the guards assembled in the garage. They immediately recognized Bob's kinky red-brown hair, and they were able to produce an identical pair of underwear. Charley accompanied the police to Santa Rosa to identify the body.

Early that morning, the guards informed Natalia, who immediately went in to tell Trotsky. He emerged in his bathrobe and slippers. "Poor Bob," they heard him say. Not long afterward, the guards saw him tending to the rabbits, his expression grave and his face streaked with tears. A telegram was sent to Harte's father, who called a few hours later and asked Trotsky to identify the body personally. Trotsky went to the morgue in San Angel and performed this disagreeable duty, struggling to contain his emotions.

One of the conspirators told the police that he had been brought to the house in Santa Rosa by the painter Luís Arenal and hired to stay with Harte—not to guard him, but rather to keep him company. Harte, in other words, was not a prisoner, although common sense could have told him that he was a doomed man. Five days later, Arenal and his brother Leopoldo returned to the house, where they paid off and dismissed the minder. The police were now looking for the brothers in connection with Harte's murder.

To Colonel Salazar, it seemed evident that Harte had been eliminated as an inconvenient co-conspirator. To Trotsky, however, Harte's corpse was definitive proof of his innocence, a refutation of all the Stalinist slander about his being an agent of the GPU. "Bob perished because he placed himself in the path of the assassins," Trotsky said in a statement released later that same day. "He died for the ideas in which he believed. His memory is spotless."

Trotsky now added another victim to the pantheon of his fallen secretaries—eight in all, all victims of Stalin and the GPU. At some level, Trotsky must have understood that the discovery of Harte's decomposing corpse was convincing proof neither of the ill-starred American's ideas nor of his loyalties. But under the circumstances, the only acceptable version of events was that Harte was an innocent victim. To honor his memory, Trotsky arranged to have a stone plaque placed upon the wall inside the patio near the entrance to the garage. Its dedication affirmed what even Colonel Salazar could now agree was a dead certainty: "In Memory of Robert Sheldon Harte, 1915–1940. Murdered by Stalin."

CHAPTER 11

Deadline

T
wo weeks after the May 24 commando raid on Trotsky's home, James Cannon and Farrell Dobbs of the Socialist Workers Party came down to Coyoacán to inspect the crime scene and consult with Trotsky on what measures needed to be taken in order to improve the defenses at Avenida Viena 19. "It was a real attack—the escape was a miracle," Cannon wrote to Trotsky's lawyer in New York, Al Goldman. "It's obvious the assailants thought they had finished the job." Another attack was certain to come, and it was believed that bombs, not bullets, now posed the greatest danger.

Cannon advised New York headquarters that several thousand dollars would be required in order to meet the threat. Concrete and steel fortifications must replace wood; steel shutters must protect the interior windows; steel nets must be raised to defend against bombs. An appeal letter went out from New York to the nineteen party branches across the country, urging comrades to do their part and reminding them that Bob Harte had "made the supreme sacrifice."

Contributing to the sense of urgency in Coyoacán was fear of the political instability and civil unrest that seemed likely to accompany Mexico's long-anticipated presidential election, set for July 7. The government enforced a "depistolization" program in the days surrounding the election in order to limit the potential for trouble in what had become a rugged contest marked by sporadic violence. President Cárdenas had refused to name a successor or throw his support behind the candidate from his own party, Manuel Ávila Camacho, his minister of defense. Camacho was a center-right candidate who ran with the support of the left, including the Communists and Lombardo Toledano's labor unions.

Camacho's opponent was Juan Almazán, an army general who had retired from the military a year earlier when he announced his candidacy for the presidency as leader of his own right-wing political party. Conservative opinion was ascendant in Mexico, and because Almazán's prospects were good, his campaign was subject to dirty tricks by his enemies on the left. At the start of the May 24 assault, Siqueiros let out a cry of "Viva Almazán!" This was intended to help obscure the identities of the assailants by drawing suspicion to Almazán's supporters, but no one was fooled. Nor did it inhibit the Mexican Communists from accusing Trotsky of conspiring with Almazán behind the scenes, part of the unceasing effort to compromise his asylum by portraying him as a meddler in Mexico's national politics.

As it happened, the election went through without any violence. Officially Camacho won an overwhelming victory, but the polling was marred by vote-rigging and intimidation of Almazán's supporters. Almazán at first refused to concede defeat, then left for the United States, where he continued to make vague threats about challenging Camacho's claim to victory. Ultimately, Almazán yielded, and fears of widespread unrest, even civil war, proved unwarranted.

At the house on Avenida Viena, where these developments were closely monitored, Camacho's victory brought a measure of relief, though it took none of the steam out of Cannon's fund-raising drive in the United States. In the two and a half months following the Siqueiros raid, the Socialist Workers Party raised over $2,250 toward improving Trotsky's security. Trotsky's household finances were a separate matter. With no further income expected from the Stalin biography until its completion, Trotsky placed his hopes on the sale of his archives, an idea that had percolated for more than two years before it was finally realized in the spring of 1940.

Trotsky initially hoped the sale of his papers might bring in upwards of $50,000, but in hard economic times this proved to be far too optimistic. The deal Goldman concluded on May 10 with Harvard University earned Trotsky a relatively modest $6,000, to be paid only after the materials were delivered and inspected. Two weeks later came the

Siqueiros raid, which appeared designed to destroy Trotsky's archives as well as to end his life. The rush was now on to organize and catalogue these voluminous papers for shipment to Cambridge as soon as possible. The precious cargo, packed in three dozen crates and boxes, left Mexico City by train on the morning of July 17.

THE TRANSFORMATION OF Trotsky's home into a fortress began on the very afternoon of the assault. By a stroke of good fortune, there happened to be a man on the scene with the necessary wherewithal. He was Hank Schultz, a comrade from Minneapolis who had come to Coyoacán on vacation with his wife and child in order to meet Trotsky. Schultz was a railway brakeman by trade who volunteered to help Local 574 during the great Minneapolis Teamsters' strike of 1934, when, as night picket dispatcher, he worked in close collaboration with Dobbs. Subsequently, he joined the Trotskyist movement and met his wife, Dorothy, who was also a party member. They arrived in Coyoacán four days before the raid but were out of town when it happened.

Schultz was a skilled mechanic and electrician, as well as an experienced organizer of men, assets that proved to be invaluable in the weeks after the assault. Had it not been for Schultz, Robins testified, they would have "all caved under the overload." Trotsky called him "indefatigable, absolutely self-less, inventive, and in spite of sickness always in a good mood. Such people will build up the party." Another enthusiast was Joe Hansen, now returned to Coyoacán to serve as a guard and help erect the fortifications. Schultz was due back at his job in Minneapolis in mid-July, so time was short.

Some of the renovations made to the house on Avenida Viena were visible from the street. The east windows were bricked in. The old wooden entrance to the garage was replaced with double iron doors: a heavy outer door that swung open and an inner folding gate, both secured by electronically controlled locks. The tower atop the roof at the northeast corner of the property was converted into a two-story bombproof redoubt, with cement floors and ceilings. Three new brick turrets appeared above the walls, each with loopholes overlooking the patio and the neighborhood. Two of these blockhouses—one at the northwest corner and another at the center of the north wall directly above the guards' quarters—looked out onto the river. The one at the southeast corner, built on the roof of the house and looking down on the police *casita* and Avenida Viena, served as the main guard station and housed the electronic switches to the garage doors.

Security measures inside the house were delayed by the police investigation of the raid. On June 24, Trotsky and Natalia, accompanied by a formidable police escort, were brought to the heavily guarded city courthouse to give sworn depositions. The judge planned to come out to the house to see the rooms as they were at the time of the attack, an inspection that was delayed until July 16. On that day, the judge and his associates took about five hours to examine the bullet holes and other evidence and to interview Seva. Also allowed to inspect the house, much to the consternation of its residents, were the Communist lawyers for the captured raiders. Trotsky's guards, their hands on their guns, kept a close watch on these unwanted visitors.

Meanwhile, the hunt for Siqueiros and his artist-accomplices continued. The trail led to midtown Manhattan and the Museum of Modern Art, where five Siqueiros paintings were on display as part of an exhibit called "Twenty Centuries of Mexican Art." It was there, at MOMA, that several witnesses spotted the Arenal brothers behaving like innocent museum-goers, a development that set the FBI on their trail and prompted the Mexican police to initiate a request for their extradition.

ONCE THE JUDGE had paid his visit, the security installations inside the house could proceed. Steel shutters went up on the interior windows. The Trotskys' bedroom was equipped with new doors, each made with two layers of heavy iron encasing sand-filled centers. Along the north wall, a second level of guards' rooms was under construction. An underground bunker was in the planning stage.

After several weeks of deliberation, it was decided not to install a photoelectric alarm system. The main reason was that the new fortifications would have necessitated an elaborate arrangement of mirrors in order to convey the light beam uninterrupted along the tops of the walls surrounding the property. And anyway, as Hansen wrote to Dobbs in New York on July 31, "the next attack will most likely be bombs." The several hundred dollars designated for the photoelectric system would be spent instead on barbed-wire entanglements and bombproof wire netting.

All of these renovations were paid for by the contributions of the American comrades, but also thanks to the generosity of a few wealthy sympathizers in the U.S. who were stirred to action by the attempted murder of Trotsky and his family. One of these benefactors, a certain "Mr. Kay" who wished to remain anonymous, was rewarded with a personal letter from the grateful beneficiary. "The only thing I know about you, through my friends Jim Cannon and Farrell Dobbs, is that you are a very sure and generous friend," Trotsky wrote on August 3. "We live here, my family and my young friends, under the permanent threat of a new 'blitzkrieg' assault on the part of the Stalinists and, as in the case of England, the material aid comes from the States."

Trotsky was referring to the intensifying Battle of Britain, the Luftwaffe's massive bombing raids designed to knock out the Royal Air Force and prepare the way for a German invasion. "During the past two months the house has been undergoing a transformation into a kind of 'fortress'; in a few more weeks we will be very well protected against new 'blitzkrieg' assaults."

One generous supporter who had the opportunity to witness the transformation was Frank Jacson, Sylvia Ageloff's "husband," who was becoming a familiar figure within Trotsky's tight circle of comrades. Around the time that Trotsky wrote to thank his American benefactor, he asked Jacson what he thought of the new fortifications. Hansen and Cornell were standing there in the patio with Trotsky, admiring their handiwork. This was all fine, they heard Jacson say, but "in the next attack the GPU will use other methods." "What methods?" he was asked. Jacson just shrugged his shoulders, but he now spoke as an authority on these matters. Thanks to the incompetence of the Siqueiros gang, the NKVD penetration agent had been promoted to assassin.

Jacson had met Trotsky for the first time four days after the assault. Alfred and Marguerite Rosmer were leaving Mexico for France and had

Trotsky in the winter of 1939–40.

booked passage on a ship sailing to New York from Veracruz on May 29. Jacson was to drive them to their ship.

Sylvia had returned to New York City eight weeks earlier, after an extended sick leave from her job as a social worker. In her absence, the Rosmers had proved to be an invaluable connection for Jacson. Marguerite Rosmer was very close to Natalia, so Jacson's friendship with the couple put him in good standing with Natalia, and thus indirectly with Trotsky. The Rosmers' departure, therefore, was a setback for the Soviet operative, but he maneuvered to take maximum advantage of their send-off. Jacson told them he traveled to Veracruz on business every two weeks and would be happy, once again, to serve as their chauffeur.

On the eve of the departure, Natalia decided that she would like to accompany the Rosmers to Veracruz. That would mean returning alone with Jacson, with an overnight stop along the way, and nobody at the house thought this was a good idea. Instead, it was agreed that Evelyn Andreas, Trotsky's American typist and the companion of one of his guards, would drive Natalia in her car.

Jacson entered the patio on May 28 at 7:58 a.m., the time recorded in the log kept by the guards. Trotsky, who was tending to the chickens, greeted him and the two men shook hands. Jacson presented Seva with a toy glider, and he was invited to the breakfast table for a cup of coffee while the Rosmers finished packing. When they appeared, he carried their luggage out to the car. Evelyn, meanwhile, had arrived for Natalia. At the moment of departure, Trotsky surprised everyone by walking several steps out into the street to see them off, the only time he had ever done this.

The travelers stopped overnight at Jalapa, a couple of hours' drive from Veracruz. In accompanying Evelyn to park the cars in the garage for the night, Jacson discovered that her car was in need of a major repair and that it would be unsafe to continue driving it. The garage attendant seconded this opinion, according to Jacson, so the next morning everyone made room in his car for the last leg of the trip. Reaching the outskirts of Veracruz, Jacson, who had claimed to be a regular visitor to the city, did not seem to know his way around and had to stop to ask directions to the ship. Natalia wondered about this, but not enough to question it.

The Rosmers sailed for New York and Jacson drove Natalia and Evelyn back to Coyoacán. The guards' log shows that Natalia was returned to the house on May 30 at 3:42 in the afternoon. Jacson then drove Evelyn to her apartment in the city. There he was introduced to Dorothy Schultz, who together with her two-year-old daughter was staying with Evelyn. For Jacson, the connection to Evelyn and Dorothy would now prove to be extremely advantageous after the departure of the Rosmers.

It was at their apartment a few days later that Jacson met Hank Schultz, who seems to have been the only one of Trotsky's associates to question him about his name, which did not seem French. Jacson clarified that it was spelled without a "k" and was French-Canadian. In fact, he was thoroughly French, he eagerly explained, for although he was born in Canada and kept a residence in Montreal, he had moved to France as a boy and was educated in Paris.

Jacson came by the apartment at least a dozen times when Dorothy was there, usually staying several hours, sometimes for lunch or supper. He played with the Schultzes' baby daughter, Ann, for whom he ex-

pressed a deep fondness. He took them on trips to the zoo at Chapultepec Park and other points of interest in and around Mexico City. He
brought them on a picnic to the foot of El Popo, which gave him
another opportunity to boast about his mountain-climbing exploits in
Europe and in Mexico.

Jacson told Hank and Dorothy that he was close to the Trotskyist
circle in Paris and had made large contributions to the organization—
for a while had even paid the entire cost of publishing its newspaper.
This is what he also told the guards. In normal times these credentials
could have been confirmed with the French comrades, but they were
in flight from the German invaders, who entered Paris on June 14, a
dizzying turn of events which, among other things, forced the Rosmers to terminate their voyage in New York. One of the casualties was
Mark Zborowski—Trotsky's indispensable Comrade Étienne and the
NKVD's agent "Tulip"—who was reportedly being held prisoner in a
German concentration camp in France.

Jacson's bragging tales about his business activities might also have
been scrutinized, except that they were the kind that seemed indiscreet
to question. To Hank and Dorothy he said his employer was a war profi-
teer from New York who was exporting supplies of food and raw materials to the Allies, mainly Britain. The business was illegal, he confided,
and as a matter of fact, he himself was in Mexico illegally. His monthly
income, he said on several occasions, was $400, on top of which he had
a very generous expense account. He often remarked on the large sums
of money he handled, and even carried on his person, intimating that
these were bribes.

It was at Evelyn's apartment, in the second week of June, that Jacson
was introduced to Cannon and Dobbs, together with other visiting
comrades from New York. Jacson proved to be a most helpful chauffeur
and tour guide, on one occasion driving the visitors out to the ancient
pyramids north of Mexico City. This was an especially valuable service
because Trotsky's Dodge remained in police custody since the assault.
At the end of a day-long excursion to Toluca, Jacson bought Natalia a
gift of sour cream and honey, which he delivered on the drive home,
making a detour to the house, where Trotsky and Natalia came into the
patio to greet everyone.

On the evening of June 11, Jacson drove Cannon and Dobbs to dinner at the Hotel Geneva, then afterward took them for a drink. He had a way of talking a lot while saying very little. It was difficult to pin him down on anything. He stayed away from political topics, although on the subject of the split among American Trotskyists, he made clear that his sympathies were with the Majority. In any case, not being a comrade, Jacson posed no threat to Cannon and Dobbs. He was, on the contrary, that relatively harmless creature in Marxist politics at that time: a capitalist.

Jacson's conversations with Dorothy were a different matter entirely. He was remarkably blunt in his opinions about Sylvia and her support for the Minority position on the dialectic and the defense of the USSR. He described in great detail the arguments they had in person and by letter. On one occasion, after sharply criticizing Sylvia's views on dialectical materialism, he expressed doubts that their "marriage" could survive. Nor was ideology the only irreconcilable difference: Jacson made disparaging remarks about Sylvia's looks, her clothes, and her refusal to have children because, as he put it, she was a coward when it came to pain.

Jacson left Mexico City for New York on June 12, explaining that he had to go away on business. He asked one of the guards to drive him to the airport in his Buick, which he arranged to leave at the house for use by the guards while he was away. This gesture was greatly appreciated by the household, although ultimately the greater beneficiary was Jacson himself: during his absence from Mexico his automobile would continue to earn him goodwill.

MERCADER-JACSON WENT TO New York to receive instructions, funds, and encouragement from his NKVD handlers. He also used the time to re-engage with Sylvia, who was surprised to learn that her "husband" had visited the house in Coyoacán and met Trotsky and Natalia. Jacson told Sylvia that he had to make one final trip to Mexico before returning to take up a permanent position in New Jersey. He left New York by train on June 30, arriving in Mexico City on July 11, on which day he phoned Evelyn to say he had business to attend to in Tampico and would return in a week. When he resurfaced, he confided to Evelyn and Dorothy that his boss had decided to form a diamond-cutting syndicate in partnership with some Dutch émigrés who had fled the Nazis with large quantities of gems.

Jacson put off reclaiming his Buick until July 29. On that day, he came by the house at 2:40 and stayed for an hour and ten minutes. Sylvia had not heard from him since July 11, when he sent a telegram upon arriving in Mexico City. She grew impatient, and then desperate, for word from him when he failed to reply to her telegrams. Finally, he sent a cable and then called to say he had been very ill in a small town near Puebla and that he hoped to return to New York before long. She wired to ask if he wanted her to come and be with him during his convalescence. This was exactly what he was counting on, though he did not immediately reply. A few days later he telephoned her. Again she asked if she should come to him, and after some hesitation, he agreed.

On July 31, Jacson stopped by the house to deliver an expensive box of chocolates to Natalia, saying it was a gift from Sylvia, who was preparing to join him in Mexico City. It was then that Leonid Eitingon, the NKVD field officer of Operation Duck, sent a coded message to his superiors: "Everything is in order."

THE DAY JACSON came by the house to retrieve his Buick, he disappointed the guards when he admitted that while in New York he had not dropped in on the headquarters of the Socialist Workers Party. All his free time had been taken up trying to convince Sylvia and her sisters of the correctness of the Majority view, he explained, while during the day he was tied up with business. The guards took this to Trotsky, who agreed that Jacson's behavior was far from exemplary. On the other hand, Trotsky now had a better appreciation of what Jacson had to endure in contending with the Ageloff sisters. For, while Jacson was away, Trotsky had experienced his first gloves-off confrontation with members of the Minority.

The opposition associated with Max Shachtman and James Burnham was still called the Minority, even though its members had broken away and formed a separate Workers Party in April. A month later, Burnham stunned his comrades by announcing his resignation from the new party and his complete break with Marxism. Burnham's astonish-

ingly candid resignation letter proved to be an endless source of amuse-
ment and ammunition to Cannon and the Majority, whose suspicions
all along that Burnham was a petty-bourgeois fraud had now been
forcefully validated.

Burnham was gone, but he remained a favorite target of abuse in
the Socialist Workers Party and also among Trotsky and his staff. This
became obvious to a group of about forty Americans who crowded
into the dining room of Trotsky's home late in the afternoon of July 17
to hear him speak, an event arranged by Professor Hubert Herring in
conjunction with his summer Latin America seminar. Among the participants was a group of seven visitors from Texas invited separately and
led by Charles Orr and his wife. The Orrs were Trotskyists who went
to Spain during the civil war; they had been arrested and imprisoned
in the crackdown on the POUM in Barcelona in June 1937. Charles
Orr taught sociology at the University of Texas, where he and his wife
attracted a small following of young partisans for the Minority. Several
of them accompanied the Orrs to Mexico City for a vacation and a
chance to meet Trotsky.

The seminar began with Trotsky making brief remarks. According to
Hansen, "The OM ripped into the democracies, their decay, and the sole
hope of humanity being socialism." A lively question-and-answer session
ensued, and Hansen reveled in the fact that Trotsky could finally get a
taste of the debates he had thus far only read about in letters from New
York. Trotsky pressed the Minority guests to defend Burnham's assessment of the USSR as the evil twin of Nazi Germany and his rejection of
dialectical materialism. Hansen says that Trotsky became quite agitated,
"and even the Old Lady, who can now follow English pretty good, began to argue, but in French, which nobody could understand."

Trotsky was so inspired by the experience that the Orrs and their
young friends were invited back a few days later for a formal debate
with the guards. Hansen, Robins, and Otto spoke for the Majority, each
taking ten minutes, while the Orrs were each given fifteen minutes to
defend the Minority position. Trotsky served as chairman. The Orrs
blamed the split on Cannon and his "Stalinist" methods, but Trotsky
had heard it all before. Jotting notes to himself in English, he wrote:
"no reason for a spleet!"—a word he spelled just as he pronounced

it—"the spleet is not accident—*inevitable*." When Chairman Trotsky rose to speak, he thanked the Orrs for confirming him in his opinion that the Minorityites were merely an inferior version of the Russian Mensheviks.

Further evidence of the Minority's feeblemindedness was provided by Sylvia Ageloff, who flew in from New York on August 8 and whom Jacson brought to the house later that afternoon for tea with Trotsky and Natalia. Jacson was still not healthy—although his gastrointestinal troubles were hardly a suitable topic for conversation at the dining table. Inevitably, the discussion centered on the Majority and Minority views on the war and the USSR. Jacson took Trotsky's side, although he barely said a word and appeared to be out of his depth. Ultimately, it was of no consequence, because Ramón Mercader had already mastered the peculiar dialectic of Jacson and Sylvia.

On August 9, with the Dodge released from police custody, Trotsky and a group from the household set off on a picnic. Trotsky called these outings "walks" for a reason, yet on this occasion, according to Hansen, "he certainly didn't act like his old dynamic self. He was scarcely interested in building the fire, hunted for nothing, tried to sleep on the ground, didn't take so much as his usual walk, dropped to sleep immediately in the car after it was over." He seemed greatly fatigued, behaving as though he needed the picnic for rest rather than to channel his energy.

This was a cause for concern because, on his doctor's orders, Trotsky was getting plenty of time for relaxation these days. He was required to take an hourlong siesta after lunch, while on Sundays he was supposed to avoid work entirely and just lie in bed. "It bores him stiff," said Hansen of Trotsky's enforced holidays. For months he had been intending to return to work on his Stalin biography. He had written to his translator in New York on March 19, "I would be really glad if I could deliver the whole during the month of August. It is possible. And I will do every-

Trotsky and Natalia on a picnic, winter 1939–40.

thing to observe this new 'deadline.'" But then came the assault and the investigation, and Trotsky was forced to defend himself against charges that he had orchestrated it himself.

In mid-June, after Trotsky accused the monthly *Futuro* and its publisher, Lombardo Toledano, of preparing the moral ground for the assault and thus serving as an arm of the GPU, the magazine responded by suing him for defamation. The suit was joined by sister publication *El Popular* and later by the Communist paper *La Voz de Mexico*. Trotsky and his staff immediately began to organize a counteroffensive, mobilizing Goldman in New York to secure depositions from Soviet defectors, such as ex-spy Walter Krivitsky, on the relationship of the Comintern and the GPU and on how Kremlin funds were distributed among pro-Soviet publications, even in distant lands like Mexico.

Trotsky appeared in court under extremely tight security for the preliminary hearing on July 2, a session that lasted most of the day. The experience seemed to energize him, giving him the chance to strike back at Lombardo Toledano and the other Stalinists who had been harassing him ever since he set foot in the country. Hansen described him as "working like a steam engine" and "still the dynamo." On the day after his court appearance, Trotsky exhorted Goldman: "It is imperative not to lose a single hour, so that I can meet my 'accusers' well armed with affidavits, concrete dates, general considerations, etc. I await with the greatest impatience your answer."

These exertions ended up taking a toll on Trotsky's health. His blood pressure was running extremely high. His lower back was giving him trouble, to the extent that he sometimes walked out into the patio in the morning doubled over. By late July, Natalia had become extremely worried about his condition. She said it was a recurrence of his old European illness, in fact one of his worst bouts yet, only without the fever. He had absolutely no stamina. A few minutes of conversation left him exhausted. Hansen alerted Dobbs on July 31 about Natalia's concerns, adding, "They have no confidence in either doctors or hospitals down here on account of the Stalinists."

Trotsky's political struggles entered a new stage in the first days of August, when David Serrano, a member of the Politburo of the Mexican Communist Party who had been arrested in connection with the commando raid, gave a deposition asserting that Siqueiros was a Trotskyist and had carried out the self-assault as Trotsky's paid agent. On August 6 Trotsky held a press conference in order to rebut this latest and most outlandish frame-up attempt. Another head of the Stalinist hydra had to be cut off, even though Trotsky understood that yet another would grow in its place.

Trotsky never appeared to lose heart. Shortly after the second Moscow trial in January 1937, he received a letter from an Old Bolshevik living in the United States whose spirit was breaking under the avalanche of false accusations manufactured by the Kremlin. To Trotsky such a reaction was unacceptable. "Indignation, anger, revulsion? Yes, even temporary weariness. All this is human, only too human. But I will not believe that you have succumbed to pessimism. History has to be taken as she is; but when she allows herself such extraordinary and filthy outrages one must fight her back with one's fists."

Now, in August 1940, as he anticipated the next eruption of lies and slander, Trotsky sounded the same defiant tone: "We await the new intrigue calmly. We don't need to invent anything. We shall only aid in elucidating the logic of facts. Against this logic the falsifiers will break their skulls!"

THE PICNIC ON August 9 was supposed to provide a respite from these battles. Hansen drove the Dodge, with Trotsky seated beside him; the Ford followed, with Robins at the wheel. They headed up to the Lagunas de Zempoala in the mountains above Cuernavaca, but there the clouds thickened and the air turned chilly, so they got back into the vehicles and descended to a lower elevation. They chose a picnic spot on a slope in the forest, the same place where Trotsky had enjoyed a picnic more than two years earlier with Diego and Frida and Van. Afterward, they drove to Cuernavaca so that the first-timers could have a look at Diego's murals. From there they continued farther southeast to the lowlands, almost to Cuautla, then headed north in the direction of Amecameca.

Trotsky slept much of the way—either that or he was lost in his thoughts. Upon rising in the morning he liked to joke to Natalia, as he opened the steel shutters on their bedroom window, that yet another night had passed without a visit from Siqueiros. Yet he worried about how Natalia would get along without him. Both their sons were gone. He was under no illusion that their younger son, Seryozha, could have survived Stalin's terror, but he did not want Natalia to believe that all hope was lost. One day not long after the raid, he said to her with emotion he could barely contain: "My death . . . may lighten Seryozha's situation." "No, no, no," Natalia refused to consider this grim calculus.

Memories of Lyova could easily make Natalia upset, and she sometimes heard her husband grieving in his study and understood the reason. An especially bittersweet memory of a private moment between father and son was one that Trotsky had described to Natalia years earlier, just after it happened. It was September 1933, shortly after they had moved from Turkey to France. He was living at the Sea Spray villa in Saint-Palais-sur-Mer, near Royan on the Atlantic Coast. Natalia had gone to Paris to see a doctor. A group of French comrades gathered at the house, and Trotsky engaged one of them in a lively discussion, with Lyova and the others looking on and then joining in. Trotsky was in fine form—he felt once again like *le Vieux*, the Old Man—and several times he caught sight of Lyova's adoring eyes locked onto him.

Afterward, Lyova came to his father's room. They exchanged words about purely mundane matters—until suddenly Lyova stepped forward, placed his head on his father's shoulder, and hugged him. "Papochka, I love you very much," he said, sounding to his father just like a little boy. Trotsky held his son close, pressing his cheek against the boy's head. Lyova could feel his father getting upset, so he turned and tiptoed out of the room.

Near Amecameca, at the foot of the snow-peaked Popocaté-petl, Trotsky was awake again. He remained alert for the ride back to Coyoacán. As they entered the neighborhood, he slid down low in the seat to hide his head from the windows facing the streets. He did not want to make himself an easy target for a Stalinist with a machine gun. "And what if Hansen were to be machine-gunned?"—the thought crossed Trotsky's mind. "After this we must have two of the best drivers in the car," he told Hansen as they pulled into the garage.

The guard now numbered seven, with two men on duty at all times through the night. Most of the guards performed secretarial tasks for some part of the day; all of them took turns cleaning in and around the chicken yard and rabbit hutches. Natalia, who before the raid had been strongly in favor of decreasing the number of guards, was now pushing for a threefold guard through the night, but they told her that the expense was pro-hibitive. Mexican comrades could not be enlisted for fear that this would leave Trotsky open to the charge of interfering in Mexican politics.

After Hank Schultz returned with his family to Minneapolis on July 24, he hit upon the idea of sending down to Coyoacán the one comrade he believed could provide the guard with the kind of military training necessary to withstand the next Siqueiros-style raid on Trotsky's home. The man he had in mind was a legendary figure in Teamster circles, a Sioux Indian known as the Rainman.

The Rainman was Ray Rainbolt, the organizer and commander of the 600-strong Defense Guard of the Minneapolis Teamsters. Rain-bolt had been one of several field organizers of the cruising pickets during the 1934 strike. He had considerable military training from multiple stints in the U.S. Army, and combat experience in the trade union struggles of the 1930s. In recommending Rainbolt to Dobbs, Schultz called him the one man with "sufficient experience, prestige, and authority to take charge down there." What Coyoacán needed was a commander to transform Trotsky's guard into a disciplined military unit. "The old man, I'm absolutely sure, is of the same opinion."

But it turned out that the Old Man was not of the same opinion—or rather, as so often in the past when it came to decisions about his body-guards, he could not make up his mind. As before, a large factor was money. Trotsky doubted the value of the Rainman coming down for six weeks, preferring instead that a permanent guard be sent to replace Hansen, who was due to leave by the end of August. Trotsky's reaction annoyed the comrades in New York and Minneapolis. Once again, on the vital matter of the guard, Trotsky was dragging his feet.

One of the considerations behind Schultz's proposal to send down someone with sufficient authority to take charge was that Trotsky was not always the most cooperative subject to guard. He was fully on board when it came to building fortifications and keeping his head down in the car, but he could also be lax about his personal safety. The fact is, he was used to coming through unscathed, whether in prison, revolution, civil war, or exile. The dumb luck of the Siqueiros raid, even though it set in motion the transformation of the villa into a fortress, may have reinforced Trotsky's fatalism. Concrete and steel replaced wood, yet he still refused to subject his visitors to the indignity of a personal search.

At a meeting of the guard shortly after the assault, Robins proposed that Trotsky always be accompanied when he was in the patio. A special source of concern was the row of tall eucalyptus trees on the far bank of the creek running along the north side of the house, which offered a perfect hiding place for a sniper. Trotsky objected and said that if the guard voted to endorse such a change he would refuse to go out to the patio at all. Robins was sympathetic to the Old Man's situation: "His life was a sort of modified prison routine; no prisoner likes guards."

On August 16, Trotsky was the recipient of two gifts from a com-rade in Los Angeles: a bulletproof vest and a siren. The chain mail vest, which suited the household's recent preference for medieval décor, had been "piously admired" by everyone, Trotsky wrote the comrade appre-ciatively, although he doubted that it was comfortable enough to sleep in. "The siren provoked even more admiration. It is wonderful enough just in appearance." They hesitated to try it out, however, lest they raise a ruckus, "for we are told that this siren can be heard from here to Los Angeles. I, personally, consider this an exaggeration."

Trotsky closed this letter on a serious note. "More than two and a half months of my time has been almost exclusively devoted to the investigation," he noted, referring to the contentious aftermath of the May 24 raid. The next day he would present the judge in the defamation suit a lengthy memorandum documenting in exhaustive detail the financial and other ties between the Mexican Stalinists and the GPU. "And now I hope to be able to go back to my book."

Ramón Mercader and Sylvia were staying in room 113 of the Hotel Montejo, on the Paseo de la Reforma, registered as Mr. and Mrs. Frank Jacson. Sylvia was troubled by the changes she observed in Ramón's health and his nerves. His light olive complexion had turned green. He was nervous and irritable. On August 15 he spent the entire day in bed with a fever. The next day, Sylvia wrote to her sister Hilda in New York that Jacson's illness, together with the "Mexican mañana," was delaying their departure for home.

Their plan was to travel to New York by plane, which would leave them enough time for a few days in Acapulco before she had to get back to work. "I'm glad I came from Jac's point of view because he really looks terrible and needs care," Sylvia wrote. "Jac has diarrhea something terrible—it just wears him out." He told her that his intestinal problem had been diagnosed long ago and required surgery. So the sooner she could get him to her doctor in New York, the better.

Sylvia indicated to her sister that beyond the vacation in Acapulco, she had no desire to stay in Mexico. Four days earlier, on August 12, she had paid Trotsky a visit. The reception was cordial, she said, "but the O.M. immediately opened up on how we were worse than the Mensheviks, etc. Argument is useless—I suppose I'll have to go there to say goodbye but I don't relish it." Trotsky had expressed concern about her husband's health. "The O.M. says that for the sake of the *majority* I should insist that Jac have an intervention chirurgical—(operation to you)—immediately—so I told him I had as much influence on him physically as I had politically."

Jacson's haggard appearance and nervous twitching were noticed by everyone at the house—although no one chose to remark on them. At this point there was only one man who could have stopped his progress: not the Sioux warrior from Minneapolis, but the Frenchman with the Dutch name who had left Coyoacán the previous autumn, after eight years by Trotsky's side.

Van sent a telegram to Trotsky after the Siqueiros raid, offering to return and take charge of his security. His intervention might have made all the difference. Frank Jacson spoke French fluently, but his accent was not quite French: not Parisian, nor Belgian, nor Canadian. This should have been detected by the Rosmers, except that they, especially Marguerite, had become infatuated with Jacson. Van's sensitive ear and his canniness about the methods of the GPU gave him the best chance of unmasking the impostor. It was Trotsky himself who sealed off this possibility. "It would be really too cruel to force you to return to this prison," he wrote to Van on August 2. "Now, after the reconstruction, it has become a genuine prison, not in the modern manner, it is true, but rather more like those in medieval days."

On August 17 at 4:35 p.m., the iron doors of the prison gave way and Jacson was admitted into the patio. He was dressed in a suit and tie, as usual, and wore his horn-rimmed glasses. He also wore a gray hat and carried a khaki raincoat over his arm. It was sunny, but being the rainy season, the weather was changeable. Natalia was off in town. Jacson met Trotsky in the patio. He brought with him a short article he had composed, a rebuttal to Burnham and Shachtman's revisionist line about the war and the USSR, and he asked Trotsky to read it.

Jacson-Mercader was invited inside. He followed Trotsky, walking from the patio toward the library at the top right of the T-shaped house. Ramón Mercader was five feet ten inches tall and wiry in build, his broad shoulders slightly stooped. The day before, Sylvia had written her sister that "he's lost so much weight his clothes flop on him like a scarecrow." Trotsky, walking ahead of him, was six feet in his shoes, broad-chested, and despite his recent bouts of fatigue and an aching back, physically powerful.

The two men made their way through the library into the dining room and then into the study. It was the first time they had been alone in the house. Trotsky sat down at his desk, strewn with books, newspapers, and manuscripts, and began reading Jacson's article. As he did so, Jacson walked over to the left side of the desk and sat on its edge, keeping his hat on and his raincoat over his arm. This was shockingly rude behavior, but Trotsky, even though he was very territorial about his desk, was too polite to say anything. Jacson sat beside the push-button switch to the alarm system. Nearby on the desk lay Trotsky's .25-caliber automatic pistol, with six cartridges in the magazine. It had been oiled and reloaded only days before. Beside it were souvenir slugs from the Siqueiros raid.

It took Trotsky only a few paragraphs to recognize that Jacson's article was unoriginal and incoherent. He suggested a few changes, and invited Jacson to bring back the article after he had revised it. Trotsky rose and escorted Jacson out to the patio. The entire visit took only eleven minutes.

The encounter left Trotsky feeling out of sorts. "I don't like him," he said to Natalia. Jacson's typewritten banalities were one thing, but it was his outrageously bad manners that unsettled Trotsky. "Yesterday he did not resemble a Frenchman at all," Trotsky brooded. "Suddenly he sat down on my desk and kept his hat on the entire time." Natalia remembered that Jacson never wore a hat, and in fact he boasted about going around without a hat or a coat, even in rainy weather. "This time he wore a hat," Trotsky stressed.

Whatever the August 17 visit was supposed to accomplish—and a failure of nerve might explain why Ramón Mercader did not attempt to carry out his deadly assignment that day—in one way it had served its purpose. With his hat, which functioned merely as a prop, Mercader had succeeded in drawing attention away from his raincoat, which concealed the tools of his trade.

THE MORNING OF August 20 was bathed in bright sunshine, although dark clouds gathered in clusters at the edges of the twin volcanic peaks to the southeast, threatening a downpour. Trotsky came outside to feed the rabbits and chickens at 7:15 a.m. as usual, then had breakfast at 9:00 a.m. He told Natalia he felt well, and they talked once again about summoning the barber because he needed a haircut.

With the morning mail came a telegram from Al Goldman in New York with important news: "HARVARD INFORMS MATERIAL ARRIVED." This came as quite a relief. Only three days earlier Trotsky had let his paranoia about his archives get the better of him, writing to Goldman of his suspicion that the FBI had intercepted the shipment and was going through his files. "If, under these conditions they really found it necessary to check the documents without my authorization, it would signify a terrible abuse of power," Trotsky complained, vowing to deliver a sharp public protest. "Disloyalty is always bad, but disloyalty of a state power toward a private person is especially despicable." These, the words of the man who helped create the first totalitarian state, which even now he championed as the world's most advanced country.

All along, Trotsky had argued that there was no fundamental distinction between Nazi Germany and the bourgeois democracies. They differed only in the way that "there is a difference in comfort between various cars in a railway train. But when the whole train is plunging into an abyss, the distinction between decaying democracy and murderous fascism disappears in the face of the collapse of the entire capitalist system." He wrote those lines in May 1940. By August, after fascism had overrun most of Europe and Hitler appeared poised to conquer Britain, Trotsky's focus had shifted. He became increasingly critical of the pacifism that was prevalent on the American left. The best way to defend "civil liberties and other good things in America," he maintained, was to aid Britain and crush Hitler.

Yet Trotsky continued to believe that fascism represented the final stage of capitalism, and he predicted that the entry of the United States into the war would give birth to an American brand of militarism that would far surpass that of Hitler's Germany. This left him open to criticism of the kind leveled at him by Dwight Macdonald in the July-August issue of *Partisan Review*, where Macdonald argued that the standard Marxist categories no longer applied. Exhibit A was Trotsky's insistence that fascism was merely a revival of Bonapartism, clear evidence that he failed to appreciate the threat. After breakfast on the morning of August 20, Trotsky began to dictate a response to the "traitor" Macdonald, which he began by calling his article "very pretentious, very muddled, and stupid."

At one o'clock, Trotsky's Mexican attorney came by to inform him that he had been accused of defamation at a banquet sponsored by *El Popular*. The charge could not go unchallenged, they decided. Another head of the hydra would have to be lopped off.

Following the afternoon meal and a brief siesta, Trotsky went back to his desk. He dictated more of his rejoinder to Macdonald and began his response to *El Popular*—all of this spoken in Russian into the Dictaphone. Through the open window of his study, his voice could be heard punctuating the end of each sentence: *"Tochka!"* He also dictated to his typist, Evelyn, two congratulatory letters to comrades in Minneapolis who had been jailed for strike activity and were to be released on August 23.

The last letter of the day was addressed to Hank Schultz. Trotsky thanked him once again for all his help, and indicated that there was still some work to be done before the fortress was finally secure. He said he was delighted with the "excellent gift" sent to him by another Minneapolis comrade: a dictionary of American slang. Now, finally, he could make headway in understanding his guards. "There is only one difficulty: at meal-times I must permanently keep this book in my hands in order to be able to understand the conversation." He continued in a humorous vein: "In the part I have already studied, which is devoted to college slang, I had hoped to find some abbreviations for the various sciences, philosophical theories, etc. but instead I found merely about 25 expressions for an attractive girl. Nothing at all about dialectics or materialism. I see that the official 'Science' is a bit unilateral." He signed off, "Fraternally yours, Old Man."

At five o'clock, Trotsky and Natalia had their usual tea. Then Trotsky went out to feed the animals. Hansen was on the roof near the block-house at the southeast corner with Charley Cornell and handyman Melquiades Benitez. They were connecting the powerful siren sent from Los Angeles. Robins was in the patio. The other guards were either running errands or had the time off. At about 5:20 p.m., Hansen looked down onto Avenida Viena and saw Jacson drive up in his car. He usually parked it facing the wall near the garage, but this time he made a full turn in the street and parked parallel to the wall, with the vehicle facing in the direction of Coyoacán.

Jacson emerged from the car, waved to the guards and shouted, "Has Sylvia arrived yet?" She had not, nor was she expected, so Hansen and Cornell assumed that Trotsky had arranged to meet the couple but had forgotten to tell them. "No," Hansen called down, "wait a moment." Cornell then operated the electronic controls on the doors to the garage, and Robins received the visitor in the patio.

Jacson wore a hat and carried a raincoat over his left arm. Hansen, Cornell, and Melquiades continued with their tasks. Jacson walked over to Trotsky, who was feeding his rabbits. He told him he had expected Sylvia to be there and that they were leaving for New York the next day. He had brought his revised article for Trotsky to have a last look. Trotsky continued to feed the rabbits, as he explained to Jacson that wet grass made their bellies swell and could be fatal.

Natalia stepped out onto the porch at the entrance to the dining room, just to the left of the stem of the T, and saw her husband standing near an open rabbit hutch. Standing next to him was an unfamiliar figure. Only when he removed his hat and began to walk toward the porch did she recognize Jacson. *"J'ai grand soif"*—"I'm very thirsty"—he said to Natalia upon greeting her. "May I have a glass of water?" "Perhaps you would like a cup of tea?" she offered. "No, no. I dined too late and feel that the food is up here," he replied, pointing at his throat. "It's choking me."

Jacson had not shaved in several days and he looked ill and seemed nervous. His face was gray-green. Natalia asked him about his hat and raincoat. "It's so sunny today," she said. "Yes," he agreed, "but you know it won't last long, it might rain." Mercader had rehearsed this response many times in his head, but Natalia's mild challenge appears to have unnerved him, so that at first he did not hear her query about Sylvia's health. "Sylvia? Sylvia?" He pulled himself together, then remarked casually: "She's fine, as always."

Natalia and Jacson walked back toward Trotsky at the rabbit hutches.

As they approached, Trotsky addressed Natalia in Russian: "You know, he is expecting Sylvia to call on us. They are leaving tomorrow." Natalia took this as a hint that she should invite them to tea, perhaps even supper. Natalia told Jacson she had no idea they were leaving so soon and that Sylvia was coming to the house that day. "Yes, yes," said Jacson, "I forgot to mention it to you." Had she known, Natalia said, she would have prepared some things to send with them to New York. Jacson said he could return the following afternoon, but Natalia declined the offer.

Turning to Trotsky, she said that she had already offered Jacson some tea, but he said he was not feeling well and asked for water instead. Trotsky looked Jacson over and said to him in a tone of mild reproach, as though admonishing a comrade, "You don't look well. That's not good."

In the momentary lull that ensued, Natalia sensed that her husband did not wish to leave his rabbits, but he managed to sound sincere nonetheless: "Well, what do you say, shall we go over your article?" He closed the doors to the hutches, brushed off his blue denim jacket, and began to walk toward the house. Natalia accompanied them to the door of the study, which Trotsky closed behind him.

Up on the roof, inside the blockhouse, Hansen was labeling the switches connecting the alarm system to the rooms of the individual guards. The wiring of this system had turned out to be extremely complicated, and after Schultz's departure the electrical work had slowed considerably. But the worst was over—although the New York comrades had recently resurrected the idea of installing a photoelectric alarm system to detect intruders.

Suddenly a terrible cry pierced the afternoon quiet—"prolonged and agonized" is how Hansen registered it, "half scream, half sob. It dragged me to my feet, chilled to the bone." He scrambled out of the blockhouse and onto the roof, searching for its source. Melquiades was aiming his rifle at the window to the study, where there were sounds of violent struggle. For a brief moment Trotsky's blue jacket became visible as he grappled with someone. "Don't shoot!" Hansen shouted to Melquiades. "You might hit the Old Man!"

Hansen switched on the general alarm and slid down the steel ladder to the library, while Melquiades and Cornell stayed on the roof, covering the exits from the house. As Hansen entered the dining room, Trotsky stumbled out of his study, blood streaming down his face. "See what they have done to me!" he moaned. Robins entered through the far door of the dining room, with Natalia close behind.

Natalia rushed over to her husband, his face now covered with blood. He had lost his glasses. His arms were hanging limp. "What happened? What happened?" she asked, flinging her arms frantically about him and walking him out onto the porch. He did not answer right away, and she thought something might have fallen on his head. "Jacson," he uttered quietly.

Hansen and Robins entered the study, which was a shambles. Chairs were overturned and broken, papers and books were scattered all about, the Dictaphone had been smashed. There were large pools of blood on the floor and blood splattered on the desk, the books, the papers. Jacson stood in the middle of the room, gasping, his face contorted, his arms hanging limp, a pistol dangling in his hand. "You take care of him," Hansen told Robins, "I'll see what's happened to the Old Man." Robins struck Jacson on the head with the butt of his revolver, sending him to the floor. Under Robins's repeated blows, Jacson kept yelling, "They made me do it!"

Natalia had brought Trotsky back inside the house, where he slumped to the floor on a small carpet by the dining table. "Natasha, I love you," she heard him say in a grave voice. This took her completely by surprise because she assumed that his wound was not serious. She wiped the blood from his face, found a pillow for his head, and held a piece of ice on the wound. "Oh, oh," he strained to speak the Russian words, "no one, no one must be allowed to see you without being searched."

Hansen returned to Trotsky's side. The wound, which appeared to be superficial, was on the top of his head, on the right side toward the back. Trotsky said he had been shot. Hansen told him this was not possible, that they had heard no noise, and that Jacson must have hit him. Trotsky looked doubtful. He brought Natalia's hand repeatedly to his lips.

As Hansen raced up to the roof to alert the police, Trotsky spoke to

Natalia slowly, forming the Russian syllables with increasing difficulty. But he was thinking clearly and urged that Seva be kept away from the house. "You know, in *there*," he said, gesturing toward the room with his eyes, "I sensed . . . understood what he wanted to do . . . He wanted to strike me . . . once more . . . but I didn't let him."

Up on the roof, Hansen shouted down to the police, "Get an ambulance!" He looked at his watch: it was ten minutes to six. Caridad Mercader and Leonid Eitingon, stationed in separate cars a few blocks away, heard the alarm and the commotion and realized that things had not gone according to plan. Ramón had not managed to do the job quietly and then slip away—or even to execute the fall-back plan and shoot his way out. They had been waiting for him to drive around the corner and transfer to his mother's car for the getaway. Now, with the police about to descend on Trotsky's house, they fled the scene.

Again Hansen was at Trotsky's side. Rather than wait for the ambulance to arrive from the city, they decided that Cornell should go for Dr. Dutren, who lived in the neighborhood and had been to the house before. Both cars were locked up in the garage, so in order to save time Cornell decided to take Mercader's car. As Cornell departed, sounds of renewed struggle came from the study where Robins was holding Mercader.

"What about *that one*?" Natalia asked her husband. "They will kill him." "No," he struggled to say the words. "He must not . . . be killed . . . he must . . . talk." Hansen entered the study, where Mercader was trying desperately to escape from Robins. "Don't kill him," he repeated Trotsky's order, although Robins was trying to beat a confession out of him. "It's the GPU who sent you! Admit it!" he threatened Mercader, who kept insisting it was not the GPU but some mystery man that made him do it. "They are keeping my mother a prisoner!"

Hansen saw Mercader's automatic pistol on Trotsky's desk. Trotsky's glasses lay there, too—one of the lenses was broken and out of the frame. On the floor Hansen's eyes took in something he had earlier missed, a blood-soaked instrument that resembled a prospector's pick: one end was pointed, like an ice pick, the other was flat and wide; the handle, about a foot long, had been cut down for concealment.

Hansen began punching Mercader, hitting him on the mouth and on the jaw below the ear until the pain in his hand forced him to stop. The urge to kill was overwhelming, and Mercader sensed this. "Kill me! Kill me!" he pleaded. "I don't deserve to live. Kill me. I did not do it on the order of the GPU, but kill me." As they were beating him, he went in and out of consciousness, moaning several times, "They have imprisoned my mother."

Suddenly Cornell burst into the room. "The keys aren't in his car." He searched Mercader's clothing for them while Hansen raced out to open the garage doors. Moments later, Cornell was driving out of the garage.

While they waited for Cornell to return with the doctor, Natalia and Hansen kneeled at Trotsky's side, holding his hands. "He hit you with a pick," Hansen told him. "He did not shoot you. I am sure it is only a surface wound," he said, this time without conviction. "No," Trotsky responded, "I feel here"—pointing to his heart—"that this time they have succeeded." Hansen again sought to reassure him, but Trotsky understood what was happening. "Take care of Natalia. She has been with me many, many years," he said, as his eyes filled with tears. Natalia began to cry over her husband, kissing his hand.

Cornell arrived with Dr. Dutren. He examined the wound and said it was not serious, although his manner said otherwise. A few moments later the ambulance arrived, and the police entered to take away the as-sailant, who was bloodied and bruised. As they dragged him out of the study, he cried, *"Ma mère! Ma mère!"*

The ambulance men brought in the stretcher. Natalia did not want her husband to be taken to a hospital: the risk of another attack was too great. Tense moments followed, as everyone waited for Trotsky to decide what to do. Hansen, Cornell, and Robins were kneeling beside him now. "We will go with you," Hansen told him. "I leave it to your decision," Trotsky said in a whisper. As they were about to place him on the stretcher, he again whispered, "I want everything I own to go to Natalia." And finally, in a voice that wrenched the hearts of the men leaning over him, he said, "You will take care of her . . ."

Natalia and Hansen rode with Trotsky in the ambulance, which jolted over the potholes and plowed through the mud of Coyoacán's near-impassable streets. The siren wailed incessantly en route to the city,

accompanied by the shrill whistles of the squadron of police motor-
cycles leading the way.

Trotsky remained conscious. His left arm was extended along the
side of his body. It was paralyzed. His right hand wandered in circles
over the white sheet, touched the water basin near his head, then
found Natalia. Bending very low she asked him how he felt. "Better
now," she heard him whisper despite the din. "Better now"—this gave
her hope.

Evening descended as the ambulance sped through the bustling
streets of Mexico City, weaving its way through traffic toward the hospital. Trotsky whispered into Hansen's ear: "He was a political assassin.
Jacson was a member of the GPU or a fascist. Most likely the GPU."

The sirens died away as the ambulance pulled up to the entrance of
the Green Cross Emergency Hospital, where a crowd had gathered. Inside, they laid him down on a narrow cot. Silently the doctors examined
the wound, as Natalia stood alongside her husband. On their instructions, a nurse began to shave Trotsky's head. With a hint of a smile, he
said to Natalia, "Look, we found a barber."

Trotsky looked over at Hansen and gestured weakly with his right
hand. "Joe, you . . . have . . . notebook?" Hansen leaned against the cot
and with his broken right hand recorded Trotsky's words. When Natalia
then asked what he had said, Hansen replied, "He wanted me to make
a note about French statistics." This surprised Natalia, who thought it
was strange for him to have his mind on French statistics at such a
moment—or perhaps he was beginning to improve.

They began to undress the patient. Using scissors they cut away
his blue jacket, then his knitted vest, then his shirt, and then they un-
strapped his wristwatch. As they began to remove his pants, Trotsky said
to Natalia, "I don't want them to undress me . . . I want you to do it."
These words, spoken in a grave and sorrowful voice, were the last he
ever spoke to Natalia. When she had finished, she bent over him and
kissed his lips. He kissed her back. Again she kissed him, and again he
responded. And then one final time.

Trotsky underwent surgery that evening. The doctors trepanned an
area of the right parietal bone. Blood and gray matter spilled out from a
wound three-quarters of an inch wide and two and three-quarters inches

deep. The direction of the pickax was from top to bottom, front to back, and right to left. Thus, it turned out, Jacson had not struck Trotsky from behind, as was initially believed, which might explain why the victim was able to prevent his assailant from striking him a second time.

The first medical bulletin stated that although the results of the operation were "very satisfactory," the prognosis was grave. Trotsky was still in a coma and was paralyzed on the left side. The chief surgeon was quoted as saying that the patient's chances were one in ten. The New York comrades made arrangements to send down Dr. Walter Dandy, director of neurosurgery at Johns Hopkins University.

Natalia remained at Trotsky's bedside through the night, despite the urging of the doctors to get some sleep. She sat beside him, dressed in a white hospital gown, holding his hand. She was waiting for him to wake up and take control of himself, as he had always done, and her hopes rose and fell with his breathing, which was alternately regular and calm, then rapid and heavy.

Hansen, Robins, and Cornell stood guard at the hospital. During the evening, Colonel Salazar arrived to question Mercader, who was being treated for his wounds in a room two doors down the corridor from Trotsky. His face was bruised and swollen, both eyes were blackened, and the gashes Robins had inflicted to the top of his head needed stitches. In Mercader's raincoat the police found a dagger nearly fourteen inches long. His gun, a .45 Star automatic, had eight bullets in the magazine and one in the firing chamber. He was also carrying a large sum of money—$890—which seemed to indicate that he had plans to flee Mexico after the attack.

A letter of confession, written in French, was also found on the assailant, presumably meant to be discovered only upon his death. From its telltale misspellings to its author's claim to be a disillusioned follower of Trotsky, it bore the fingerprints of the NKVD. Jacson had been driven to murder, the letter alleged, because Trotsky had been pressuring him to break with Sylvia, one of the "Minority rabble," and go to the Soviet Union to engage in acts of sabotage and organize the assassination of Stalin.

Under questioning by the police, the assailant began to spin a web of tangled lies about his background, his contacts in Mexico City, and his movements before the attack. "It was a veritable maze," said Colonel Salazar. Yet Mercader's account of the details of his crime had the ring of authenticity. He said he closed his eyes before striking the blow, which could explain why it failed to knock Trotsky unconscious. "The man cried out in a way that I shall never forget as long as I live," said Mercader. "His cry was 'aaaaaah . . .' very long. Infinitely long. And it appears to me still in these moments that this cry penetrates my brain."

Trotsky rose up like a madman, Mercader said, threw himself on him and bit his hand. "You see, here, I still have the marks of his teeth." Mercader pushed him away and he fell to the floor but managed to get up and leave the room. "I remained like one demented, without knowing what to do. At this time people entered and beat me." He begged Trotsky's guards to kill him, he said, but they refused. "I want to die."

This was the fate that Sylvia now wished for him. Upon hearing that Jacson had attacked Trotsky, she rushed over to the house as a horrible truth began to sink in. She informed the police that the Canadian Frank Jacson was really a Belgian named Jacques Mornard—though of course this was nowhere close to the bottom of things. She was placed under arrest and transferred to the Green Cross hospital in a state of nervous collapse. Every time the detectives mentioned Trotsky's name, she sobbed uncontrollably. She cursed Mornard as a Stalinist agent and kept screaming at her interrogators, "Kill him! Kill him!"

During the day after the attack, Trotsky remained in a coma. His blood pressure and pulse approached normal, but the doctors offered little hope. A press release composed by Hansen for distribution to the dozens of reporters gathered at the hospital revealed Trotsky's last words, as recorded by Hansen, before he slipped into unconsciousness: "I am close to death from the blow of a political assassin, who struck me down in my room. I struggled with him. He had entered the room to talk about French statistics. He struck me. Please say to our friends that I am sure of the victory of the Fourth International. Go Forward!"

Hansen telephoned Dobbs at noon New York time to say that Trotsky's condition was getting worse. He had lost all his reflexes, including control of his eyelids. An hour later Hansen reported a sharp rise in blood pressure. Dobbs sent a telegram to each branch of the

Socialist Workers Party warning that the outlook was bleak. "We will all do everything possible to help preserve the flickering life of our Old Man."

In Los Angeles, Jan Frankel recognized the hand of the GPU and puzzled over how it was that Trotsky had been left alone with his assailant. In Minneapolis, Hank and Dorothy Schultz searched their memories for missed clues about Jacson and cursed themselves for having been played as his pawns. In Baltimore, Van was out taking a walk on the morning of August 21 when he glanced down at a sidewalk stack of sale copies of *The New York Times* and spotted the headline in the middle of the front page: "Trotsky, Wounded by 'Friend' in Home, Is Believed Dying." He headed home to listen for reports on the radio, racked with guilt for having left Trotsky's side before the real danger set in.

TROTSKY'S BREATHING HAD become more rapid now, early in the evening of August 21, alarmingly so. Natalia, losing her composure, asked the doctors what it meant. For the next twenty minutes they worked to save the patient, but at 7:25 Trotsky's last struggle ended.

Colonel Salazar stepped out the front door of Green Cross Hospital and uttered the words that became the next day's headline around the world: "Gentlemen! Trotsky is dead!" This set off a mad scramble for the telephones by the hundred or so Mexican and foreign correspondents on the scene.

When it was over, Natalia knelt down and pressed her face against the soles of her husband's feet. Until the very end she had waited for him to awaken and decide matters for himself. She would see this happen, though only several months later, in a dream. She had moved out of their bedroom, and into the adjacent room that once belonged to Seva, at the base of the T. Trotsky came out of his study, passed through their bedroom, and entered her room. He appeared vibrant and was immaculately dressed. His white hair was thick and full. His eyes were a piercing blue. He walked over to her, stood there a moment, then said calmly, "Everything is finished."

Shipwreck

Trotsky's ashes are buried in the patio of his home in Coyoacán, beneath a monolith engraved with a large hammer and sickle. The Mexican government bought the house from Natalia in November 1940 and arranged for her to reside there as its caretaker. Over the next twenty years, she maintained the house just as it was when Trotsky lived there, leaving untouched even the remaining bullet holes from the May 1940 assault.

Natalia also kept a close eye on Trotsky's political legacy. In the United States, the Red Decade of the 1930s ended with a veritable stampede from Marxism. This exodus was abetted by an outpouring of patriotism during the Second World War, despite the wartime alliance between the United States and "Uncle Joe" Stalin's USSR. Trotsky's Fourth International persisted nonetheless, led from New York by James Cannon and the Socialist Workers Party. Into the postwar years, the Fourth International continued to adhere to Trotsky's old position that Stalin's USSR was a degenerated workers' state, a designation it also applied to the Soviet bloc countries of Eastern Europe, which had been occupied by the Red Army at the end of the war.

Natalia was increasingly skeptical of this assessment. She believed that, were Trotsky alive, he would not consider the postwar USSR to be any kind of workers' state. Privately she told the leaders of the Fourth International that Stalinism had by now completely destroyed the Revolution and that the so-called people's democracies of Eastern Europe were nothing more than Soviet vassal states.

The last straw for Natalia was the Korean War, which broke out in June 1950 and which Trotsky's disciples portrayed as the mortal struggle of the East's colonial peoples against American imperialism. Natalia resigned from the Fourth International, a decision she explained at great length in an open letter to its Executive Committee. In it she recalled the greeting sent to her by a recent congress of the Socialist Workers Party, which had assured her that the party continued to be guided by Trotsky's ideas. "I must tell you that I read these lines with much bitterness," Natalia said. "As you will see from what I am writing you, I do not see his ideas in your politics."

Stalin's death in March 1953 was a major turning point in Soviet history. Beria was executed later that year, as the Terror came to an end and the camps began to release their prisoners. Emerging as Stalin's successor, Nikita Khrushchev delivered a momentous speech to the 20th Party Congress in February 1956 denouncing Stalin's crimes and his cult of personality. Khrushchev limited his criticism to the wickedness of Stalin and his henchmen, carefully avoiding a wholesale indictment of the Soviet system.

Khrushchev's de-Stalinization led to the political rehabilitation of selected victims of Stalin's Terror—though not the defendants in the major purge trials, and certainly not Trotsky, whose name and image had been thoroughly erased in Stalin's time from books, museums, and films. Trotsky remained useful as a Soviet bogeyman, however. When Mao Zedong accused Khrushchev of "revisionism" and challenged his leadership of the world communist movement, the Kremlin condemned the Chinese Communists for their "neo-Trotskyist deviation."

After Khrushchev's speech, Natalia addressed a letter to the Soviet government, in the person of Kliment Voroshilov, her husband's old antagonist going back to the days of the Russian civil war. As chairman of the Presidium of the Supreme Soviet, Voroshilov was now formally head of state. Natalia wrote to ask for information about her son Seryozha. The last news she heard about him was of his arrest in January 1937 as a mass poisoner. Natalia never received a reply to her letter.

VOROSHILOV STEPPED DOWN as head of the Supreme Soviet on May 7, 1960. One day earlier, Ramón Mercader walked out of a Mexico City prison after serving a twenty-year sentence for Trotsky's murder. Mercader's stay in Mexico might have been much shorter. In 1944, Soviet intelligence hatched an escape plan for Mercader, who was informed of the plot. The conspiracy was frustrated by the unexpected return to Mexico of Caridad Mercader—this time not as an operative but as a rogue mother, guilt-ridden over the fate of her son. Caridad's careless behavior in Mexico City may have thwarted the operation to liberate Trotsky's assassin.

Caridad seems to have lost her bearings after the success of Operation Duck. In a private Kremlin ceremony on June 17, 1941, she and Leonid Eitingon were awarded the Order of Lenin for their role in Trotsky's assassination, while Pavel Sudoplatov received the Order of the Red Banner. After the war, Iosif Grigulevich—the elusive "Felipe"— was presented with the Order of the Red Star. Ramón's recognition would have to wait until his release and, as Stalin made clear, it would depend on how well he acquitted himself while behind bars.

In prison, Mercader stuck to his story about being a disgruntled Belgian follower of Trotsky by the name of Jacques Mornard. It was only thanks to Mexican detective work that his true identity was discovered in 1950, although he never confessed his ties to Soviet intelligence. When he was set free in May 1960—three months early, for reasons of secrecy—he went first to Cuba, then to Czechoslovakia, then to the Soviet Union. After his release, he rarely saw his mother. His brother Luis claimed that Ramón never forgave her for her recklessness in returning to Mexico City. "Thanks to her, I had to spend an extra sixteen years in prison," he said.

On June 8, 1961, Leonid Brezhnev, who succeeded Voroshilov as chairman of the Presidium of the Supreme Soviet, received Mercader in the Kremlin and, in a secret ceremony, awarded him the title Hero of the Soviet Union, the Order of Lenin, and the Gold Star medal. The award citation praised him for displaying "heroism and bravery" in carrying out a "special task."

NATALIA MOVED TO Paris in 1960 and died there two years later. Her ashes are buried alongside her husband's in the patio of the house in Coyoacán, which is now part of Mexico City. Seva and his wife and children occupied the house until 1990, when it became a public museum financed and administered by the Mexican government, with Seva's participation.

A few blocks away, the Blue House also became a favorite tourist attraction. After Trotsky's death, Diego Rivera and Frida Kahlo became ardent supporters of Stalin's USSR, and both painters contributed their talents to the burgeoning Stalin cult during the postwar years. Rivera's long-term quest to rejoin the Mexican Communist Party finally succeeded in 1954, four years after Frida's own admission to the party and two months after her death.

When Frida died, an unfinished portrait of Stalin stood on her easel in the Blue House, where she lived out her final days. It still stands in the same place in what is now the Museo Frida Kahlo. For many years, a small bust of Stalin adorned the bedroom where Trotsky and Natalia used to spend their nights. Visitors to the Blue House are not told that Trotsky once lived there—although as Frida has achieved cult status, more and more people have heard the story of how she had an affair with old man Trotsky, right under Diego's nose.

At Frida's funeral, Rivera was accompanied in the procession through the city streets by former president Lázaro Cárdenas and by Diego's onetime nemesis David Alfaro Siqueiros. Siqueiros arrived there by a circuitous route. Six weeks after Trotsky's murder, Colonel Salazar finally caught up with the fugitive painter, in Hostotipaquilla, a village in the state of Jalisco. Siqueiros was held for six months before his trial, during which time Mexican and Spanish intellectuals petitioned President Manuel Ávila Camacho to allow this national treasure to go free.

At the trial, Siqueiros spoke passionately in his own defense, justifying his extreme actions by pointing to the counterrevolutionary activity Trotsky had conducted from his Coyoacán redoubt. The purpose of the raid, he claimed, was not to murder Trotsky but rather "to help expose the treason of a political center of espionage and provocation" that violated Mexico's independence and, moreover, undermined the defense of Republican Spain. Siqueiros criticized President Cárdenas and the Communist Party for tolerating Trotsky's perfidy and forcing him to take matters into his own hands.

Siqueiros was acquitted of all the serious charges: homicide in the death of Robert Sheldon Harte, attempted homicide of Trotsky, criminal conspiracy, and the illegal use of firearms. He was still on the hook for the lesser charges of trespassing and breaking and entering, but he was released on bail. President Camacho summoned Siqueiros and told him he could have his liberty only if he left the country, which he did in April 1941, moving to Chile.

Siqueiros was allowed to return to Mexico in November 1943, although he did not steer clear of political controversy nor manage to stay out of jail. He went on to create some of his greatest works of art, perhaps none greater than the massive interior mural he painted in Chapultepec Castle on the history of Mexico: *From the Dictatorship of Porfirio Díaz to the Revolution*, which he completed in 1965. In 1966, the Mexican government awarded him the National Art Prize, and the following year the Soviet government honored him with the Lenin Peace Prize.

In his last years, Siqueiros was revered as a national institution, and when he died in 1974 he was given a hero's burial. Early in his career his political activism had so often pulled him away from his art that as a muralist he was far less prolific than either Rivera or Orozco. In the end, however, he left behind an imposing body of work. Visitors to the Leon Trotsky Museum in Coyoacán can still view the scars left by his most politically inspired endeavor, on the walls of Trotsky's old bedroom.

On the afternoon of June 27, 1941, a team of FBI agents and U.S. marshals raided the branch headquarters of the Socialist Workers Party in Minneapolis and St. Paul. They seized a large amount of radical literature, two red flags, and an autographed photo of Trotsky.

On the basis of this and other evidence, the federal authorities charged twenty-nine Trotskyist militants with conspiracy against the U.S. government. Four of the accused were national leaders of the Socialist Workers Party: James Cannon, Farrell Dobbs, Al Goldman, and

Felix Morrow. Fourteen others were connected to the Teamsters, including Ray Rainbolt, head of the union's Defense Guard, and Dorothy Schultz, Hank's wife and Twin Cities secretary of the Workers Defense League, as well as former Coyoacán guards Jake Cooper and Emil Hansen, the better half of the disgruntled duo of Bill and Emil.

The indictment contained two counts, the first of which charged the defendants with conspiracy to overthrow the government of the United States through armed revolution. The prosecution sought to demonstrate that the accused took their inspiration from the Russian Revolution and their ideas from Lenin and Trotsky: "and accordingly, certain of the defendants would, and they did, go . . . to Mexico City, Mexico, there to advise with and to receive the advice, counsel, guidance, and directions of the said Leon Trotsky." The second count invoked the Smith Act of 1940, which made it a criminal offense to advocate the overthrow of the U.S. government. Among the evidence presented in court was Trotsky's photograph and his writings, along with the published works of Marx, Engels, and Lenin.

All the defendants were cleared of the conspiracy charge, but eighteen of them were found guilty of violating the Smith Act, including Cannon, Dobbs, Goldman, Cooper, and Hansen. Their prison sentences ranged from a year and a day to sixteen months, which they began serving on January 1, 1944.

The Communist Party aided the Justice Department in prosecuting these Trotskyists by providing incriminating documents, some of which had been collected by Sylvia Caldwell, Cannon's secretary, who was an informant for the NKVD. After she was exposed by a Communist defector in 1954, she was brought before a grand jury but refused to testify, invoking her Fifth Amendment right against self-incrimination. She finally confessed to her NKVD past before a second grand jury in 1958.

That was the year that the law also caught up with Mark Zborowski, aka Étienne and "Tulip," who was Lyova's shadow and his successor as Trotsky's right-hand man in Paris. Zborowski came to the United States in December 1941. In New York, where he moved in Russian émigré and Trotskyist circles, he reported on the activities of Al Goldman and Jean van Heijenoort, among others. He was also put on the trail of

Victor Kravchenko, a trade official whose defection to the United States in 1944 greatly embarrassed the Moscow government.

In New York, Zborowski continued his research in cultural anthropology, at first in association with the American Jewish Committee, then as a consultant to a Columbia University research project on contemporary cultures, sponsored by the U.S. Navy. This work led to his co-authorship in 1952 of *Life Is with People*, a groundbreaking study of the Jewish shtetls of Eastern Europe before World War II, with an introduction by Margaret Mead. In 1954 he began research at the Veterans' Administration Hospital in the Bronx on the rehabilitation of disabled people. It was at that time that he came under scrutiny from the FBI.

The noose tightened around Zborowski when Alexander Orlov, the Soviet defector who had gone underground in 1938, surfaced in New York. Orlov and his wife had been living on the NKVD funds he absconded with at the time of his defection and flight to the United States. As these funds ran low, he decided he would support himself by publishing his memoirs—although a carefully crafted version that would honor his pledge of secrecy to Moscow and at the same time not make him look like a man with blood on his hands. By a stroke of good luck, he completed his book manuscript on the eve of Stalin's death, in March 1953, which generated enormous interest in what was published later that year as *The Secret History of Stalin's Crimes*.

Orlov made his first big splash a month after Stalin's death with a series of articles in *Life* magazine. According to Orlov, during his time as NKVD station chief in Spain, he functioned more or less as a political attaché, and played no role in the murder of Andrés Nin and others on the non-Communist left during the Spanish civil war. Orlov had known for years that he would have some explaining to do about his actions in Spain. Soviet defector Walter Krivitsky, in a 1939 memoir, had identified Orlov as the NKVD's chief terrorist there. The FBI assumed that Orlov was keeping secrets, although the agency's interrogators failed to break him. Orlov dismissed Krivitsky's claims about him as a "Trotskyist invention." Krivitsky was not there to back up his assertions: in February 1941 he was found dead in a Washington, D.C., hotel room, a victim of suicide.

During a casual conversation with a Russian émigré, Orlov discovered that Zborowski was living in New York. He seems to have been surprised that his warning letter to Trotsky in Mexico had failed to expose the agent provocateur he identified as "Mark." Orlov took his story about Zborowski to the FBI, and then publicly unmasked him in his testimony before the U.S. Senate's Internal Security Subcommittee in September 1955.

Zborowski was then summoned to appear before the same committee in February 1956. The senators questioned the witness about his spying activities in France, from his alleged role in the theft of Trotsky's archives in November 1936, to the circumstances surrounding the death of Lyova in a Paris clinic in February 1938. Zborowski skillfully ducked and weaved, often pleading a faulty memory. In any case, he was not legally culpable for espionage work carried out in France, and he told the Senate panel the same thing he had told the FBI: that he had emphatically refused to work for Soviet intelligence in the United States, despite being pressured to do so.

On February 20, 1957, Zborowski appeared in court to testify at the trial of an American accused of spying for the Soviets. Under oath, Zborowski told the grand jury that he did not know the defendant, although in fact the man in question was his former Soviet handler in New York—and the FBI soon had the evidence to prove it. In April 1958, Zborowski was arrested for perjury. At the time, he was a researcher at the Harvard School of Public Health, working on the same campus that housed Trotsky's archives.

Zborowski was convicted and sentenced to five years in prison. The conviction was then overturned on a technicality, but he was retried in 1962 and again convicted, and he served a sentence of three years and eleven months.

VAN HAD AGREED to appear at Zborowski's trial in 1958, but it turned out that his testimony was not needed. By then Van had been out of the Trotskyist movement for over a decade. The Socialist Workers Party expelled him in 1947 for asserting that Marx's predictions about the revolutionary capacity of the working class had been mistaken. In an article devoted to the 100th anniversary of the *Communist Manifesto* which appeared in the March 1948 issue of *Partisan Review*, Van called for a rethinking of the fundamental assumptions of Marxism. *Partisan*

Review, the literary haven of *Trotskysant* intellectuals in the late 1930s, was becoming a voice of Cold War liberalism. Among those leading the anti-Communist charge were Trotsky's old ideological antagonists Max Eastman and Sidney Hook. Max Shachtman joined this chorus somewhat later, from his position on the social democratic left.

James Burnham, once Trotsky's most formidable adversary and bête noire inside the Socialist Workers Party, turned his thesis about a brave new world of "bureaucratic collectivism" into a hugely successful book, *The Managerial Revolution*, published in 1941. The book stirred enormous controversy, not least because of the author's apparent coldbloodedness about the dawning age of authoritarianism pioneered by the Soviet Union, Nazi Germany, and New Deal America. George Orwell accused Burnham of looking forward to a German victory in the war and to a totalitarian future, a charge Burnham denied. Orwell later drew upon Burnham's dark vision to create his dystopian masterpiece, *1984*.

After the war, the fiercely anti-communist Burnham fell out with Cold War liberals over their objections to the red-baiting crusade of Senator Joseph McCarthy, and he moved toward the political right. In 1955, the year after McCarthy's downfall, Burnham helped William F. Buckley Jr. launch the conservative weekly *National Review*, contributing as a columnist and a senior editor. Burnham and his fellow editors were early and unwavering champions of Ronald Reagan, whom they helped get elected to the White House in 1980. In 1982, in a speech he gave to the British House of Commons, President Reagan famously declared, "The march of freedom and democracy will leave Marxism and Leninism on the ash heap of history." Few were aware at the time that Reagan's words were an ironic echo of Trotsky's banishment of the Mensheviks to the dustbin of history back in 1917. In 1983, the year Reagan characterized the Soviet Union as an "evil empire," he awarded Burnham the Presidential Medal of Freedom.

Even then, Trotskyist sects endured throughout the non-Communist world. In the United States alone, there were numerous splinter groups, factions, and tendencies—remnants of the party that Cannon and Shachtman built. Joe Hansen, Trotsky's favorite American secretary-guard-driver, remained in the thick of these obscure Trotskyist politics until his death in 1979.

The year 1979 was the 100th anniversary of Trotsky's birth, an occasion marked by a ceremony held at Columbia University. In attendance that day was a middle-aged Russian Jewish émigré newly arrived from Moscow by the name of Yulia Akselrod. Yulia was the daughter of Seryozha, Trotsky and Natalia's son—she was a granddaughter they never knew they had.

Seryozha was shot in October 1937, although all that his family in the Soviet Union knew was that he was last seen in prison in Krasnoyarsk, Siberia. He had been exiled there after his 1935 arrest in Moscow. Yulia's mother was allowed to join her husband in Krasnoyarsk, but when she was six months pregnant with Yulia, Seryozha was arrested again and disappeared. Yulia's mother returned to Moscow, where she was arrested two years later and sent to Kolyma, in northeastern Siberia, the site of the most infamous of the gulag camps. Yulia remained in Moscow with her grandparents, until they were arrested in 1951 and all three of them were deported to Siberia.

Yulia eventually moved back to Moscow, where she kept her family background a carefully guarded secret. As a Jew, she was allowed to emigrate to the United States in 1979. A few months after her arrival in New York, she saw a poster advertising the Columbia University event in honor of her grandfather, and out of curiosity she decided to attend. It was an unforgettable experience. The Iran hostage crisis had just begun, yet the hall was decorated with a huge black-and-green banner bearing the slogan "Long live the revolution in Iran!" Yulia's command of English was not good enough to enable her to understand the particular substance of the speeches, but she grasped their tenor.

When the meeting ended, she could not resist disclosing her identity to the organizers of the event. She had the feeling that they did not quite believe her story, but nonetheless they introduced her to another special member of the audience, a man who had lived with her grandfather in Mexico and had subdued his assassin. This was Harold Robins, a hardened veteran of the Trotskyist factional struggles in the years since Trotsky's murder. Yulia and Robins became good friends, even though, as she later said, "I sometimes had to suppress the urge to kill the dear fellow. He was a true believer—a man who had never lost faith in Trotsky's ideas and his dream of a world revolution—and we never stopped arguing."

Robins kept urging her to read the Marxist classics, and she could not make him understand that for years she and her college classmates in Moscow had been force-fed their Marxism, and that it had left a distinctly bad taste in their mouths. "Telling this to Harold was like talking to the wall. I had nothing but my experience to go on, after all, whereas he had a vision."

Robins died a true believer in 1986, the same year that Van met a violent end. After abandoning Trotskyist politics, he received a doctorate in mathematics from New York University in 1949 then pursued a distinguished academic career in the fields of mathematics and formal logic. He also continued to serve informally as Trotsky's archivist, first at Harvard, where he helped catalog Trotsky's papers, and later at Stanford, after a section of Trotsky's Paris archive long presumed lost or stolen by the NKVD came to light at the Hoover Archives in the early 1980s. This included a large cache of letters between Trotsky and Lyova.

On one of his trips to Mexico to negotiate the purchase of Natalia's papers for Harvard, Van started up a romantic liaison with the daughter of Adolfo Zamora, one of Trotsky's Mexican friends and sometime legal adviser. The romance led to marriage, and Van found himself in a tempestuous relationship with an increasingly unstable woman. The marriage kept Van connected to Mexico City, which is where he was, asleep in his study, when his wife fired three bullets from a Colt .38 into his head before turning the gun on herself.

Van died just as General Secretary Mikhail Gorbachev's efforts to reform the Soviet system were gathering momentum and making household words of the terms *perestroika* and *glasnost*. Gorbachev hoped to salvage the original Bolshevik project, and he understood that this would require filling in the many "blank spots" of Soviet history.

Yet from Gorbachev's point of view, there was no room for Trotsky in the pantheon of honorable Bolshevik victims of Stalin. In a speech he delivered in November 1987 to mark the seventieth anniversary of the Bolshevik Revolution, Gorbachev stated that Trotsky "had, after Lenin's death, displayed excessive pretensions to top leadership in the Party, thus fully confirming Lenin's opinion of him as an excessively self-assured politician who always vacillated and cheated." Trotsky's ideas, said Gorbachev, were "essentially an attack on Leninism all down the line."

Bukharin, Zinoviev, Kamenev, and many other fallen Bolsheviks were soon legally rehabilitated, as were many non-Bolsheviks, including Trotsky's son Seryozha—but not Trotsky himself. In the new atmosphere of openness, however, he was no longer taboo, and Soviet journalists and historians began to publish articles about his role as Lenin's essential comrade in 1917 and as the organizer of the Red Army. In January 1989, a Soviet publication told its readers for the first time that the Kremlin had ordered Trotsky's murder. Later, Trotsky's articles began to appear in print, and in post-Soviet Russia his books became available, by which point they were completely harmless.

In 1988, the Soviet government granted Seva a visa so that he could be reunited with his half-sister Alexandra after a separation of sixty years. Their mother, Zina, had been forced to leave Alexandra behind when she went to live with Trotsky in Turkey. In the postwar Stalin years, during a wave of arrests of the children of "enemies of the people," Alexandra was sentenced to ten years of exile in Kazakhstan, a deportation that was cut short by Khrushchev's thaw in 1956. When they met again, Seva and Alexandra shared no common language and had to communicate through an interpreter and by sign language. After his visit, Seva said of the experience, "It was a little like people from a shipwreck who meet safe and sound on the beach."

Gorbachev had unleashed forces he could not control, and by 1990 the *glasnost* indictment of Stalin began to move on to Lenin and the Bolshevik Revolution. The unintended effect was to undermine the legitimacy of Marxist-Leninist ideology and the Bolshevik Revolution, and therefore of the entire Soviet system, preparing the way for the collapse of the USSR in 1991.

One of those observing these historic developments with intense interest from the sidelines was Albert Glotzer. Glotzer, a Chicago native, had joined the American Trotskyist movement at its creation in

1928 and served as Trotsky's bodyguard in Turkey, where he fished and hunted with the Old Man, listened to him dictate his *History of the Russian Revolution*, and gave a beleaguered young Seva an abortive lesson in boxing. At the Dewey hearings in Coyoacán in 1937, Glotzer served as court reporter. He sided with Shachtman and the Minority in the factional split inside the Socialist Workers Party in 1939–40. He drifted away from Trotskyism, but he stayed in contact with Natalia and became friends with Seva, whom he visited over the years in Mexico City.

Looking back on Trotsky and Trotskyism from the perspective of August 1991, as history turned a corner, Glotzer could not overcome a profound sense of waste. "Many things we wrote and said in the Thirties were simply bullshit," he wrote to novelist Saul Bellow. Bellow was a student Trotskyist in Chicago in the 1930s. He and a former classmate had an appointment to meet with Trotsky in August 1940. They were in Taxco when they heard about the attack and rushed to Mexico City and to Green Cross Hospital. Passing themselves off as reporters, the two young men were led into a room where, as Bellow described the scene, "Trotsky was lying dead with a bloody turban of bandages, and his face streaked with iridescent iodine."

Bellow was never as ideologically invested as the party cadres like Glotzer, and his Trotskyism died at the same time Trotsky did. A half-century later, as the Soviet Union was disintegrating, he learned for the first time by reading Glotzer's own memoir of Trotsky how the man he once revered had been an outsider to Bolshevism until 1917, how his actions as a Bolshevik leader turned him into a prisoner of the myth of October as a workers' revolution, and how in his last exile he made his disciples its prisoners as well.

"The Soviet Union will live and develop as the new social basis created by the October Revolution," Trotsky had declared after arriving in Mexico in 1937, when he predicted that the birthplace of socialism "will produce a regime of true democracy and will become the greatest factor for peace and for the social emancipation of humanity." Although doubts crept into his later writings, through the Kremlin's purges and the Nazi-Soviet pact and its bloody aftermath he refused to surrender this utopian vision. The fact is, as Glotzer elucidated for Bellow, Trotsky could not disavow the USSR without also repudiating Red October, which would have meant renouncing his life's work. Instead, as his prospects grew dim and as Stalin's assassins closed in, he kept reaffirming his absolute faith in the dogma of Marxism and pointing toward a glorious Soviet future. "Optimism was all he really had."

ACKNOWLEDGMENTS

—————————————————

The idea for this book came from Donald Lamm, my agent and my friend, who has been an endless source of inspiration for me as a writer. I am enormously grateful to him, as well as to his colleagues Christy Fletcher, Emma Parry, and Melissa Chinchillo.

I have been fortunate to have the collaboration of two outstanding editors. Tim Duggan, at HarperCollins, helped me shape the book with his unerring sense of narrative and his relentless pursuit of clear and readable prose. Neil Belton, at Faber and Faber, enforced rigor and precision by challenging me with his skeptical queries and acute insights at every step along the way. I am grateful to both men for their confidence and their support.

Terence Emmons and Donald Sommerville vetted the manuscript for accuracy and readability. Allison Lorentzen was a kind and efficient facilitator. The staff of the Hoover Institution Library and Archives provided expert and courteous assistance throughout.

I could not have written this book without the generosity and encouragement of family and friends, especially Inga Weiss, Kristin Engel, Jack Morton, Austin Hoyt, John Brande, William Free, Chris Roberge, my parents, Bertrand and Muriel Patenaude, and my wife, Christina Patenaude.

SOURCES AND NOTES

This book draws extensively on two main Trotsky archives, abbreviated in the notes as follows:

TEP — Trotsky Exile Papers, The Houghton Library, Harvard University, Cambridge, Massachusetts.

TC — Trotsky Collection, 1917–1980, Hoover Institution Archives, Stanford University, Stanford, California.

Other archival collections frequently cited are abbreviated as follows:

Buchman papers — Alexander H. Buchman Papers, Hoover Institution Archives, Stanford University, Stanford, California.

Glotzer papers — Albert Glotzer Papers, Hoover Institution Archives, Stanford University, Stanford, California.

Hansen papers — Joseph Hansen Papers, Hoover Institution Archives, Stanford University, Stanford, California.

Solow papers — Herbert Solow Papers, Hoover Institution Archives, Stanford University, Stanford, California.

Volkogonov papers — Dmitri Antonovich Volkogonov Papers, Hoover Institution Archives, Stanford University, Stanford, California.

A number of works are cited throughout the Notes section. They are listed here, in alphabetical order, according to the abbreviations used for them in the notes:

Andrew & Mitrokhin — Christopher Andrew and Vasili Mitrokhin, *The Sword and the Shield: The Mitrokhin Archive and the Secret History of the KGB* (Basic Books, 2001).

Brenner — Anita Brenner, *Idols Behind Altars* (Payson & Clarke, 1929).

Broué — Pierre Broué, *Trotsky* (Fayard, 1988).

Broué, *Léon Sedov* — Pierre Broué, *Léon Sedov, fils de Trotsky, victime de Stalin* (*Les Éditions Ouvrières*, 1993).

Cambridge History	Alan Knight, "Mexico, c. 1930–46," *The Cambridge History of Latin America*, vol. VII, Leslie Bethell, ed. (Cambridge University Press, 1984), 1–82.
Case	*The Case of Leon Trotsky: Report of Hearings on the Charges Made Against Him in the Moscow Trials* (Merit Publishers, 1968).
Craig I	Gordon A. Craig, *Europe, 1815–1914*, 3rd ed. (Harcourt Brace Jovanovich College Publishers, 1971).
Craig II	Gordon A. Craig, *Europe Since 1914*, 3rd ed. (Harcourt Brace Jovanovich College Publishers, 1972).
Deadly Illusions	John Costello and Oleg Tsarev, *Deadly Illusions* (Crown, 1993).
Deutscher I	Isaac Deutscher, *The Prophet Armed: Trotsky, 1879–1921* (Verso, 2003).
Deutscher II	Isaac Deutscher, *The Prophet Unarmed: Trotsky, 1921–1929* (Verso, 2003).
Deutscher III	Isaac Deutscher, *The Prophet Outcast: Trotsky, 1929–1940* (Verso, 2003).
Diary	Leon Trotsky, *Trotsky's Diary in Exile, 1935*, Elena Zarudnaya, trans. (Harvard University Press, 1958).
Dugrand	Alain Dugrand, *Trotsky in Mexico*, Stephen Romer, trans. (Carcanet, 1992).
Eastman	Max Eastman, *Love and Revolution: My Journey Through an Epoch* (Random House, 1964).
Eastman, *Companions*	Max Eastman, "Problems of Friendship with Trotsky," *Great Companions* (Museum Press Limited, 1959).
Eastman, *Heroes*	Max Eastman, "Great in a Time of Storm: The Character and Fate of Leon Trotsky," *Heroes I Have Known: Twelve Who Lived Great Lives* (Simon and Schuster, 1942).
FBI	Leon Trotsky's Federal Bureau of Investigation file, made available through the Freedom of Information Act.
Feferman	Anita Burdman Feferman, *Politics, Logic, and Love: The Life of Jean van Heijenoort* (A K Peters, 1993).
Glotzer	Albert Glotzer, *Trotsky: Memoir & Critique* (Prometheus Books, 1989).
Hansen, "With Trotsky in Coyoacan"	Hansen, "With Trotsky in Coyoacan," introduction to Leon Trotsky, *My Life* (Pathfinder Press, 1970).
Herrera	Hayden Herrera, *Frida: A Biography of Frida Kahlo* (Bloomsbury, 1998).
Howe	Irving Howe, *Leon Trotsky* (Penguin Books, 1979).
In Defense of Marxism	Leon Trotsky, *In Defense of Marxism (against the petty-bourgeois opposition)* (Pathfinder Press, 1970).
Kelly	Daniel Kelly, *James Burnham and the Struggle for the World: A Life* (ISI Books, 2002).
Kern	Gary Kern, *A Death in Washington: Walter G. Krivitsky and the Stalin Terror* (Enigma Books, 2003).
Knei-Paz	Baruch Knei-Paz, *The Social and Political Thought of Leon Trotsky* (Clarendon Press, 1978).
Kolpakidi	Aleksandr Kolpakidi and Dmitrii Prokhorov, *KGB: Spetsoperatsii sovetskoi razvedki* (Olimp, Astrel, 2000).

Legacy	*The Legacy of Alexander Orlov: Prepared by the Subcommittee to Investigate the Administration of the Internal Security Act and Other Internal Security Laws of the Committee on the Judiciary of the United States Senate, Ninety-Third Congress, First Session* (U.S. Government Printing Office, 1973).
Leon Trotsky	Joseph Hansen et al., *Leon Trotsky: The Man and His Work* (Merit Publishers, 1969).
Levine	Isaac Don Levine, *The Mind of an Assassin* (Farrar, Straus and Cudahy, 1959).
Montefiore	Simon Sebag Montefiore, *Stalin: The Court of the Red Tsar* (Alfred A. Knopf, 2004).
Mosley	Nicholas Mosley, *The Assassination of Trotsky* (Michael Joseph, 1972).
My Life	Leon Trotsky, *My Life: An Attempt at an Autobiography* (Dover Publications, 2007).
Natalia	Victor Serge and Natalia Sedova Trotsky, *The Life and Death of Leon Trotsky*, Arnold J. Pomerans, trans. (Basic Books, 1975).
Nikandrov	Nil Nikandrov, *Grigulevich: Razvedchik, "kotoromu vezlo"* (Molodaya Gvardia, 2005).
Ocherki	*Ocherki istorii rossiiskoi vneshnei razvedki*, vol. 3 (Mezhdunarodnye otnosheniia, 1997).
Polizzotti	Mark Polizzotti, *Revolution of the Mind: The Life of André Breton* (Farrar, Straus and Giroux, 1995).
Poretsky	Elizabeth K. Poretsky, *Our Own People: A Memoir of "Ignace Reiss" and His Friends* (The University of Michigan Press, 1970).
Rochfort	Desmond Rochfort, *Mexican Muralists: Orozco, Rivera, Siqueiros* (Chronicle Books, 1993).
Salazar	General Leandro A. Sanchez Salazar, with the collaboration of Julian Gorkin, *Murder in Mexico: The Assassination of Leon Trotsky*, Phyllis Hawley, trans. (Secker & Warburg, 1950).
Stein	Philip Stein, *Siqueiros: His Life and Works* (International Publishers, 1994).
Sudoplatov	Pavel Sudoplatov and Anatoli Sudoplatov, with Jerrold L. and Leona P. Schecter, *Special Tasks: The Memoirs of an Unwanted Witness—a Soviet Spymaster* (Little, Brown and Company, 1994).
Trotsky, "The Comintern and the GPU"	Leon Trotsky, "The Comintern and the GPU," *Fourth International*, vol. 1, no. 6, November 1940, 148–63; available at the Marxist Internet Archive, http://www.marxistsfr.org/archive/trotsky/1940/08/gpu.htm.
Tucker, *Stalin as Revolutionary*	Robert C. Tucker, *Stalin as Revolutionary, 1879–1929: A Study in History and Personality* (W.W. Norton & Company, 1974).
Tucker, *Stalin in Power*	Robert C. Tucker, *Stalin in Power: The Revolution from Above, 1928–1941* (W.W. Norton & Company, 1990).
Ulam	Adam Ulam, *The Bolsheviks* (Collier Books, 1965).
Van	Jean van Heijenoort, *With Trotsky in Exile: From Prinkipo to Coyoacán* (Harvard University Press, 1978).

Venona	John Earl Haynes and Harvey Klehr, *Venona: Decoding Soviet Espionage in America* (Yale University Press, 1999).
Venona Secrets	Herbert Romerstein and Eric Breindel, *The Venona Secrets: Exposing Soviet Espionage and America's Traitors* (Regnery Publishing, 2000).
Volkogonov	Dmitri Volkogonov, *Trotsky: The Eternal Revolutionary*, Harold Shukman, trans., ed. (The Free Press, 1996).
Wald	Alan M. Wald, *The New York Intellectuals: The Rise and Decline of the Anti-Stalinist Left from the 1930s to the 1980s* (The University of North Carolina Press, 1987).
Wolfe	Bertram D. Wolfe, *The Fabulous Life of Diego Rivera* (Cooper Square Press, 2000).
Writings	*Writings of Leon Trotsky*, 2nd ed., Naomi Allen and George Breitman, eds., vols. 9–12 (Pathfinder Press, 1973–1978).

Prologue: A Miraculous Escape

2   he needed the money: Van to Frankel, February 27, 1938, TC 23:14; Natalia, "Father and Son," in *Leon Trotsky*, 42–43.

2   Trotsky often said to his wife: Natalia, 252.

2   Trotsky's editors in New York: Alan Collins to Charles Walker, September 22, 1938, TEP 13957.

2   a boon to the Soviet caricaturists: for example, *Moscow News*, February 3–10, 1937.

2   Trotsky confidently predicted: *Writings*, 12:290–91.

3   "Death to Trotsky!": Hansen, "The Attempted Assassination of Leon Trotsky," and Alfred Rosmer, "A Fictionalized Version of the Murder," in *Leon Trotsky*, 5–12, 77–79.

3   a meeting of his guards: Harold Robins, unpublished memoir, TC 30:1.

4   sound of automatic gunfire: my account of the "miraculous escape" is drawn from *Writings*, 12:233–35; Natalia, 256–61; Adam [Hank Schultz] to Farrell Dobbs, May 25, 1940, TC 23:12; Harold Robins's account, TEP 17193; Jake Cooper's account, TEP 10725.

7   "assassination failed" . . . pretended to be dead: *Writings*, 12:235.

7   Mexican detectives . . . not a miracle but a hoax: Salazar, 3–26.

7   Harte was a victim: Trotsky's most elaborate statement is "False Suspicions about Robert Sheldon Harte" (unpublished manuscript in Russian), July 15, 1940, Hansen papers, 69:57.

9   Moscow radio: Deutscher III, 270.

9   interned in a large house: Deutscher III, 278–79; Natalia, 206–9; *Writings*, 9:21–36.

9   "gravedigger of the revolution": Deutscher II, 248.

9   "enemy number one": *Writings*, 12:241.

9   Trotsky was predicting . . . revolutionary shock wave: for example, "Hitler and Stalin: How Long Will It Last?" *Liberty*, January 27, 1940.

10   training ground . . . took refuge in Mexico: *Cambridge History*, 46.

10   gathering danger: Trotsky to John Glenner [Jan Frankel], April 12, 1939, TC 10:56.

Chapter One: Armored Train

13   two sirens: *Writings*, 9:56.

13   aged him five years: Trotsky to Tamada Knudsen, January 20, 1937, TEP 8696.

13   forests and fjords . . . shrouded in secrecy . . . "mysterious Mexico": *Writings*, 9:37–41.

14   apprehension rose . . . Baku, on the Caspian Sea . . . disembark voluntarily . . . Max Shachtman . . . more than two months: *Writings*, 9:75–79; Natalia, 210; George Novack to Felix Morrow, January 13/15, 1937, TC 23:2.

14   two hours straight: George Novack to Felix Morrow, January 13/15, 1937, TC 23:2.

14   "the whole New World": Natalia, 210.

15   General Beltrán: George Novack to Felix Morrow, January 13/15, 1937, TC 23:2.

15   by airplane or by train: George Novack to Felix Morrow, January 13/15, 1937, TC 23:2.

15   culture shock: *Writings*, 9:79.

15   Novack arrived . . . the train . . . was armored: George Novack to Felix Morrow, January 13/15, 1937, TC 23:2.

16   shattering Rubio's jaw: *Time*, February 17, 1930.

16   sun-baked landscape . . . huddled in a compartment: *Writings*, 9:80; Natalia, 210.

16   formed a committee: Dewey et al. to Dear Friend, October 22, 1936, TC 25:4; "American Committee for the Defense of Leon Trotsky: Declaration of Principles," n.d., TC 25:4.

17   Dewey, the famous philosopher: George Novack to Felix Morrow, January 13/15, 1937, TC 23:2.

17   series of ballads . . . Mexican folk songs: Dugrand, 17.

17   Rivera was livid . . . Anita Brenner: Herrera, 204–5.

18   Rivera's great surprise: Wolfe, 238.

18   announced the good news . . . "splendid decision": Suzanne La Follette to Dear Friend, December 11, 1936, TC 25:4.

18   Cárdenas rose to prominence . . . To establish his authority: Don M. Coerver, Suzanne B. Pasztor, and Robert M. Buffington, *Mexico: An Encyclopedia of Contemporary Culture and History* (ABC-CLIO, 2004), 64–68.

19   sympathetic to Marxist ideology . . . "the revolution itself!": George Novack to Felix Morrow, January 7, 1937, Hansen papers, 69:64; Van, 106.

19   anti-Trotsky posters . . . independent liberal class: George Novack to Felix Morrow, January 5, 1937, Hansen papers, 69:64.

19   Cárdenas summoned Rivera . . . not land secretly . . . freedom of movement: Max Shachtman to Felix Morrow, January 5, 1937, TC 23:2.

20   change in the political atmosphere: George Novack to Felix Morrow, January 5, 1937, and January 13/15, 1937, TC 23:2.

20   serve Trotsky as a bridge . . . "one chance in a hundred" . . . attempt on his life . . . Thompson submachine gun: Max Shachtman to Felix Morrow, January 5, 1937, TC 23:2.

21   town of Cárdenas: George Novack to Felix Morrow, January 13/15, 1937, TC 23:2.

21   "china-blue eyes": *Writings*, 9:41.

21   an additional locomotive: *Writings*, 9:80.

21   his glory days: for accounts of Trotsky's armored train, see *My Life*, 411–22, and Volkogonov, 163–73.

21   Russia's time of troubles: Nicholas V. Riasanovsky, *A History of Russia*, 4th ed. (Oxford University Press, 1984), 453–61.

22   Treaty of Brest-Litovsk . . . "breathing spell": Ulam, 382–410.

22   Czechoslovak soldiers: George F. Kennan, *Russia and the West under Lenin and Stalin* (Little, Brown and Company, 1961), 97–99.

22   "the shape of a noose": *My Life*, 396.

23   "flying administrative apparatus": *My Life*, 413.

23   "Cowards, scoundrels, and traitors": Volkogonov, 349.

23   "a real army": *My Life*, 408.

23   political commissars: Deutscher I, 344, 356.

23   internal lines of operation: Deutscher I, 358–59.

24   125,000 miles: Volkogonov, 165.

24   "Pullman wheels": *My Life*, 413.

24   "war of movement": *My Life*, 419.

24   "leather-coated detachment": *My Life*, 420.

24   supplies and gifts: *My Life*, 414–15.

24   questions of strategy: Volkogonov, 143.

25   every tenth deserter: Volkogonov, 137.

25   "gangrenous wound": *My Life*, 401–2.

25   "Masses of men": *My Life*, 411.

25   Lev Davidovich Bronstein: Deutscher I, 1–47.

25   virulent form of anti-Semitism: Riasanovsky, *A History of Russia*, 394–95; Richard Pipes, *The Russian Revolution* (Vintage Books, 1991), 70–71.

26   Many Bolsheviks had assumed: Deutscher I, 337, 345.

26   awards for bravery: Deutscher I, 349.

26   running feud with Stalin: Volkogonov, 132, 140–43; *My Life*, 440–44; Deutscher I, 352–53.

27   Stalin's intrigues: Deutscher I, 361–65; Volkogonov, 193.

27   heroic defense of Petrograd: *My Life*, 423–35.

27   Order of the Red Banner: Volkogonov, 169.

28   "despised fascist hireling": Volkogonov, 128.

28   "northerner's fear of the tropics": *Writings*, 9:80.

28   Cárdenas himself typically arrived . . . "fat and smoldering": *Time*, January 25, 1937.

29   separated from Natalia: Natalia, 211.

29   "mad dash": George Novack to Felix Morrow, January 13/15, 1937, TC 23:2.

30   occupy the Blue House temporarily: Max Shachtman to Felix Morrow, January 5, 1937, TC 23:2; George Novack to Felix Morrow, January 13/15, 1937, TC 23:2.

30   patio filled with plants and flowers: Natalia, 211.

30   orange tree: *Writings*, 9:80.

30   "wild confusion" . . . retreat into private life: George Novack to Felix Morrow, January 13/15, 1937, TC 23:2.

30   *La Venida de Trotsky*: George Novack to Felix Morrow, January 5, 1937, Hansen papers, 69:64.

30   "Out with Trotsky": Novack to Morrow, January 13/15, 1937, TC 23:2.

30   "Down with Trotsky": *Time*, January 25, 1937.

30   "polemic with flunkeys": *Writings*, 9:82.

31   bloodthirsty chorus: Natalia, 210.

31   secretarial staff: Max Shachtman to Felix Morrow, January 5, 1937, TC 23:2.

31   balance of power: *Writings*, 9:80.

31   "ideal country for an assassination": *Time*, January 25, 1937.

Chapter Two: Mastermind

32   "mad dogs be shot" . . . screaming headline: Tucker, *Stalin in Power*, 370.

32   "Anti-Soviet Trotskyite Center": Robert Conquest, *The Great Terror: A Reassessment* (Oxford University Press, 1990), 147–49.

32   mastermind: Leonard Schapiro, *The Communist Party of the Soviet Union*, 2nd ed. (Vintage, 1971), 415–17.

32   Yuri Pyatakov . . . breakneck speed: Schapiro, 415; Montefiore, 211.

33   "semi-Trotskyites": *Moscow News*, February 3–10, 1937.

33   strenuous time: Van, 104–5.

33   Nikita Khrushchev: Montefiore, 210–11; Conquest, 167.

33   confessions . . . endless fascination . . . hardened Old Bolsheviks: for example, "The Trial of the Trotskyites in Russia," *The New Republic*, September 2, 1936; "The Moscow Trials," *The Nation*, October 10, 1936.

34   Kingsley Martin: *New Statesman*, April 10, 1937.

34   "One can only be right with the Party": Deutscher II, 114–15.

35   "The Pit and the Pendulum": *Writings*, 9:94.

35   Hippodrome: George Novack to Trotsky, February 4, 1937, TEP 3651; Elinor Rice to Trotsky, February 10, 1937, TEP 4250.

35   "one of the most dramatic events": Harold Isaacs to Trotsky, February 2, 1937, TEP 2041.

35   telephone exchange: Van, 106–7.

36   atmosphere inside the Hippodrome: *The New York Herald Tribune*, February 10, 1937; *The New York Daily News*, February 10, 1937; *The New York Times*, February 10, 1937. Trotsky's Hippodrome speech published as *I Stake My Life* (Pioneer Publishers [1937]).

37   series of resignations: Minutes of meeting of Trotsky defense committee, March 1, 1937, TC 25:5; Novack to "Committee Member," March 16, 1937, TC 25:5.

37   signed a petition: Harvey Klehr, *The Heyday of American Communism: The Depression Decade* (Basic Books, 1984), 360; George Novack, "Radical Intellectuals in the 1930s," *International Socialist Review*, Vol. 29, No. 2, March–April 1968, 21–34; Deutscher III, 299.

37   morally responsible . . . positive achievements: Mauritz A. Hallgren to Hortense Alden, February 11, 1937, TC 25:5; Mauritz A. Hallgren, *Why I Resigned From the Trotsky Defense Committee* (International Publishers [1937]); James T. Farrell to Trotsky, February 8, 1937, TEP 936.

38   John Dewey viewed the matter: TC 25:5.

38   Dreyfus affair: Craig I, 331–35.

38   Dewey was by reputation: David C. Engerman, *Modernization from the Other Shore: American Intellectuals and the Romance of Russian Development* (Harvard University Press, 2003), 174–84.

38   Dewey's reluctance: George Novack to Trotsky, March 22, 1937, TC 13:62.

39   Sidney Hook: Wald, 130, 132.

39   Trotsky himself was enlisted: Trotsky to Suzanne La Follette, March 15, 1937, TEP 8741.

39   Dewey relented: James Cannon to Bernard Wolfe, March 19, 1937, TEP 480.

39   "a great holiday in my life": *Case*, 584.

40   overdrive: Van, 108–9.

40   calls for Trotsky's expulsion: George Novack to American Committee, April 28, 1937, TC 25:6.

40   hearings in a public hall: Press release for May 10, 1937, TC 25:7.

40   magenta blossoms: Dewey to Robbie Lowitz, undated [April 11, 1937], Glotzer papers, box 11.

40   six-foot barricades . . . atmosphere inside the Blue House: James T. Farrell, "Dewey in Mexico," in *John Dewey: Philosopher of Science and Freedom* (The Dial Press, 1950), 361; Glotzer, 259; Van, 108.

41   turned away: Glotzer, 259.

41   Klieg lights: Herbert Solow to Margaret de Silver, April 10, 1937, Solow papers, box 1.

41   pitch of his voice: Albert Glotzer to Alan Wald, March 16, 1977, Glotzer papers, box 35.

42   "expulsed": Farrell, "Dewey in Mexico," 361.

42   the fate of his children: Solow to Margaret de Silver, April 10, 1937, Solow papers, box 1; *The New York Times*, April 18, 1937; *Case*, 41–42.

43   Hotel Bristol: *Case*, 167–73.

43   Berlin to Oslo, *Case*, 204–26.

43   invited representatives: *Case*, 64–65.

44   "Truth, justice, humanity": Dewey to Robbie Lowitz, undated [April 15, 1937], Glotzer papers, box 11.

44   "dictatorship *for* the proletariat": *Case*, 357.

44   Dewey remained skeptical: *Case*, 437.

45   Trotsky's prophetic formulation: Deutscher I, 74, 79.

45   "permanent revolution": Howe, 25–33; Knei-Paz, 108–74.

46   New Economic Policy: Stephen F. Cohen, *Bukharin and the Bolshevik Revolution: A Political Biography, 1888–1938* (Vintage Books, 1975), 123–59.

46   "socialism in one country": Tucker, *Stalin as Revolutionary*, 368–94.

46   "super-industrializer": *Case*, 245.

46   slaughtered by the millions: Robert Conquest, *The Harvest of Sorrow: Soviet Collectivization and the Terror-Famine* (Oxford University Press, 1986).

47   "brute force" . . . "successes" . . . unnecessary brutality: *Case*, 248–51.

47   "degenerated": *Case*, 282.

47   "under the Iron Heel": Trotsky to Margaret de Silver, October 25, 1937, TEP 7672.

47   Carleton Beals . . . Frank Kluckhohn: Glotzer 266–69; *Case*, 411–18; *The New York Times*, April 18 and 19, 1937; *News Bulletin* of the American Committee

for the Defense of Leon Trotsky, May 3, 1937, Glotzer papers, box 4; Dewey to Robbie Lowitz, April 20, 1937, Glotzer papers, box 11; Felix Morrow to Edwin L. James, April 5, 1937, TC 25:6; Wolfe, Van, and Frankel to Felix Morrow, April 22, 1937, Glotzer papers, box 4; Bernard Wolfe to Edwin James, May 21, 1937, Glotzer papers, box 4.

48   Dewey called it "a book": Dewey to Robbie Lowitz, undated [April 15, 1937], Glotzer papers, box 11; Trotsky's closing statement is in *Case*, 459–585.

48   "And when he finished": Glotzer to Alan Wald, March 16, 1977, Glotzer papers, box 35.

48   Dewey avoided stepping: *Case*, 585.

49   "lion in a circus": Dewey to Robbie Lowitz, undated [April 15, 1937], Glotzer papers, box 11.

49   Dewey said to Trotsky: Glotzer, 271; Van, 110.

49   "You were right about one thing": Dewey to Max Eastman, May 12, 1937, Glotzer papers, box 11.

50   American press coverage: Pearl Kluger to Bernard Wolfe, May 12, 1937, TEP 6778.

50   Dewey came out fighting: text of Dewey's speech is in TC 25:7.

50   best speech of his career ... Hook told Dewey: Pearl Kluger to Bernard Wolfe, May 12, 1937, TEP 6778; Harold Isaacs to Cdes, May 10, 1937, TEP 6481.

51   thunderclap out of Moscow: Tucker, *Stalin in Power*, 435–38; Volkogonov, 316–29; Schapiro, *The Communist Party*, 423–24.

51   Kronstadt rebellion: Paul Avrich, *Kronstadt 1921* (W. W. Norton & Company, 1974); Ulam, 472–73; Schapiro, *The Communist Party*, 205–8.

52   memory of Kronstadt: Wendelin Thomas to Trotsky, June 24, 1937, TEP 5504; Knei-Paz, 556–57; Volkogonov, 393–94.

52   "One would think": Deutscher III, 353–54; Trotsky to Wendelin Thomas, July 6, 1937, TEP 10569.

52   "shot like partridges": Avrich, 146.

52   special source of concern: Trotsky to Goldman, September 5, 1937, TEP 8289; Trotsky to Dear Friend [Jan Frankel], January 26, 1938, TEP 8158.

52   "That you seek vindication": Thomas to Trotsky, December 7, 1937, TEP 5506.

53   announced its verdict: press release of December 12, 1937, TC 25:8.

53   "our first great victory": Hansen to Harold Isaacs, December 16, 1937, TEP 11535.

53   "tremendous": Trotsky to Albert Goldman, December 21, 1937, TEP 8291.

53   "great moral shock": Trotsky to Suzanne La Follette, December 22, 1937, TEP 8765.

53   Dewey made a radio broadcast: *The New York Times*, December 14, 1937.

53   Dewey expanded: *The Washington Post*, December 19, 1937, quoted in Glotzer, 137–38.

54   Trotsky ... was indignant: Trotsky to Walker, January 12, 1938, TEP 10766; Van, 110.

54   one long essay: "Their Morals and Ours," *New International*, June 1938; Knei-Paz, 556–67; Howe, 165–173.

54   "Idealists and pacifists": Knei-Paz, 557.

54   "the end is justified": Knei-Paz, 559.

54   "Means and Ends": published in *New International*, August 1938.
54   "He was tragic": Farrell, "Dewey in Mexico," 374; Louis Menand, "The Real John Dewey," *The New York Review of Books*, Vol. 39, No. 12 (June 25, 1992).

Chapter Three: Man of October

55   "The old man relaxed": Bernard Wolfe to James Cannon, May 26, 1937, TC 23:2.
56   "Trotsky displayed all his amiability" ... "like an object" ... "most brusque": Van, 26.
57   "slammed the door": Van, 109–10.
57   "all my predictions" ... "real prisoner" ... "it would be a catastrophe": Jan Frankel to Charles Walker, June 8, 1937, Glotzer papers, box 2.
58   experienced philanderer: Van, 114.
58   "richest vocabulary of obscenities": Wolfe, 240–41.
59   considerable hardship: Herrera, chs. 4, 5; Wolfe, 242–43.
59   best-known work of art: Herrera, 109–11; Wolfe, 395.
59   *Henry Ford Hospital*: Herrera, 143–45.
59   the fantastic and the grotesque: Wolfe, 394.
59   *Fulang-Chang and I*: Herrera, 209–10.
60   "Frida did not hesitate": Van, 110–12.
60   Diego's own brazen philandering: Wolfe, 357–58; Herrera, 181, 199, 209.
61   "Make love, take a bath": Herrera, 199.
61   "little goatee": Herrera, 209.
61   Cristina's house: Herrera, 210.
61   "Natalia was suffering": Van, 112.
61   "one of the saddest faces": Farrell, "A Memoir on Leon Trotsky," *University of Kansas City Review* 23 (1957), 293–98.
61   athletic Seryozha: *Diary*, 58–60, 69–70; Deutscher II, 311–12.
62   Kirov's murder: Robert Conquest, *Stalin and the Kirov Murder* (Oxford University Press, 1989).
62   "proving extremely difficult": Seryozha to Natalia, December 9, 1934, TEP 13521; *Diary*, 108.
62   brutally interrogated ... blamed themselves ... "they will torture him": *Diary*, 60, 62–63, 69–70, 108, 129–30, 134–35.
62   Natalia issued an open letter: Deutscher III, 233.
62   "N. is haunted": *Diary*, 70.
63   mass poisoning of workers: Deutscher III, 294.
63   "To the Conscience of the World": February 4, 1937, TEP 17310.
63   the "poisoner": *Case*, 40.
63   "drive Sergei to insanity": Deutscher III, 294.
63   "ventured to speak to Trotsky" ... "morbidly jealous": Van, 112.
64   "Diego came by with a gun": Herrera, 200–1.
64   habit of threatening people: Feferman, 144–45.
64   temporary separation: Van, 112.
64   Frida paid him a visit ... "The stakes were too high": Van, 112.
65   "very tired of the old man": Herrera, 212.

65   Trotsky and Natalia's correspondence, July 11–22, 1937: TEP 5573–75, 10613–27; published in French translation as *Correspondance 1933–1938* (Gallimard, 1980), translated, annotated, and with an introduction by Jean van Heijenoort.

66   They had met in Paris . . . "exceeded all expectations" . . . "Resembles Odessa": Joel Carmichael, *Trotsky: An Appreciation of His Life* (Hodder and Stoughton, 1975), 70–73; *My Life*, 142–49; Natalia, 11–12.

67   "face to face with real art": *My Life*, 147.

67   "He never forgave me": Van, 113.

69   "Old age is the most unexpected": *Diary*, 106.

70   "A distance had been established" . . . "the hands of the G.P.U.": Van, 114.

71   blow hot and cold: Herrera, 468–69; Hansen, "With Trotsky in Coyoacan," xx–xxi.

71   self-portrait: Herrera, 213–14.

71   more prolific . . . best thing that had ever happened: Herrera, 215.

72   Trotsky's plan of escape: Van, 118–19.

Chapter Four: Day of the Dead

73   the scene of a fiesta: description of the November 7 fiesta draws on Hansen, "With Trotsky in Coyoacan," xiv–xv; Hansen to James Cannon, November 7, 1937, TC 23:3; Hansen to Pearl Kluger, November 9, 1937, TEP 11828.

75   "whirl of mass meetings": *My Life*, 294.

75   "speaking simultaneously": Nikolai Sukhanov, quoted in Volkogonov, 84.

75   "hidden reserve of nervous energy": *My Life*, 294–95.

75   "bare, gloomy amphitheatre": John Reed, *Ten Days That Shook the World* (Boni & Liveright, 1919), 21.

75   a human tinderbox . . . "nipples of the revolution" . . . "like a sleepwalker": *My Life*, 295–96.

76   revolutionary oath: Volkogonov, 88.

77   "willingly die fighting": R. H. Bruce Lockhart, *Memoirs of a British Agent* (Putnam, 1932), 26–27.

77   "float on countless arms": *My Life*, 296.

77   Trotsky managed to find his voice: Hansen, "With Trotsky in Coyoacan," xv.

77   "great and swelling stream": Hansen to Pearl Kluger, November 9, 1937, TEP 11828.

77   Seryozha . . . was executed: Volkogonov, 354–55.

78   October's "greatest interpreter" . . . "breadth and profundity": Trotsky letter to *Partisan Review*, June 17, 1938, published as "Art and Politics in Our Epoch," *Partisan Review*, August–September 1938.

78   tumultuous revolutionary decade: Rochfort, 11–21.

78   Rivera . . . settled in Paris: Wolfe, 64–75.

79   Mexican Renaissance . . . Three major figures: Rochfort, 24–33; Wolfe, 118, 141–49, 154–57, 159–61.

79   Ministry of Education building: Rochfort, 51–67; Wolfe, 167–81; Brenner, 277–87.

80   "frog-faced man" . . . "frog or a housefly": Wolfe, 179.

80   "Frog-toad": Herrera, 109.

81   "so-called easel art": Herrera, 82–83; Rochfort, 38–39; Brenner, 244–59.

81   party of radical painters: Wolfe, 151, 384.

81   "passionate dilettante" ... "commonplace slogans": Wolfe, 384–85, 419.

81   tenth-anniversary celebrations: Wolfe, 214–24.

82   anniversary demonstrations: Deutscher II, 312–18; Volkogonov, 300–1.

82   Sergei Eisenstein: Wolfe, 215.

82   "Look at your icon painters": Wolfe, 221.

82   Moscow's Red Army Club: Wolfe, 217–20.

83   "millionaire artist for the establishment": Rochfort, 123; Wolfe, 259; Herrera, 201–2.

83   unmask the "Right Danger": Bertram D. Wolfe, *A Life in Two Centuries: An Autobiography* (Stein and Day, 1981), 305.

83   Rivera fit the description: Wolfe, 230–36.

83   popular one-man show: Wolfe, 300–2.

83   San Francisco . . . Detroit Institute of Arts: Rochfort, 121–30; Wolfe, 280–96, 302–16.

84   Battle of Rockefeller Center: Wolfe, 317–41; Rochfort, 130–37.

84   Trotsky expressed his admiration: Trotsky to Rivera, June 7, 1933, TEP 9790.

85   "Rockefeller exploiters" . . . two minor fresco panels: Wolfe, 333–38; Diego Rivera, *Portrait of America* (Covici, Friede, 1934), 31, 178–79, 228–29.

85   "lifeless faces" . . . "rhythmic dance": Wolfe, 425.

86   attacks on Rivera became an onslaught: Wolfe, 238.

86   "Do you wish to see with your own eyes the hidden springs . . . ?": Trotsky, "Art and Politics in Our Epoch," *Partisan Review*, August–September 1938.

87   "a bit of an anarchist" . . . Trotsky reproached him: Van, 134.

87   Ilya Ehrenburg: Wolfe, 65.

87   "Rivera was the one": Van, 134.

88   risqué jokes: Curtiss memoir, in Buchman papers, box 3, folder "Mexico 1987."

88   telling of tall tales: Wolfe, 6.

88   not exotic enough . . . "white, red, and black": Wolfe, 13–14.

88   Frida used hand signals: Herrera, 362.

88   "stupid or banal": Wolfe, 398.

88   bathed irregularly: Wolfe, 26.

88   "enemy of clocks and calendars": Wolfe, 226, 255.

89   "the considerable subtlety": [Unidentified] to University of Chicago, August 31, 1938, TEP 17402.

89   "spontaneity was kept in check": Farrell, "A Memoir on Leon Trotsky."

89   "haughty and arrogant" . . . Karl Radek: Natalia, 120–21.

90   "gift of personal friendship": Eastman, *Heroes*, 247–49.

90   "servants to an aim": Farrell, "A Memoir on Leon Trotsky."

90   "anniversary smoker": Eastman, *Heroes*, 247.

91   "I can't stand it": Natalia, 120–21.

91   "we drove in the Dodge": Hansen to Sara Weber, November 3, 1937, TEP 12491.

91   "few days relaxing": Hansen to James Cannon, November 22, 1937, TEP 11049.

91   "The Old Man enjoyed . . . mud holes": Hansen to Pearl Kluger, July 21, 1938, TEP 11857.

92 tiny scratch pad ... "funeral was in process": Hansen to Reba Hansen, November 1, 1937, Hansen papers, 18:5.

92 "*LD lost patience*": "Joe's notes on Trotsky," Hansen papers, 40:7.

92 retrieve a Thompson submachine gun: Dugrand, 18.

92 Restaurant Acapulco: Wolfe, 360–61.

92 perpetual disorder ... had not made him rich: Wolfe, 202, 354; Max Shachtman to Felix Morrow, January 5, 1937, TC 23:2; Van to Jan Frankel, February 16, 1938, and March 3, 1938, TC 23:4.

93 man arrived at the door: Van to Jan Frankel, February 4, 1938, TC 23:4; Trotsky to James Cannon, February 15, 1938, TC 9:54; Hansen to James Cannon, February 3, 1938, TC 23:4.

93 Trotsky was irate ... "criminal lightmindedness": Van to Jan Frankel, February 4, 1938, TC 23:4.

93 comings and goings: Van, 118–19.

93 Diego mortgaged his home: Van to Jan Frankel, January 16, 1938, TC 23:4.

94 slid into the backseat ... arranged pillows: Van, 119.

94 "house is in an uproar": Hansen to Reba Hansen, February 14, 1938, Hansen papers, 18:8.

94 angry clashes: Van to Jan Frankel, February 16, 1938, TC 23:4.

94 "the blackest day": *Writings*, 10:177.

94 thunderstruck ... "Does Natalia know?": Van, 119–20.

95 "Lyova is ill": Natalia, 228.

Chapter Five: The Trouble with Father

96 "Goodbye, Leon" ... moving tribute: "Leon Sedov—Son, Friend, Fighter," *Writings*, 10:166–79.

96 stormy scene erupted . . . "gravedigger" . . . "I have smelled gunpowder": Deutscher II, 247–49; *Diary*, 69; Natalia, 149.

97 "children and grandchildren": *Diary*, 69.

97 Whether Lyova died a natural death: Broué, *Léon Sedov*, 219–64.

98 Lyova's appendicitis became acute . . . the patient died: Volkogonov, 357–61; Deutscher III, 320–23; Gérard Rosenthal, *Avocat de Trotsky* (Éditions Robert Laffont, 1975), 229–35.

98 no sign of poisoning: the documentation in Trotsky's archives of the investigation and speculations surrounding Lyova's death is substantial; among the key documents are Henri Molinier to Trotsky, February 22 and 25, 1938, TEP 3188, 3190; Gérard Rosenthal to Trotsky, February 23, 1938, TEP 4336; Pierre Naville to Trotsky, February 22, 1938, TEP 3522; reports of Elsa Reiss (February 23, 1938), Lelia Estrin (February 24, 1938), and Mark Zborowski (February 25, 1938), TEP 17131, 15949, 17388; report of Dr. Marcel Thalheimer, February 18, 1938, TEP 15532; Trotsky to Monsieur le Juge d'Instruction, March 14, 1938, TEP 2995; Trotsky to Examining Magistrate of the Lower Court, Department of the Seine, July 19 and August 24, 1938, *Writings*, 10:386–91, 421–25.

99 "terrible cry": "Joe's notes on Trotsky," Hansen papers, 40:7.

99 mere sight of them: Hansen to Pearl Kluger, February 26, 1938, TEP 11843; Hansen to Rose Karsner, February 17, 1938, TEP 11760; Van, 120.

99   "my best friend": *Writings*, 10:163.

99   idolized his father: Volkogonov, 357; Deutscher II, 311; Deutscher, III, 115–17.

99   "politics in his blood": *Diary*, 69–70.

99   On the evening of January 16, 1928: *My Life*, 539–42; Natalia, 155–57.

101   "He will remind Seryozha": *Diary*, 62–63.

101· "We called him our minister of foreign affairs": *Writings*, 10:168.

102   Lyova became homesick: Volkogonov, 324–25; Van, 26–27.

102   Relations between Trotsky and Lyova: Deutscher III, 116; Van, 102.

102   Lyova's involvement with a woman: Van, 24, 85.

102   gone to live in Berlin: Deutscher III, 117.

103   Stalin said with a sneer: Volkogonov, 325.

103   "singsong Moscow accent": *Diary*, 58–59.

103   "Lev Davidovich!": Glotzer, 50.

103   Zina was already mentally unstable . . . worshipped her father: Deutscher III, 117–21; Volkogonov, 348–53.

104   "I am a good-for-nothing" . . . "You are an astonishing person": Deutscher III, 120–21, 142.

104   Soviet government deprived Trotsky: Volkogonov, 350–51; Deutscher III, 142.

105   "Zina is terribly oppressed": Volkogonov, 350.

105   "Mama is tied down" . . . "gentle, quiet little boy": Van, 35.

105   "usual little cruelties": Glotzer, 50.

105   "I expect a letter from you": Van, 37.

105   turned on the gas taps: Deutscher III, 157–58; Volkogonov, 350–51.

105   left instructions: Zina's note, TEP 17339.

105   "Poor, poor, poor child": TEP 17340.

106   something terrible had happened . . . "Two deep wrinkles": Van, 35.

106   open letter: Van, 35.

106   "I will go mad myself": Alexandra Sokolovskaya to Trotsky, January 31, 1933, TEP 12608.

106   radical young activist . . . lovers had married: Deutscher I, 35–36, 47; Volkogonov, 8–16; Trotsky to Gérard Rosenthal, April 10, 1939, TEP 9828.

107   Trotsky's response: Van, 40; published in part for the first time in *Istoricheskii arkhiv*, No. 1, 1992, 36.

107   "The two furrows": Van, 41.

107   "they should be shot": Van, 42.

107   Alexandra was arrested . . . They disappeared without a trace: fate of Trotsky's family members: Volkogonov, 352–54, 366–67; Deutscher III, 228; *Diary*, 70, 160–61; on Trotsky's brother, *The New York Herald Tribune*, February 26, 1938; "Genealogy of Trotsky's Family," Lubitz Trotskyana Net, http://www.trotskyana. net/Leon_Trotsky/Genealogy/genealogy.htm.

108   tremendous load: Deutscher III, 144–45.

108   despairing letters: Deutscher III, 145.

109   volatile and often contentious: Deutscher III, 145; Poretsky, 261.

109   "all Papa's deficiencies": quoted in Feferman, 310.

109   "trouble with father": Deutscher III, 146.

109   Lyova remembered these Old Bolsheviks: Deutscher III, 281–83, 319–20.

109   he became hysterical: *Legacy*, 21–22.

110 "I became completely engrossed": *Writings*, 10:174.

111 "a labyrinth of sheer madness": Deutscher III, 319.

111 statement in a Paris newspaper: *Writings*, 10:387–88.

111 reproaches of his son for delays: Deutscher III, 116, 144–45, 295–96.

111 "I am a beast of burden": Deutscher III, 310–11.

112 "a ridiculously transparent pose": Van, 92.

112 Mark Zborowski: Volkogonov, 334–36; Deutscher III, 283–84; Broué, *Léon Sedov*, 126 *et passim*; Poretsky, 261–62.

112 "sullen, frowning face": Van, 99–100.

112 waves lasting five or six days: *Legacy*, 22.

113 "Étienne can be trusted absolutely": Volkogonov, 336.

113 "I never had any special suspicions": Van, 99.

113 Zborowski kept Moscow thoroughly acquainted: Volkogonov papers, reel 4; Broué, *Léon Sedov*, 210–11.

113 "Stalin must be killed": Zborowski reports, February 8, 1937, and February 11, 1938, Volkogonov papers, reel 2; *Deadly Illusions*, 282–84.

113 defection of Ignace Reiss: see his widow's memoir, Poretsky, *Our Own People*.

114 upper reaches of the GPU: *Deadly Illusions*, 293–302.

114 "Long live Trotsky!": Deutscher III, 315.

114 bullet-ridden body: Deutscher III, 315–16; Rosenthal, *Avocat de Trotsky*, 205–20; Broué, *Léon Sedov*, 184–93; Sneevliet to Trotsky, September 25 and 30, 1937, TEP 5204, 5206; Elsa Reiss [Poretsky] to Trotsky, September 30, 1937, TEP 4242.

114 "He is able, brave, and energetic": Deutscher III, 318; Broué, *Léon Sedov*, 211.

114 *"le dernier refuge"* . . . *"Ton Vieux"*: Trotsky to Lyova, November 18, 1937, TEP 10237; Trotsky to Chers amis [Lelia Estrin and Mark Zborowski], November 18, 1937, TEP 7710.

115 alcoholic and depressed . . . "lost all faith": Zborowski report, July 23, 1937, Volkogonov papers, reel 2.

115 he and his colleagues were stumped: Sudoplatov, 82–83.

115 "Both of them have aged terribly": Hansen to Rose Karsner, February 17, 1938, TEP 11760.

116 small automatic pistol: Hansen, "With Trotsky in Coyoacan," xxxiii.

116 "Slovenliness bordering on treachery" . . . "excuses and promises": Trotsky to Lyova, February 15, 1937, TEP 10198.

116 "money to buy postage stamps": Deutscher III, 297.

116 "outright crime": Trotsky to Lyova, January 21, 1938, TEP 10244.

116 Lyova's last letter: Deutscher III, 320.

116 brick, plaster, lime, and sand: Hansen to Reba Hansen, February 14 and 17, 1938, Hansen papers, 18:8.

117 The OM was seated at a small table: Hansen, "With Trotsky in Coyoacan," xxxiii.

117 his affecting tribute: *Writings*, 10:166–79.

118 rejoined the rest of the household: Hansen to Reba Hansen, May 23, 1938, Hansen papers, 18:11.

118 "our beloved daughter": Trotsky to Jeanne Martin, March 10, 1938, *Writings*, 10:257–58.

118 testament he produced in great haste: Camille [Rudolf Klement] to Trotsky, February 18, 1938, TEP 2035; Alfred Rosmer to Trotsky, February 16, 1938, TEP 4487.

118 Jeanne abducted Seva: Broué, 876–77; Deutscher III, 326–28.

118 "his banging window shattering glass": "Joe's notes on Trotsky," Hansen papers, 40:7.

119 "You are with my enemies": Van, 120–21.

119 take Lyova's place: Deutscher III, 329–30.

119 penetrate Trotsky's household: Volkogonov, 445.

Chapter Six: Prisoners and Provocateurs

120 twenty trained men: Van, 18–19. On Blumkin, see *Legacy*, 111; Christopher Andrew and Oleg Gordievsky, *KGB: The Inside Story* (HarperPerennial, 1991), 155; Kolpakidi, 119; Andrew & Mitrokhin, 40; Poretsky, 146–47.

120 "hire an assassin for a few dollars": Lyova to Trotsky and Natalia, December 7, 1936, TEP 4863.

121 "question of life and death" . . . "they use in American banks": Lyova to Trotsky [January 1937], TEP 4870.

121 authorized selected local Trotskyists: Van, 105, 133; Broué, 848.

122 Bernard Wolfe: Alan Wald, "Bernard Wolfe (1915–1985)," Glotzer papers, box 40.

122 disassembling and then reassembling the Luger: Bernard Wolfe, *Memoirs of a Not Altogether Shy Pornographer* (Doubleday & Company, 1972), 33–35.

122 events in Spain: Craig II, 639–45; Antony Beevor, *The Spanish Civil War* (Peter Bedrick Books, 1983).

123 Stalin had a complicated political agenda: Kolpakidi, 130–41; Andy Durgan, *The Spanish Civil War* (Palgrave, 2007), 66–70, 91–92.

123 May Days, Barcelona: Beevor, 187–91; Durgan, 92–97.

123 Erwin Wolf: Trotsky to George Novack and Felix Morrow, September 25, 1937, TEP 9431; *Writings*, 9:508–12.

124 George Mink: Mink profile in Solow papers, box 11, "Spies"; FBI, 2:33–34; Vernon L. Pedersen, "George Mink, the Marine Workers Industrial Union, and the Comintern in America," *Labor History*, Vol. 41, No. 3, 2000; Kern, 57; *Venona Secrets*, 106–10; Broué, 926; Whittaker Chambers, *Witness* (Random House, 1952), 302–3; Hansen to Reba Hansen, May 4, 1938, Hansen papers, 18:11.

124 Blue House was on high alert: Hansen to James Cannon, October 26, 1937, TC 23:3; Trotsky to Jack Weber, December 1, 1937, TEP 10800; Harry Milton to Trotsky, October 11, 1937, TEP 3161.

124 GPU defector Ignace Reiss: [Jan Frankel] to Friend, October 25, 1937, TC 23:3.

125 Harry Milton: Trotsky to James Cannon, October 3, 1937, TEP 7510; Albert Glotzer to Trotsky, November 3, 1937, Glotzer papers, box 3. "Gosh! Are you hit?" George Orwell, *Homage to Catalonia* (Beacon Press, 1955), 185–86.

125 organizing a hunger strike: Milton to Martin Abern, May 19 and May 21, 1937, TEP 15057, 15058.

125 generous contribution to the legend: *Time*, May 2, 1938.

125   "gangsterist activity": Trotsky to Milton, November 6, 1937, Glotzer papers, box 2.

125   preempted Milton's appointment: Jan Frankel to Natalia Trotsky, November 15, 1937, TC 26:14; also James Cannon to Trotsky, November 10, 1937, TEP 489; Trotsky to Cannon, November 14, 1937, TEP 7514; Hansen to Cannon, November 14, 1937, TC 23:3; Jack Weber to Hansen, November 28, 1937, TEP 7166.

125   "the O.M. is extremely restive": Harold Isaacs to Comrades, July 3, 1937, Hansen papers 69:64.

126   "Milton, Stone matter": Trotsky to Jack Weber, December 1, 1937, TEP 10800.

126   the new man be an experienced driver: Jan Frankel to Harold Isaacs, June 30, 1937, attached to Isaacs to Comrades, July 3, 1937, Hansen papers, 69:64.

126   "Fifty million Americans drive autos": Cannon to Bernard Wolfe, August 11, 1937, TEP 6237.

126   Cannon failed to understand: Trotsky to Sara Weber, August 17, 1937, TEP 10822.

126   Hansen was born in the farming town: Hansen bio online on Lubitz Trotskyana Net, http://www.trotskyana.net/Trotskyists/Bio-Bibliographies/bio-bibliographies.html; Sara Weber to Trotsky, August 25, 1937, TEP 5887.

127   The Dodge got a hearty reception: Hansen to Cannon, September 30, 1937, TC 23:3; Hansen, "With Trotsky in Coyoacan," viii-ix.

127   the Old Man could be very difficult: Hansen to Flo, October 18, 1937, Hansen papers, 5:26.

127   towering historical figure . . . "friendly terms with a volcano": Hansen to Sara and Jack Weber, October 21, 1937, TEP 12490.

127   Fernández family in the suburb of Tacuba . . . "He will never learn!": Van, 116.

127   "gave me hell" . . . "The driver is good": Hansen to Reba Hansen, October 11, 1937, Hansen papers, 18:5.

128   "hairskin shaves from death": Hansen to Reba Hansen, June 27, 1938, Hansen papers, 18:12.

129   "She kept crying as we drove along": Hansen to Reba Hansen, September 16, 1940, Hansen papers, 19:4.

129   Trotsky behaved like a revisionist: Hansen to Reba Hansen, October 1, 1937, Hansen papers, 18:4.

129   "I never heard him make a remark about the food": Van, 16.

129   "To dress up, to eat": Van, 61.

129   two types of meals . . . "lost somewhere in the clouds": Hansen to Reba Hansen, May 23, 1938, Hansen papers, 18:11.

129   "coldness, silence, oppression": Hansen to Reba Hansen, November 8, 1937, Hansen papers, 18:5.

130   visit to the doctor: Hansen to Reba Hansen, January 21, 1938, Hansen papers, 18:7.

130   Trotsky's friendly jesting: Van, 17; Hansen to Reba Hansen, January 21, 1938, Hansen papers, 18:7.

130   "no laughter but of mockery": Eastman, *Heroes*, 249.

130   "a good deal like a prison": Hansen to Reba, January 21, 1938, Hansen papers, 18:7.

130   Natalia, who was high-strung: Hansen to Reba Hansen, February 6, 1938, Hansen papers, 18:8.

131   warmer climes of her native Tampico: Hansen to Reba Hansen, December 8, 1937, Hansen papers, 18:6.

131 incident involved Van's wife: Van, 116–17; Hansen to Reba Hansen, December 8, 1937, Hansen papers, 18:6; Feferman, 151–53.

131 Fernández family in Tacuba ... "Damn, can they dance": Hansen to Reba Hansen, December 13, 1937; and May 17 and June 12, 1938, Hansen papers, 18:6, 18:11, 18:12.

132 "heavy, powerful, accurate, sure action": Hansen to Reba Hansen, October 13, 1937, Hansen papers, 18:4.

132 barking and howling of the neighborhood dogs: Hansen to Reba Hansen, March 2, 1938, Hansen papers, 18:9.

132 recruitment of a full-time guard: Van to Jan Frankel, February 4, 1938, TC 23:4; Hansen to Cannon, February 5, 1938 (twice), TC 23:4; Rose Karsner to Hansen, February 8, 1938, TEP 6653; Cannon to Hansen, February 14, 1938, TEP 6211; Van to Jan Frankel, February 16, 1938, TC 23:4.

132 authorized the hiring of the new guard: Van to Jan Frankel, February 22, 1938, TC 23:4.

132 garrison of three comrades: Jan Frankel to Van, February 22, 1938, TC 23:4.

133 gas masks: Van to Jan Frankel, February 22, 1938, TC 23:4.

133 The "Trial of the 21": Cohen, *Bukharin and the Bolshevik Revolution*, 372–81; Deutscher III, 332–33; Jean-Jacques Marie, *Trotsky: Révolutionnaire sans frontières* (Payot & Rivages, 2006), 506–8, 549.

133 war room: Hansen to Reba Hansen, March 2, 1938, Hansen papers, 18:9.

134 "It all seems like a delirious dream": *Writings*, 10:201.

134 Dewey himself now denounced: *The New York Times*, March 4, 1938.

134 "a too easy victory to the G.P.U.": Trotsky to Margaret de Silver, March 31, 1938, TEP 7673.

134 Hank Stone, the first chief of the guard: "Henry Malter dit Hank Stone (1908–1986)," *Cahiers Léon Trotsky*, No. 28, December 1986; Cannon to Trotsky, November 10, 1937, TEP 489; Jan Frankel to Natalia, November 15, 1937, TC 26:14.

134 hammer and nails: Stone to Jan Frankel, April 7 and April 16, 1938, TC 23:5.

134 "cobwebs inside the barrel" ... "buying bananas": Stone to Jan Frankel, April 7, 1938, TC 23:5.

135 $100 per month: Stone to Jan Frankel, May 17, 1938, TC 23:5.

135 Minneapolis became a Trotskyist stronghold: Farrell Dobbs, *Teamster Rebellion* (Monad Press, 1972); Charles Rumford Walker, *American City: A Rank-and-File History* (Farrar & Rinehart, 1937).

135 "big meaty fellow": Hansen to Reba Hansen, May 15, 1938, Hansen papers, 18:11.

135 absence of six teeth: Hansen to Reba Hansen, March 30, 1938, Hansen papers, 18:9.

135 first indication of trouble: Stone to Jan Frankel, March 16, 1938, TC 23:5.

136 "women's work" ... "go jump in a lake": Stone to Jan Frankel, April 8, 1938, TC 23:5.

136 bread without butter: Hansen to Rose Karsner, April 4, 1938, TEP 11764.

136 "potatoes and gravy": Hansen to Reba Hansen, May 23, 1938, Hansen papers, 18:11.

136 Edith offered to cook: Hansen to Reba Hansen, April 21 and 22, 1938, Hansen papers 18:10; Stone to Jan Frankel, May 17 and May 21, 1938, TC 23:5.

137 "The Mink": *Time*, May 2, 1938. Mink's photo: Van to Pearl Kluger, April 21, 1938, TC 23:5.

137 *"une maison de bourgeois"*: Van to Jan Frankel, May 30, 1938, TC 23:5.

137 Trotsky would erupt: Van to Jan Frankel, May 2, 1938, TC 23:5.

137 Hank's demoralization was now complete: Stone to Jan Frankel, May 17, 1938, TC 23:5.

138 the goodbye was sad . . . Chris Moustakis: Hansen to Reba Hansen, June 12, 1938, Hansen papers, 18:12.

138 elaborate alarm system: Stone to Jan Frankel, May 31, 1938, TC 23:5.

138 effect of the floodlights: Hansen to Reba Hansen, May 11, 1938, Hansen papers, 18:11.

138 two cedars and a pine: Hansen to Reba Hansen, February 6, 1938, Hansen papers, 18:8.

138 more permanent structure made of bricks: Hansen to Reba Hansen, May 11, 1938, Hansen papers, 18:11.

138 "getting fed up with the entire matter": Sara Weber to Rose Karsner, August 8, 1938, TC 23:6.

139 Rudolf Klement: Deutscher III, 330–31; Sara Weber to John G. Wright [Joseph Vanzler], July 17, 1938, TEP 12568; Hansen to Rose Karsner, July 21, 1938, TC 23:5; *Writings*, 11:24–25, 137.

139 theft of Trotsky's archives: Broué, *Léon Sedov*, 172–74; Deutscher III, 283–85.

139 a tense meeting to sort the matter out: Volkogonov, 426; Poretsky, 250–54.

139 "I am sending you 103 letters": Dmitri Volkogonov, *Trotskii: Politicheskii portret*, Vol. 2 (Novosti, 1994), 274–75.

139 Zborowski regularly supplied Moscow: Volkogonov, 358–61, 370–73, 378–80; Volkogonov papers, reel 2; Broué, *Léon Sedov*, 210–11.

140 "we have dreamed about getting hold of it": Volkogonov, 448.

140 September 6 in Reims, France: Étienne and Paulsen to International Secretariat, February 22, 1938, TEP 15642.

140 Trotsky was incensed at Sneevliet: Trotsky to International Secretariat, September 30, 1937, TEP 8052; Trotsky to Comrades, September 30, 1937, TC 12:18; Trotsky to Elsa Reiss [Poretsky], October 13, 1937, TEP 9783; Trotsky to Sneevliet, December 2, 1937, TEP 10422; *Writings*, 9:448–51, 459–60, 492–95; 10:146–47, 150–52.

140 staff member at the Soviet embassy: Walter Krivitsky, *In Stalin's Secret Service*, 2nd ed. (Harper & Brothers, 1939), 253.

140 Sneevliet's misgivings came to focus on Zborowski: Deutscher III, 316.

140 Walter Krivitsky: Kern, 147 *et passim*; Poretsky, 220 *et passim*.

141 "There is a dangerous agent in your party": Poretsky, 252–53.

141 Krivitsky was wary of the Trotskyists: Deutscher III, 317.

141 Trotsky, the man, was a formidable figure: Jan Frankel to Trotsky, July 12, 1939, TEP 1279.

141 Père Lachaise cemetery: Kern, 155–56.

142 He used his position as "Sonny's" successor: Étienne and Paulsen to International Secretariat, February 22 and June 24, 1938, TEP 15642, 15643; Étienne and Paulsen to Trotsky, November 11, 1937, TEP 879; Deutscher III, 329–31.

142   outraged at the "slanderer": Deutscher III, 329; Trotsky to International Secretariat, March 12, 1938, TEP 8058.

142   most devoted comrade: Étienne to Natalia [February 1938], TEP 13396.

142   "This is to your credit": Volkogonov, 445.

142   "to get to the OLD MAN": Volkogonov, 444; Volkogonov, *Trotskii*, 307.

142   his letter to Van: Marie, *Trotsky*, 503.

142   Trotsky received a letter: TEP 2321, TC 13:40.

142   helped confirm Sneevliet in his suspicion: Étienne and Paulsen to Trotsky, November 11, 1937, TEP, 879; Deutscher III, 330–31.

143   suspicions about Serge: Elsa Reiss [Poretsky] to Trotsky, November 7, 1938, TEP 4245.

143   "The sooner, the more decisively": Trotsky to Dear Friends, December 2, 1938, TEP 7729.

143   a letter arrived at the Blue House: TEP 6137; TC 13:63; Deutscher III, 331–32.

144   "extremely confidential, extremely important": TEP 8105; TC 12:25.

144   Alexander Orlov: *Deadly Illusions*; on the NKVD purges in Spain: Andrew & Mitrokhin, 73; Kolpakidi, 138; *Deadly Illusions*, 268, 279–80, 287–89; Andrew and Gordievsky, *KGB: The Inside Story*, 158–60.

144   The fatal summons: *Deadly Illusions*, 301.

145   listed all the secrets he could reveal: *Deadly Illusions*, 308–12, 430.

145   Trotsky published the ad: *Socialist Appeal*, January 14 and February 4, 1939.

146   mysterious correspondent was Krivitsky: Trotsky to Vanzler [John G. Wright], January 21, 1939, TEP 10927.

146   Cannon's secretary was an informant for the GPU: *Venona*, 262–63.

146   "Long live Trotsky!": *Legacy*, 112.

146   Ramón Mercader: Kolpakidi, 156–57.

Chapter Seven: Fellow Travelers

147   "Looking as mischievous as an art student": Van, 132.

147   Breton was the leader of Surrealism: Polizzotti, *Revolution of the Mind*.

148   published a laudatory review . . . *Planet without a Visa*: Polizzotti, 245–47, 399–400; Broué, 898.

148   European branch of the Dewey Commission: Polizzotti, 435–36; Gérard Roche, "La rencontre de l'aigle et du lion: Trotsky, Breton et le manifeste de México," *Cahiers Léon Trotsky*, No. 25, March 1986, 25 [hereafter: Roche].

148   Van arranged . . . Meyer Schapiro: Roche, 26; Van, 121.

148   He wrote extensively about literary fiction: Knei-Paz, 454–75; Deutscher I, 39–46; Deutscher, II, 150–168; Howe, 94–102.

148   "The novel is our daily bread": Dugrand, 34.

148   "miracle of reincarnation": Knei-Paz, 460.

149   a group called Proletcult: Edward J. Brown, *Russian Literature Since the Revolution* (Collier Books, 1969), 136–40; Knei-Paz, 289–96; Deutscher II, 139–42, 150–51.

149   "We Marxists have always lived in tradition": Leon Trotsky, *Literature and Revolution* (Haymarket Books, 2005), 115; for this particular quotation I have used the translation in Deutscher III, 154.

149   "What the worker will take from Shakespeare": Trotsky, *Literature and Revolution*, 184–85.

149   "Art must make its own way": Trotsky, *Literature and Revolution*, 178.

150   Symbolist poet Andrei Bely: Trotsky, *Literature and Revolution*, 54–60.

150   "fellow travelers": Trotsky, *Literature and Revolution*, 62.

150   the champions of proletarian culture had their day: Brown, *Russian Literature Since the Revolution*, 27–34.

150   idealized depictions of Soviet life: Brown, 31; Max Eastman, *Artists in Uniform: A Study of Literature and Bureaucratism* (Columbia University Press, 1953).

151   "It is impossible to read Soviet verse and prose" . . . Alexis Tolstoy . . . "manufacturer of 'myths' to order!": Trotsky, "Art and Politics in Our Epoch," *Partisan Review*, August–September 1938.

152   "New York intellectuals": Wald, *The New York Intellectuals*; Terry Cooney, *The Rise of the New York Intellectuals: Partisan Review and Its Circle, 1934–1945* (The University of Wisconsin Press, 1986); Alexander Bloom, *Prodigal Sons: The New York Intellectuals & Their World* (Oxford University Press, 1986).

152   William Phillips and Philip Rahv: Wald, 76–77; Cooney, 39–41.

152   "vulgarizers of Marxism": Cooney, 90.

152   "continuum of sensibility": Cooney, 62.

153   "There is now a line of blood": Cooney, 99–100.

153   the *Partisan Review* circle: Cooney, 107–9.

154   City College in upper Manhattan: Wald, 277, 313, 350; Web site for the 1997 PBS documentary film *Arguing the World*, http://www.pbs.org/arguing/ny intellectuals_geneology.html.

154   Farrell moved there from Chicago . . . the road toward open anti-Stalinism: Cooney, 102–3; Wald, 82–85; Alan Wald, "Farrell and Trotskyism," *Twentieth-Century Literature*, Vol. 22, No. 1, February 1976.

154   "started out more or less as Trotskyism": Wald, 5.

155   Farrell . . . developed sinus trouble: Wald, 136.

155   "he read French novels": Bloom, 112; on Macdonald, see Michael Wreszin, *A Rebel in Defense of Tradition: The Life and Politics of Dwight Macdonald* (Basic Books, 1994).

155   children of immigrants: Wald, 76–77; Cooney, 6.

155   Macdonald . . . wrote to invite Trotsky: Macdonald to Trotsky, July 7, 1937, TEP 2836.

155   "very happy to collaborate": Trotsky to Macdonald, July 15, 1937, TEP 8951; also Trotsky to Rahv, March 21, 1938, TEP 9765.

156   Trotsky found it too vague: Trotsky to Macdonald, September 11, 1937, TEP 8952.

156   "What is Living and What is Dead in Marxism?": *Partisan Review* editors to Trotsky, January 14, 1938, TEP 3714.

156   "extremely pretentious and at the same time confused": Trotsky to Macdonald, January 20, 1937, TEP 8953.

156   His defection was total: Max Eastman, "The End of Socialism in Russia," *Harper's*, February 1937.

156   "retreat of the intellectuals": James Burnham and Max Shachtman, "Intellectuals in Retreat," *New International*, Vol. 5, No. 1 (January 1939).

157 Rahv . . . fired a respectful blast: Rahv to Trotsky, March 1, 1938, TEP 4211.

158 Desperate for money: André Breton, "Visite à Léon Trotsky," *Cahiers Léon Trotsky*, No. 12 (December 1982), 105–6 [hereafter: Breton, "Visite"]; Polizzotti, 446.

158 Surrealism preached the virtues of poetry . . . "pure psychic automatism" . . . "convulsive beauty" . . . autobiographical adventure journals: Polizzotti, 209–12, 264–73, 432–33; Roger Shattuck, "The Dada-Surrealist Expedition," *The New York Review of Books*, Vol. 18, No. 9 (May 18, 1972), and No. 10 (June 1, 1972).

159 he confided to Stefan Zweig: Polizzotti, 466–67.

159 leonine and noble in appearance: Herrera, 226.

159 "the pope of Surrealism": Polizzotti, 215.

159 Jacqueline—blond, lithe, and birdlike: Polizzotti, 403.

159 Breton, who was moved to tears: Breton, "Visite," 110.

159 "blossomed forth . . . into pure surreality": Herrera, 228.

159 Breton later described his state of excitement . . . "something electrifying": Breton, "Visite," 110–11.

160 no major topics were discussed . . . Trotsky asked Van: Van, 121–24.

160 Their next meeting . . . was more memorable: Van, 122.

160 "experiments with the inner life": Shattuck, "The Dada-Surrealist Expedition," *The New York Review of Books*, Vol. 18, No. 9 (May 18, 1972).

160 "Degenerate Art": Wolfgang Benz, *A Concise History of the Third Reich*, Thomas Dunlap, trans. (University of California Press, 2006), 67–68.

160 "line of demarcation between art and the GPU": Trotsky, "Art and Politics in Our Epoch," *Partisan Review*, August–September 1938.

161 agreed to draft the founding manifesto: Van, 122.

161 the excursions and the road trips: Breton, "Visite," 112.

161 "land of convulsive beauty": Roche, 24.

161 pre-Columbian sculptures from Chupicuaro: Van, 127.

161 "*Cette séduction est extrême*": Breton, "Visite," 112.

161 "skirmishes" between them: Breton, "Visite," 115–16.

161 "keep open a little window": Breton, "Visite," 116.

161 stopped to visit a church: Van, 124–25.

162 "Have you something to show me?" . . . Van wisely declined: Van, 125.

162 headed for Guadalajara: Van, 124–25.

162 José Clemente Orozco: Van, 126; Rochfort, 99–119, 137–45; Brenner, 268–76; Wolfe, 159–61.

163 "the Mexican Goya": Brenner, 268; Wolfe, 161.

163 bulwark mustache . . . perfunctory smile: Brenner, 269.

163 "He is a Dostoevsky!": Van, 126.

163 *Creative Man . . . The Rebellion of Man*: Rochfort, 139–41.

164 Orozco assured Trotsky: "Joe's notes on Trotsky," Hansen papers, 40:7.

164 like schoolkids playing hooky: Polizzotti, 461.

164 a trip to Pátzcuaro: Van, 127–28; Herrera, 227.

165 "an Aristotle, a Goethe, or a Marx": Trotsky, *Literature and Revolution*, 207.

165 "a small square of canvas?" . . . an attack of aphasia: Van, 128.

165 giving Trotsky a couple of pages of text . . . Trotsky decided to bow out: Van, 128–29; "Manifesto: Toward a Free Revolutionary Art," in Paul N. Siegel, ed., *Leon Trotsky on Literature and Art* (Pathfinder Press, 1970), 115–21.

165 Breton's initial draft: Breton, "Visite," 116; Roche, 39.

165 International Federation of Independent Revolutionary Artists: Polizzotti, 468–71.

166 "a resounding flop": Cooney, 142.

166 Trotsky and Breton parted: Van, 129; Polizzotti, 464.

166 "boundless admiration" . . . "Cordelia complex": Breton to Trotsky, August 9, 1938, TEP 369.

166 "I am sincerely touched": Trotsky to Breton, August 31, 1938, TEP 7428.

167 this was the start of the trouble: Van, 134–37; Broué, 903–7.

167 seemed to disorient Diego: Herrera, 247; Van, 136.

167 Julien Levy Gallery: Herrera, 230–33.

167 The Paris show, called "Mexique": Herrera, 250–51.

167 "all that junk": Herrera, 250.

167 Rivera's fame, money, and force of personality: Van, 132–33.

167 "incomparable political intuition and insight": Hansen to Reba, November 16, 1937, Hansen papers, 18:5.

168 "passion, courage, and imagination": Trotsky to Frida Kahlo, January 12, 1939, TC 11:37.

168 "You are a painter. You have your work": Herrera, 473, n. 247.

168 Trotsky later lamented this choice of language: Trotsky to Cannon, October 30, 1938, TEP 7536.

168 "gift for alienating people": Eastman, *Companions*, 119.

168 "The idea of my wanting to be rid of Diego": Trotsky to Frida Kahlo, January 12, 1939, TC 11:37.

168 the O'Gorman affair: Trotsky to Jan Frankel, March 27, 1939, TEP 8178; Broué, 903–7.

169 Cárdenas had nationalized Mexico's petroleum reserves: *Cambridge History*, 44.

169 "vandalism": Broué, *Trotsky*, 904.

170 "reactionary bootlicker of Hitler and Mussolini" . . . "Mexico is an oppressed country": Trotsky to Jan Frankel, March 27, 1939, TEP 8178.

170 a letter to Breton in Paris: Van, 136–37; Natalia's declaration, TEP 17313.

170 launching a number of initiatives: Trotsky to Curtiss, February 15, 1939, TC 23:7; Van, 137.

171 "purely personal adventures": Trotsky, "A Necessary Statement," January 4, 1939, *Writings*, 11:269–74.

171 "very, very good hour": Trotsky to Frida Kahlo, January 12, 1939, TC 11:37.

171 Rivera sent a letter of resignation: Rivera to Pan-American Bureau, March 19, 1939, TEP 15303.

171 "Now, dear Frida, you know the situation here": Trotsky to Frida Kahlo, January 12, 1939, TC 11:37.

171 *"Diego is completely right"*: Herrera, 246–47.

171 she and Diego divorced: Herrera, 272–73.

171 Mexico's presidential politics: Van, 138; Charles Curtiss memoir, in Buchman papers, box 3, folder: "Mexico, 1987."

171 "series of incredible zigzags": Trotsky to Pan-American Committee, March 22, 1939, *Writings*, 11:283–90.

171 he had to separate himself from the painter: Trotsky to Curtiss, February 15, 1939, TEP 7636.

172   "morally and politically impossible": Trotsky to Curtiss, February 14, 1939, TEP 7635.

172   about 1,500 former foreign volunteers in Spain: Trotsky to Goldman, January 7, 1939, TEP 8303; *Cambridge History*, 46.

172   "He wishes to impose his generosity on me" ... donated to the local comrades: Trotsky to Curtiss, February 14, 1939, TEP 7635.

172   Diego made public his break with Trotsky: Trotsky to Jan Frankel, April 12, 1939, TC 10:56; *The New York Times*, April 15, 1939.

172   the methods of the GPU: report of Curtiss meeting with Rivera, March 11, 1939, TC 23:7.

172   Diego's promiscuous application ... "A tremendous impulsiveness": Trotsky to Pan-American Committee, March 22, 1939, *Writings*, 11:283–90.

173   "You warned us many times": Trotsky to Jan Frankel, March 27, 1939, TEP 8178.

173   *"the painter's case is a part of the retreat of the intellectuals"*: Trotsky to Breton, January 11, 1939, TEP 11027.

173   fantastic political U-turn: Wolfe, 385–86; Herrera, 341–42, 435.

173   for the purpose of having him assassinated: Herrera, 297.

173   At the moment of his departure: Van, 138.

Chapter Eight: The Great Dictator

174   an important meeting in the Kremlin: Sudoplatov, 65–69; Andrew & Mitrokhin, 41, 69, 76, 85–86; Montefiore, 4–5, 115–16.

175   let the "chatterbox" out of his grasp: Montefiore, 33.

175   "Stalin would now give a great deal": *Diary*, 26–27.

176   "treacherous infiltrations": Sudoplatov, 67.

176   "The greatest delight is to mark one's enemy": *Diary*, 63–64.

176   "operetta commander": Montefiore, 33.

176   "outstanding mediocrity" ... "gravedigger" of the Revolution: *My Life*, 512; Volkogonov, 322; *Diary*, 69; *Writings*, 10: 202.

176   "He is thinking of how to destroy you": *Diary*, 23–24.

177   "His craving for revenge on me is completely unsatisfied" ... "Stalin would not hesitate a moment": *Diary*, 63–64.

177   "If Trotsky is finished": Sudoplatov, 67.

177   Sudoplatov, an experienced killer: Sudoplatov, 25–27.

178   "Trotsky should be eliminated within a year": Sudoplatov, 67.

178   "But they cannot tear me away from history!": Volkogonov, 299.

178   When Stalin found out, he was incredulous: Volkogonov, 315–16.

178   listed him as a writer: Van, "Lev Davidovich," in *Leon Trotsky*, 44–47.

178   he dreamed of becoming a writer: *My Life*, 339; Bertram D. Wolfe, "Leon Trotsky as Historian," *Slavic Review*, Vol. 20, No. 3 (October 1961), 495–502.

179   took a toll on his nerves and his health: Natalia, "Father and Son," in *Leon Trotsky*, 44.

179   "How could you lose power?": Van, 58; *My Life*, xiv, 504–5.

179   library books that were shuttled back and forth: Glotzer, 38.

179   a demonstration of 2,500 Petrograd workers: L. D. Trotskii, *Istoriia russkoi revoliutsii* (two volumes in three, Respublika, 1997), Vol. 1, 124.

180   George Bernard Shaw once remarked: Wolfe, "Leon Trotsky as Historian," 496.

180   he did not pretend to be impartial: Knei-Paz, 497; Howe, 155.

180   deliberately places himself in Lenin's shadow: Deutscher III, 185, 203, 204; Knei-Paz, 499.

180   lifting himself onto the pedestal: Volkogonov, 433.

180   "gray blur" in 1917: the famous phrase is that of Nikolai Sukhanov; see Tucker, *Stalin as Revolutionary*, 178–79.

181   During a sedate autumn and winter of 1933–34: Van, 60.

181   a writer's paradise: Volkogonov, 429.

181   *Time* magazine gave its readers the impression: *Time*, January 25, 1937.

181   his financial situation was extremely precarious: Christiania Bank og Kreditkasse to Trotsky, January 20, 1937, TEP 568; Trotsky to Jan Frankel, December 19, 1937, TEP 8154; Holm og Rode to Trotsky, January 21, 1938, TEP 1990.

181   Max Lieber . . . was behaving like a "counter-agent": Trotsky to Sara Weber, February 3 and February 12, 1937, TEP 10812, 10813.

181   "What is the matter with Lieber?": Trotsky to Shachtman, Novack, and Weber, January 31, 1937, TEP 10323; Trotsky to Sara Weber, February 12, 1937, TEP 5883.

181   a front for Soviet espionage activity: Chambers, *Witness*, 44, 355, 365–66, 376–77, 394–95, 397, 408–10; Albert Halper, *Good-bye, Union Square: A Writer's Memoir of the Thirties* (Quadrangle Books, 1970).

181   Doubleday was insisting: H. E. Maule to Trotsky, January 28, 1937, TEP 3007; Trotsky to Maule, February 2, 1937, TEP 9030.

181   a book on the Moscow trials could be the best-seller: Trotsky to Sara Weber, January 15, 1937, TEP 12499.

182   a counterindictment he called "Stalin's Crimes": Trotsky to Harper & Brothers, February 20, 1937, TC 11:1.

182   Trotsky felt compelled to abandon his project: Jan Frankel to Charles Walker, June 9, 1937, TC 23:2.

182   "the average man on the New York street": Trotsky to Walker, November 5, 1937, TEP 10763.

182   updating *My Life*: Trotsky to Walker, August 25, 1937, TEP 10755.

182   departure of his Russian typist: Trotsky to Walker, September 30, 1937, TEP 10761; Hansen to Sara and Jack Weber, October 21, 1937, TEP 12490; Trotsky to Walker, November 5, 1937, TEP 10763.

182   his financial position was "extremely acute": Trotsky to Walker, December 19, 1937, TEP 10765.

182   a breakthrough occurred in New York . . . worked out an arrangement: Walker to Trotsky, February 16, 1938, TEP 5811; Harper & Brothers to Curtis Brown Ltd, April 12, 1938, TEP 636.

182   Natalia was borrowing funds: Hansen to Rose Karsner, March 2, 1938, TEP 11763.

182   an anxious letter to Van: Jan Frankel to Van, February 22, 1938, TC 23:4.

182   "totally acceptable": Trotsky to Walker, February 26, 1938, TEP 10769.

182   *"à contre-coeur"*: Van to Jan Frankel, February 27, 1938, TC 23:4.

182   warned by the Doubleday editors: Walker to Trotsky, March 10, 1938, TEP 5813; Trotsky to Walker, March 15, 1938, TEP 10771; Walker to Trotsky, March 30, 1938, TEP 5814; Walker to Trotsky, April 9, 1938, TEP 5817.

182   a hardheaded arrangement: Walker to Trotsky, April 9, 1938, TEP 5817.

183 first advance check arrived: Curtis Brown to Trotsky, April 27, 1938, TEP 637.

183 "I have waged the fight chiefly with a pen in my hand": *My Life*, xvi.

183 he was nicknamed Pero: *My Life*, 135.

183 Listening to Trotsky's resonant voice: Van, 13–14; Van, "Lev Davidovich," in *Leon Trotsky*, 42; Glotzer, 38.

183 pace the floor of his study . . . lose his train of thought and his patience: Sara Weber, "Recollections of Trotsky," *Modern Occasions*, Spring 1972, 182; Hansen to Reba Hansen, November 1, 1937, Hansen papers, 18:5; Hansen, "With Trotsky in Coyoacan," xxiii.

184 "At least one-third of my working time": Trotsky to Canfield, September 25, 1938, TEP 7483; Trotsky to Malamuth, October 12, 1938, TEP 8971.

184 freedom afforded by Russian syntax: Van, "Lev Davidovich," in *Leon Trotsky*, 45.

184 He complained that Max Eastman's translation: Trotsky to Jan Frankel, February 3, 1938, TEP 8160.

184 a scholar of Russian literature: Charles Malamuth.

184 a comrade in New York would serve as his researcher: Joseph Vanzler, pseudonyms John G. Wright and Usick.

184 he had lost the habit of writing by hand: Trotsky to Kopp, June 1, 1938, TEP 8711; Trotsky to Jan Frankel, March 27, 1939, TEP 8179; Marie, *Trotsky*, 498, 550.

184 Trotsky chafed at the slow pace: Trotsky to Sara Weber, October 4, 1937, TEP 10829.

184 Sara Weber . . . family illness: Rae Spiegel to Jan Frankel, April 9, 1938, TC 23:5.

185 "Let her come! We shall win her over!"; Van, 101.

185 "She is a quite young girl": Trotsky to Jan Frankel, May 14 and May 31, 1938, TEP 8167, 8168.

185 "A girl of eighteen cannot make conspiracies": Trotsky to Jan Frankel, June 18, 1938, TEP 8171.

185 Trotsky discovered the Dictaphone: Sara Weber to Rose Karsner, August 8 and August 23, 1938, TC 23:6.

185 "like a peasant shying away from an optician" . . . his enthusiasm for his recording machine: Hansen to Rose Karsner, December 7, 1938, TC 23:6.

185 Russian émigré living in Mexico: Sara Weber to Rose Karsner, August 17, 1938, TC 23:6.

185 a fresh supply of wax cylinders: Hansen to Rose Karsner, December 7, 1938, TC 23:6.

186 "guerrilla polemics": Trotsky to Martin Abern, August 4, 1938, TEP 7254.

186 The man who became Stalin: Tucker, *Stalin as Revolutionary*, 64–114; Montefiore, 26–32.

187 Lenin's promotion of Stalin . . . "wonderful Georgian": Tucker, *Stalin as Revolutionary*, 152.

187 he sent off the first chapter: Trotsky to Collins, July 1, 1938, TEP 7607.

187 "far more than 80,000 words": Trotsky to Walker, August 20, 1938, TEP 10782.

187 Trotsky had changed the conception of the book: Collins to Walker, September 22, 1938, TEP 13957; Walker to Trotsky, September 25, 1938, TEP 5825.

187 he hoped to finish the book by February 1: Trotsky to Collins, September 25 and 27, 1938, TEP 7612, 7613.

187   Trotsky responded with a vigorous defense . . . "*My* book on *Stalin* must be unattackable": Trotsky to Walker, October 3, 1938, TEP 10783; Malamuth to Trotsky, December 22, 1938, TEP 2876; Cass Canfield to Trotsky, October 7 and November 26, 1938, TEP 452, 454; Trotsky to Canfield, November 29, 1938, TEP 7485.

188   The writing would proceed more quickly now: Trotsky to Collins, November 28, 1938, TEP 7615.

188   "Health is revolutionary capital": Van, "Lev Davidovich," in *Leon Trotsky*, 44.

188   His obsession with matters of health and fitness: Ulam, 515–18.

188   passionate about hunting and fishing: Van, 11–13; Van, "Lev Davidovich," in *Leon Trotsky,* 44; Deutscher II, 335–36.

189   he often paced the patio: "Joe's notes on Trotsky," Hansen papers, 40:7.

189   He became a gardening addict: Hank Stone to Jan Frankel, May 31, 1938, TC 23:5; Hansen to Reba Hansen, June 19, 1938, Hansen papers, 18:12; Hansen to Rose Karsner, August 21, 1938, TC 23:6; Trotsky to Rae Spiegel, November 3, 1938, TC 10:3; Trotsky to Jan Frankel, November 3, 1938, TEP 8174.

189   Cactus expeditions: Charles Cornell, "With Trotsky in Mexico," in *Leon Trotsky*, 64–67.

189   Old Man Cactus: Dugrand, 38.

189   Natalia made jokes: Deutscher III, 363.

189   hunting stories . . . "We flushed a covey of mourning doves" . . . This gave Trotsky an opening to poke fun: Hansen to Reba Hansen, January 2, 1938, Hansen papers, 18:7.

190   a Sunday in October 1923: *My Life*, 495–98.

191   "L.D.'s temperature mounted": Natalia quoted in *My Life*, 499–500.

191   Trotsky's illness continued to plague him: "Auszug aus der Krankengeschichte von Herrn Leon Sedoff," 1935, TEP 15749.

191   "my high temperature paralyzed me": *My Life*, 522.

192   Trotsky headed south for the Black Sea resort of Sukhumi: *My Life*, 508.

192   "Lenin is no more": Nina Tumarkin, *Lenin Lives! The Lenin Cult in Soviet Russia* (Harvard University Press, 1983), 158.

192   Lenin's body lay in state for four days: Tumarkin, *Lenin Lives!*, 138–62; Deutscher II, 110–11.

193   Walter Duranty described a series of false rumors: *The New York Times*, January 8, 1924.

193   No one informed Trotsky of the postponement . . . "I had no choice": *My Life*, 508–11; Trotsky to Malamuth, October 21, October 29, and November 17, 1939, TEP 8979, 8980, 8981.

193   In Eastman's view, Trotsky did have a choice: Eastman, *Love and Revolution*, 408.

193   The funeral was held on Sunday . . . "like a smoke sacrifice": Tumarkin, *Lenin Lives!*, 162; Volkogonov, 266.

194   bright, warm January sun: *My Life*, 509.

194   At 3:55 p.m. . . . The effect was deafening: Tumarkin, 162.

194   "It is the moment of Lenin's burial": *My Life*, 509; Trotsky to Malamuth, October 21, 1939, TEP 8979.

194   The mail brought disconsolate letters . . . seventeen-year-old Lyova: *My Life*, 511.

194   "I should have come at any price!": Leon Trotsky, *Stalin: An Appraisal of the Man and His Influence* (Harper & Brothers, 1941), 381.

194   cryptogenic fever: Van, 57; *Diary*, 118–19, 145–46, 148, 159; Trotsky to Sara Weber, December 11, 1938, TEP 10836.

195   He lost ten pounds . . . no signs of a heart problem: "Health Report by Harry Fishler, M.D.," TEP 15751; Dr. Alfred Zollinger to Dr. Hartmann, March 5 and April 16, 1938, TEP 15646, 14740.

195   He did not leave the house for more than two months: Irish O'Brien to Rose Karsner, February 23, 1939, TC 23:7; Lillian to Rose Karsner, March 29, 1939, TC 23:7.

195   "The general name of my illness is 'the sixties' ": Trotsky to Jan Frankel, March 31, 1939, TEP 8180.

195   the loss, yet again, of his Russian typist: Trotsky to Hansen, March 17, 1939, TEP 8436; Trotsky to Jan Frankel, March 27, 1939, TEP 8179.

195   Trotsky pleaded his hard luck case: Trotsky to Collins, January 3 and April 8, 1939, TEP 7616, 7619.

195   "I am almost desperate": Trotsky to Hansen, March 17, 1939, TEP 8436.

195   The new residence . . . in dilapidated condition: Van, 138; Natalia, 251; Irish O'Brien to Usik, May 14, 1939, TEP 12537.

196   he found a Russian typist: Trotsky to Collins, May 9, 1939, TEP 7621.

196   less detailed and more "synthetic" . . . "if nothing extraordinary happens": Trotsky to Collins, June 3, 1939, TEP 7622.

196   Trotsky was thoroughly disgusted: Natalia, "Father and Son," in *Leon Trotsky*, 42–43.

196   "Stalinism is counterrevolutionary banditry": Volkogonov, 421–22.

196   an aggressive prosecuting attorney: Knei-Paz, 528–32; Deutscher III, 361–67.

196   "never-slumbering envy" . . . "in the full panoply of power": Trotsky, *Stalin*, 336.

196   Stalin had hastened Lenin's death: Deutscher III, 367–68.

197   death of Lenin's widow, Krupskaya: Trotsky, "Kroupskaia est morte," March 4, 1939, TEP 15726.

197   "They proved to be not only peppery but poisoned": Trotsky, *Stalin*, 372–73.

197   "the monstrosity of such suspicion": Trotsky, *Stalin*, 372, 376–80.

198   Trotsky accused *Life* of caving in to "the Stalinist machine": quote is in Trotsky to Bush, January 8, 1940, TEP 8925; also see Bush to Trotsky, October 3, 1939, TEP 2790; Trotsky to Bush, October 15, 1939, TEP 8919; Bush to Trotsky, November 22, 1939, TEP 2793; Trotsky to Editorial Board, *Life*, November 23, 1939, TEP 8922; Trotsky to *Saturday Evening Post*, January 27, 1940, TEP 10018.

198   the text of an anonymous letter: TC 13:40.

199   Trotsky . . . decided that they deserved to be taken seriously: Trotsky to Jan Frankel, May 10, 1939, TEP 8185.

199   Orlov was now living in Los Angeles: *Deadly Illusions*, 322.

199   he tried to reach Trotsky by telephone: Trotsky to Jan Frankel, May 10, 1939, TEP 8185; *Legacy*, 20.

200   room 735 of the Lubyanka: Sudoplatov, 68–69.

200   Operation Utka . . . Stalin authorized the operation: Kolpakidi, 153–54; *Ocherki*, 93.

Chapter Nine: To the Finland Station

201   Alfred and Marguerite Rosmer: Van, 143.

201   they were rejuvenated: Van to Rose Karsner, August 10, 1939, TC 23:9; Hansen to Reba Hansen, November 7, 1939, Hansen papers, 19:2; Hansen to Rose Karsner, November 9, 1939, TC 23:10.

201   Hubert Herring: Van, 131; Herring to Trotsky, March 9, 1937, TEP 1970.

201   the Nazi-Soviet pact: Craig II, 654–57.

202   Trotsky insisted that the pact was of secondary importance . . . "the OM refused to be disturbed": O'Brien to Rose Karsner, August 31, 1939, TC 23:9.

202   predicting a rapprochement between Stalin and Hitler: Glotzer, 287, 314.

203   keenly sensitive to the danger posed by Hitler: Van, 2; Glotzer, 57.

203   the Red Army should immediately be mobilized: Eastman, *Heroes*, 254–55.

203   Trotsky changed his mind about remaining inside the Comintern: Knei-Paz, 414–15; Glotzer, 200–1; Jean van Heijenoort, "How the Fourth International Was Conceived," in *Leon Trotsky*, 61–64.

203   The moment was hardly propitious: Deutscher III, 342; Volkogonov, 400–1.

204   many Trotskyists were skeptical: Glotzer, 309–10; Deutscher III, 340–41; Knei-Paz, 417.

204   the voice of supreme optimism: Deutscher III, 345–46; Trotsky to Cannon, June 16, 1939, TEP 7546.

204   "The Death Agony of Capitalism": Knei-Paz, 413.

204   scene of the founding congress . . . At the end of the day: Deutscher, III, 340–41; Volkogonov, 400–7.

204   Zborowski protested that the Russian section: Deutscher III, 342; Volkogonov, 401.

205   Sylvia Ageloff . . . Ruby Weil: Levine, 43–47; FBI, 1:22, 7:58–67; *American Aspects of Assassination of Leon Trotsky: Hearings Before the Committee on Un-American Activities, House of Representatives, Eighty-First Congress, Second Session* (United States Government, 1951), December 4, 1950, 3401–17; Weil profile, TC 24:1.

205   Jacques took the ladies sightseeing . . . perfect dilettante: "Statement of Walta Karsner," August 21, 1940, TC 24:4; "Memorandum of talk with Hilda and Ruth Ageloff," August 24, 1940, TC 24:2.

205   Ramón Mercader . . . Ramón's flamboyant mother: *Ocherki*, 94; Levine, 14–21, 35–37, 43, 61–65; Kolpakidi, 155–57.

206   Spanish civil war became the NKVD's training ground . . . Leonid Eitingon: Levine, 32–35.

206   she was asked to serve as a translator . . . she was worried that her Trotskyism . . . things about Jacques that did not add up: "Memorandum of talk with Hilda and Ruth Ageloff," August 24, 1940, TC 24:2; "Statement of Walta Karsner," August 21, 1940, TC 24:4.

207   Trotsky was especially fond of Sylvia's sister Ruth: Van, 146.

207   the American Trotskyists gathered: Cannon to Trotsky, October 13, 1938, TEP 499; Rose Karsner to Lillian, October 30, 1938, TEP 6637.

208   "I hope that this time my voice will reach you" . . . "stinking cadaver" . . . "Long live the Fourth International!": *Writings*, 11:85–87.

209   Socialist Workers Party: Constance Ashton Myers, *The Prophet's Army: Trotskyists in America, 1928–1941* (Greenwood Press, 1977); Glotzer, 21–24.

209   James Cannon: Kelly, 47–49; Wald, 169–70.

209   Max Shachtman: Kelly, 48; Wald 165, 172–75.

209   James Burnham: Kelly, *James Burnham*; Wald, 176–78.

210   Cannon was wary: Cannon to Trotsky, December 16, 1937, TEP 491.

210   Burnham objected to Cannon's authoritarian management style . . . "The tendency in your letters": Burnham to Cannon, June 15, 1937, TEP 13825.

210   "degenerated workers' state" . . . "unconditional defense": Knei-Paz, 410–18; Glotzer, 294–95; Eastman, *Heroes*, 244.

210   regarded as "treason": Trotsky quoted in Glotzer, 283.

211   Burnham and Carter described the Soviet system as "bureaucratic collectivism": Kelly, 63–64; Glotzer, 284.

211   "a little epidemic of revisionism": Cannon to Trotsky, November 15, 1937, TEP 490; Trotsky to Burnham and Carter, December 6, 1937, TEP 7456; Trotsky to Burnham, December 15, 1937, Glotzer papers, box 3, folder: "Trotsky, Leon"; Trotsky to Cannon, December 21, 1937, TEP 7516.

211   especially strong among the youth: Jack Weber to Trotsky, November 22, 1937, Glotzer papers, box 4; Jan Frankel to Trotsky, December 23, 1937, TEP 1263; Israel Kugler to Trotsky, November 17, 1938, TEP 2422.

211   four out of seventy-five votes: Jan Frankel to Trotsky, January 2, 1938, TEP 1265.

211   "totalitarian" twins bearing a "deadly similarity": Leon Trotsky, *The Revolution Betrayed*, Max Eastman, trans. (Dover, 2004), 208, 210.

211   This confounded the Trotskyists: Shachtman, "The Soviet Union and the World War," *New International*, April 1940.

211   The Germans had launched their blitzkrieg: Craig II, 659–61.

212   the Soviets arrested and deported hundreds of thousands of Poles . . . Katyn Forest Massacre: Allen Paul, *Katyń: The Untold Story of Stalin's Polish Massacre* (Scribner's, 1991).

212   "The USSR in War": *In Defense of Marxism*, 3–21.

212   "nothing else would remain": *In Defense of Marxism*, 9, 14–15.

212   took his followers by surprise: Glotzer, 315; *In Defense of Marxism*, 30.

213   In Trotsky's view, the Red Army . . . was serving as a vehicle for progress in Poland: *In Defense of Marxism*, 18.

213   Shachtman had now joined forces with Burnham: *In Defense of Marxism*, ix–x; Deutscher III, 382.

213   Anatoly Lunacharsky . . . wrote a profile of him: Anatoly Vasilievich Lunacharsky, *Revolutionary Silhouettes*, Michael Glenny, trans. (Penguin, 1967), 61–62, 66.

214   "a man of exceptional abilities": Volkogonov, 18.

214   second congress of the Russian Social Democrats: Deutscher I, 60–69; Volkogonov, 25–28.

215   *Our Political Tasks*: Knei-Paz, 176–199; Deutscher I, 73–77; Volkogonov, 30–31.

215   his break with the Mensheviks: Knei-Paz, 206–14.

215   His ineptitude as a conciliator: Deutscher I, 160–64; Knei-Paz, 180.

215   "the poisonous seeds of its own destruction": Trotsky quoted in Volkogonov, 31.

215   Trotsky turned down Lenin's offer: Knei-Paz, 225.

216   "Go where you belong from now on: into the dustbin of history!": Ulam, 363–73; Deutscher I, 259; Knei-Paz, 509; Glotzer, 125.

216   Trotsky's passivity in the struggle to succeed Lenin: Glotzer, 149–53.

216   "In the time of revolutionary storm": Eastman, *Heroes*, 258–59.

217   "sharing the bitter fate": Ulam, 373.

217   Trotsky's account of the October events: *Istoriia russkoi revoliutsii*, Vol. 2/2, 277–78.

217   Eastman . . . and his wife visited Prinkipo: Eastman, *Companions*, 114–15.

217   Eastman was strikingly handsome: John P. Diggins, "Getting Hegel out of History: Max Eastman's Quarrel with Marxism," *American Historical Review*, Vol. 79, No. 1 (February 1974), 38–39 [hereafter: Diggins].

217   pale blue color of his eyes . . . kept insisting were black: Eastman, *Love and Revolution*, 557–58.

217   "Trotsky's throat was throbbing and his face was red": Eastman, *Companions*, 114.

218   twice tried and twice acquitted: Eastman, *Love and Revolution*, 85–99, 118–24.

218   an invitation from Lenin and Trotsky: Eastman, *Love and Revolution*, 78.

218   Lenin's still-secret political testament: Eastman, *Love and Revolution*, 442–55.

218   the dialectic, a principle of change . . . "historical materialism": Walter Kaufmann, *Hegel: A Reinterpretation* (University of Notre Dame Press, 1978), 153–62; Edmund Wilson, *To the Finland Station: A Study in the Writing and Acting of History* (Farrar, Straus and Giroux, 1972), 210–30; Max Eastman, "Russia and the Socialist Ideal," *Harper's*, March 1938; George Novack, "Trotsky's Views on Dialectical Materialism," in *Leon Trotsky*, 94–102.

219   Eastman was puzzled by the connection . . . the library of the Marx-Engels Institute: Eastman, *Love and Revolution*, 125–32, 416–18; Diggins, "Getting Hegel out of History."

219   *Marx and Lenin: The Science of Revolution*: Eastman, *Love and Revolution*, 460–63.

220   Hook and Eastman were Dewey's "bright boys": Eastman, *Love and Revolution*, 499–500.

220   "he became almost hysterical": Eastman, *Companions*, 115.

220   "to trim Marx's beard": Van, 63; Lunacharsky, *Revolutionary Silhouettes*, 66.

220   Eastman's "petty-bourgeois revisionism": Eastman, *Love and Revolution*, 593–95.

220   fixated on the subject of Eastman's heresy . . . "Pragmatism, empiricism is the greatest curse": George Novack, "Trotsky's Views on Dialectical Materialism," in *Leon Trotsky*, 94–102; *In Defense of Marxism*, 44–47; Shachtman to Trotsky, March 5, 1939, TEP 5107.

221   an article in *Harper's* in March 1938: Eastman, "Russia and the Socialist Idea."

221   that Eastman be dealt with "mercilessly": Trotsky to Burnham, March 22, 1939, TEP 7458.

221'   Burnham was prepared to defend the October Revolution . . . but not dialectical materialism: Shachtman to Trotsky, March 5, 1939, TEP 5107.

221   Hook . . . Dewey . . . Edmund Wilson: Diggins, 59–60.

222   the "greatest blow" . . . "the best of gifts to the Eastmans of all kinds": Trotsky to Shachtman, January 20, 1939, TEP 10337; Trotsky to Shachtman, March 9, 1939, Glotzer papers, box 3.

222   the intellectual equivalent of an appendectomy: Kelly, 77.

222   "Trotsky does not write on the dialectic": Hansen to Trotsky, June 23, 1939, TC 18:12.

222   a defense of Marxism's core principles: Trotsky to Cannon, January 9, 1940, TEP 7558; Knei-Paz, 485–86; Glotzer, 285–86.

222 Events in Europe in the autumn of 1939: Craig II, 659–61.

222 the Trotskyist Minority . . . proposed a referendum: Cannon to Trotsky, September 8 and October 26, 1939, TEP, 13874, 6222; Hansen to Reba Hansen, October 19, 1939, Hansen papers, 19:1; *In Defense of Marxism*, 33.

223 the vote was eight to four: Cannon to Trotsky, November 8, 1939, TEP 523.

223 the Soviet invasion of Finland: Craig II, 661–62.

223 "A Petty-Bourgeois Opposition in the Socialist Workers Party": *In Defense of Marxism*, 43–62.

223 "The ABC of Materialist Dialectics": *In Defense of Marxism*, 48–52.

223 not sure how it related to current debates: Hansen to Trotsky, January 1, 1940, TEP 1814.

223 "Cannon represents the proletarian party": *In Defense of Marxism*, 61.

224 his analysis of the Finnish events: *In Defense of Marxism*, 56–59.

225 "moral and material support": Stanley [Stanley Plastrik] to Trotsky, December 23, 1939, TEP 5379.

225 utterly fantastic: Manny Garrett [Geltman] to Bob, December 26, 1939, Glotzer papers, box 2; Burnham, "The Politics of Desperation," *New International*, January 1940; Glotzer, 305–6.

225 the Old Man had gone "completely haywire": Glotzer to John [Jan Frankel], January 21, 1940, Glotzer papers, box 12; Shachtman, "The Crisis in the American Party," *New International*, March 1940.

225 Sherman Stanley . . . secretary-guard: Young to Charles Cornell, May 3, 1940, TEP 7239.

225 "the most monstrous and shameful *non-sequitur*": Stanley [Stanley Plastrik] to Trotsky, December 23, 1939, TEP 5379.

225 "petty-bourgeois," a time-honored Bolshevik term of abuse: Van, 130.

225 "L.D. has laid the gauntlet": Manny Garrett [Geltman] to Bob, December 26, 1939, Glotzer papers, box 2.

225 "enraged petty-bourgeois": Trotsky to Friends, December 27, 1939, TC 13:31.

225 "Stalinist agents working in our midst": Trotsky to Cannon, December 29, 1939, TEP 7555.

226 laying the basis for a split: Hansen to Trotsky, January 1, 1940, TEP 1814.

226 "wrong side of the barricades": Trotsky to Shachtman, December 20, 1939, TC 12:14.

226 Hansen's . . . reputation for heavy-handed sarcasm: Stanley [Stanley Plastrik] to Trotsky, December 23, 1939, TEP 5379; Hansen to Trotsky, March 15, 1940, TEP 1820.

226 "declassed kibitzers" and "petty-bourgeois smart alecks": Cannon to Trotsky, January 11, January 18, and February 20, 1940, TEP 530, 532, 6203.

226 a "madhouse": Hansen to Trotsky, January 15, 1940, Hansen papers, 34:3.

226 "Where's the civil war in Finland?": Hansen to Trotsky, March 15, 1940, TEP 1820.

226 Howls of laughter: Hansen to Paul Anderson, March 7, 1940, Hansen papers, 18:6.

226 "provincials, blockheads, stupid yokels": Hansen to Trotsky, January 15, 1940, Hansen papers, 34:3.

226 Trotsky gritted his teeth: Trotsky to Friends, January 3, 1940, TC 12:32.

226   He had lived in the Bronx: Trotsky obituary, *The New York Times*, August 22, 1940.

226   "The oppositionists, I am informed": *In Defense of Marxism*, 104.

227   "the Jewish petty-bourgeois elements": *In Defense of Marxism*, 109; also, Trotsky to Cannon, October 10, 1937, and March 27, 1939, TEP 7511, 8108.

227   "petty-bourgeois disdain": *In Defense of Marxism*, 145.

227   Burnham's "brutal challenge": Trotsky to Friends, January 3, 1940, TEP 7556.

227   "each contribution by the OM": Cannon, "On the Party," undated [spring 1940] manuscript, TEP 6238.

227   "The Finnish events were absolutely decisive": Hansen to Trotsky, April 20, 1940, TEP 1823.

227   "petty-bourgeois windbags": Cannon, "Measures to Combat a Split," January 24, 1940, TEP 13879.

227   "a vigorous intervention in favor of unity": Trotsky to Albert Goldman, February 19, 1940, TC 10:66.

227   "Back to the Party!": *In Defense of Marxism*, 153–55.

227   "enemies and traitors" . . . "war of political extermination": Cannon to Trotsky, February 20, 1940, TEP 6203.

227   special convention of the Socialist Workers Party: Farrell Dobbs to Trotsky, April 10, 1940, TEP 799.

228   Shachtman announced . . . to form a separate party: Hansen to Trotsky, April 20, 1940, TEP 1823.

228   "The OM did nothing": Stanley [Stanley Plastrik] quoted in Young to Trotsky, May 3, 1940, TEP 7239.

228   Frankel on the other side of the barricades . . . "the old *Iskra* days": Young to Trotsky, May 3, 1940, TEP 7239; Young to Trotsky, July 6, 1940, TEP 6081; Van to Trotsky, March 4, 1940, TEP 5664; Dobbs to Trotsky, February 29, 1940, TEP 795.

228   Trotsky was shaken by the loss: Trotsky to Young, July 29, 1940, TEP 10953; Trotsky to Van, January 7 and February 27, 1940, TEP 10702, 10203.

228   a note to Trotsky: Sylvia Ageloff to Trotsky, January 25, 1940, TEP 122.

228   Sylvia was invited to come to the house: Trotsky to Sylvia Ageloff, January 26, 1940, TEP 11000.

228   "petty bourgeois Menshevism of the minority": Robins to Rose Karsner, February 3, 1940, TC 24:16.

229   "the factional struggle provides a *perfect* cover": Wright to Walter O'Rourke, March 28, 1940, TEP 7204.

229   As Sylvia left Trotsky's home: "Memorandum of talk with Rosmers," August 23, 1940, TC 24:6.

Chapter Ten: Lucky Strike

230   last will and testament: TC 22:4.

230   recent examination by his doctor: Hansen to Usick [Wright], September 21, 1940, TC 22:4.

231   "This magnetism is colossal": Tucker, *Stalin as Revolutionary*, 35–36.

231   "He was my master": *My Life*, 394.

231 London in October 1902: *My Life*, 142–43; Deutscher I, 48–49; Ulam, 174–75.

232 "We were lying side by side": *My Life*, 327–28.

232 "It's a bowl of mush we have": *Diary*, 83–84.

232 an assassination attempt: Ulam, 428–30.

232 "He had a way of *falling in love* with people": *Diary*, 84.

232 "Lenin and I had several sharp clashes": *Diary*, 85.

233 Krupskaya writes to say: *My Life*, 511.

233 Lenin's testament: Deutscher II, 57–58; Ulam, 562–63.

233 Trotsky was pressured by Stalin . . . avoid a premature clash: Deutscher II, 169–70, 247–48; Eastman, *Love and Revolution*, 442–55, 510–16.

234 Trotsky's shabby treatment of Eastman: Cannon to Trotsky, February 20, 1940, TEP 6203; Trotsky to Hansen, February 29, 1940, TEP 8444.

234 Max Eastman, accompanied by his wife . . . "more mellow": Eastman, *Love and Revolution*, 596; on Trotsky's mellowing, see also Van, 27; Hansen, "With Trotsky in Coyoacan," xxiii.

235 March 1940 . . . Veracruz harbor: Buchman describes his films in "Black and White Roll," TC 32:12.

235 Young was born Alexander Buchman: Buchman biography on Lubitz Trotskyana Net, http://www.trotskyana.net/Trotskyists/Bio-Bibliographies/bio-bibliographies.html; Suzi Weissman remembrance, Buchman, box 1.

236 arranged for him to visit Trotsky in Mexico: Frank Glass to Trotsky, August 30, 1939, TEP 1429; Cannon to Trotsky, October 27, 1939, TEP 520.

236 Trotsky's cactus-hunting picnics: "Black and White Roll"; Robins memoir, TC 30:1.

237 Avenida Viena 19: Irish O'Brien to Usick, May 14, 1939, TEP 12537; Hansen, "The Attempted Assassination of Leon Trotsky," in *Leon Trotsky*, 5–12; Natalia, 251; Mosley, 37; Julius H. Klyman, "Revolutionist in Exile," *St. Louis Post-Dispatch*, March 26, 1940.

238 feeding his rabbits and chickens . . . began at the Blue House: Lillian to Sara Weber, December 30, 1938, and February 4, 1939, TEP 12487, 12488.

238 later arranged to purchase: Trotsky to Dear Friends, March 20 and April 16, 1940, TC 9:75, 9:77.

238 Rhode Island Reds . . . fifty in all: Hansen to Reba Hansen, October 13, 1939, Hansen papers, 19:1.

238 new three-decker cages: Al Goldman to Hansen, February 29, 1940, Hansen papers, 15:4; "Black and White Roll."

238 the chief buck take a hard bite: Hansen to Reba Hansen, October 13, 1939, Hansen papers, 19:1.

238 "a flock of rabbits": Eastman, "Political Murder à Outrance," *The New Leader*, December 14, 1959.

238 quiz the guards: Klyman, "Revolutionist in Exile."

239 "Well, that's all there is": "Black and White Roll."

239 inspecting the alarm system: Buchman to Glotzer, January 5 and August 27, 1990, Glotzer papers, box 48.

239 Van . . . departed for the United States: Van, 145; Feferman, 177–78.

239 "You treat me as though I were an object": Van, "Lev Davidovich," in *Leon Trotsky*, 45; Van, 136.

240   "messy and complicated": Buchman to Glotzer, July 19, 1991, Glotzer papers, box 48.

240   a Hollywood version of Sing Sing: Hansen to Reba Hansen, October 13, 1939, Hansen papers, 19:1.

240   *St. Louis Post-Dispatch*: Klyman, "Revolutionist in Exile"; Buchman to Glotzer, January 5, 1990, Glotzer papers, box 48.

241   wanted more time to take photographs and films: Hansen to Rose Karsner, December 14, 1939, TC 23:10.

241   guard and secretariat were overstretched: Robins to Rose Karsner, May 6, 1940, TC 24:10.

241   increased from four to five: Cornell to Rose Karsner, February 20, 1940, TC 24:10.

241   Otto Schüssler . . . Charley Cornell: Irish O'Brien to Rose Karsner, April 6 and July 2, 1939, TC 23:8, 23:9; Lillian to Rose Karsner, April 6, 1939, TEP 11743.

241   Robins was born Harold Rappaport: Robins statement on May 24 raid, TEP 17193; Glotzer to Ralph Kessler, August 31, 1988, Glotzer papers, box 40; Ralph Kessler memorial tribute to Robins, February 1, 1988, Buchman papers, box 1.

241   a worker-intellectual . . . "one of the coolest militants": Hansen to Van, August 19, 1939, TEP 6424.

242   "the most important condition": Trotsky to Rose Karsner, September 1, 1939, TEP 11717; Rose Karsner to Irish O'Brien, September 7, 1939, TEP 2167.

242   This was too much for Trotsky: Trotsky to Sara Weber, January 2, 1940, Trotsky's letters to Sara Weber, Houghton Library, Harvard University, Cambridge, Massachusetts.

243   training in small-arms fire: Buchman to Glotzer, January 5, 1990, Glotzer papers, box 48.

243   submachine gun, which had a tendency to jam: Goldman to Hansen, February 29, 1940, Hansen papers, 15:4; Cornell to Rose Karsner, March 21, 1940, TEP 11723.

243   a strain in relations . . . "the height of folly": Young to Dobbs, April 26, 1940, TC 24:10.

243   "the cream of the earth": Robins to Rose, May 13, 1940, TC 24:10.

243   Bob Shields, a twenty-five-year-old New Yorker: Bridget Booher, "Death of a Romantic Revolutionary: Sheldon Robert Harte, Duke 1937," unpublished manuscript, Spring 1991, Duke University Archives, Durham, North Carolina; Dobbs to Cornell, March 8, 1940, TC 24:10; Rose Karsner to Cornell, March 10, 1940, TEP 6631; Dobbs to Cornell, March 22, 1940, TEP 6277.

244   To the NKVD . . . known by the code name "Amur": *Ocherki*, 100–101.

244   now posing as a Canadian businessman named Frank Jacson: *Ocherki*, 96–97; Kolpakidi, 158–59.

244   he explained his change of identity to Sylvia . . . Ramón said goodbye to Sylvia . . . she suffered from a sinus condition: FBI, 7:58–67.

245   Sylvia encountered Alfred Rosmer . . . they were introduced to Jacson . . . picnic to Mount Toluca . . . political lightweight . . . he spoke Parisian French . . . her Jac was fine, just busy . . . Alfred was admitted to the French hospital . . . he understood why this would be impossible: "Memorandum of talk with Rosmers about Frank Jacson," August 23, 1940, TC 24:6; Rosmers, "We never saw 'Jacson' in Paris," TC 24:6; Salazar, 137.

246   NKVD established a second and much larger network: Kolpakidi, 160; Andrew & Mitrokhin, 86.

246   David Alfaro Siqueiros: Stein, *Siqueiros*; Rochfort, 28–31, 38–39, 50; Wolfe, 154–57; Brenner, 260–67; Nikandrov, 90–91; Salazar, 60, 205–9; Siqueiros obituary, *The New York Times*, January 7, 1974.

246   the paper's famous masthead: Stein, 41.

246   Siqueiros was a swashbuckler: Levine, 69–70; Wolfe, 154–57, 426.

246   friends called him Caballo: Nikandrov, 101.

247   a union organizer among the silver miners: Stein, 56–57.

247   Moscow in March 1928: Stein, 59–60.

247   expelled from the party . . . seized in a police sweep: Stein, 66–67.

247   house arrest in Taxco: Stein, 70.

247   moved to Los Angeles to teach and to paint . . . *Tropical America*: Rochfort, 145–47; Stein, 72–80.

247   One of the workshop participants was Jackson Pollock: Rochfort, 150–51; Stein, 98.

247   Siqueiros published a savage attack on Rivera: Rochfort, 149; Stein, 93.

247   Siqueiros sailed to Spain: Stein, 102.

247   Carlos Contreras: Stein, 102–3; Nikandrov, 52.

248   considerable energy lobbying President Cárdenas: Stein, 111.

248   a mural for the new headquarters of the Mexican Electricians' Union: Rochfort, 149–59; Jennifer Jolly, "Art and the Collective: David Alfaro Siqueiros, Josep Renau and Their Collaboration at the Mexican Electricians' Syndicate," *Oxford Art Journal*, Vol. 31, No. 1, 2008; Stein, 112–14.

248   Pierre Matisse Gallery: Jolly, 137; Stein, 111.

248   Luís Arenal and Antonio Pujol: Stein, 76, 96; Nikandrov, 81–82.

248   the mural's theme was antifascist . . . a more generic anticapitalism: Jolly, 144.

248   "one of the great moments in twentieth-century mural art": Rochfort, 151.

248   "difference between a brush and a gun": Brenner, 242–43.

249   Iosif Grigulevich: Nikandrov, *Grigulevich*; Andrew & Mitrokhin, 86–87; Kolpakidi, 138, 147, 153–54; *Ocherki*, 94.

249   Orlov and Grigulevich were part of a mobile group: Nikandrov, 63–72; Kolpakidi, 138.

249   Grigulevich and a colleague were sent to Mexico City . . . taken to meet Beria: Nikandrov, 74–78, 91; *Ocherki*, 95; Kolpakidi, 160.

250   booby-trapped potted cactus: Nikandrov, 94.

250   Dies Committee: HUAC to Trotsky [October 12, 1939], TC 13:73; Trotsky to HUAC, October 12, 1939, TC 12:53; Hansen, "Report on Invitation of Dies Committee," December 14, 1939, TEP 16906.

250   use the reactionary Dies Committee as a tribune: Trotsky to Dear Friends, November 28, 1939, TEP 8112.

251   Stories in the American and Mexican press . . . an agent of Yankee imperialism: Broué, 928; Levine, 81.

251   "Death to Trotsky!": Trotsky, "The Comintern and the GPU"; Rosmer in *Leon Trotsky*, 78.

251   sweeping purge of its top leadership: Broué, 928–29; Levine, 78–79.

251   Trotsky understood that such a purge: Trotsky, "The Comintern and the GPU."

251 "Throw out the most ominous and dangerous traitor Trotsky": Levine, 81; Robins to Rose Karsner, May 6, 1940, TC 24:10.

251 concentration in the city of Stalinist killers: Trotsky, "The Comintern and the GPU"; Levine 70.

251 Trotsky called a meeting: Robins memoir, TC 30:1.

251 Robert Sheldon Harte: Booher, "Death of a Romantic Revolutionary."

251 harbored literary aspirations: for example, Robert Sheldon Harte, "Strike Scenes," TC 30:1.

252 forged documents supplied by Hitler: Nikandrov, 103–4.

252 The rain fell heavily at times during the night of May 23–24: general accounts of the Siqueiros raid: Salazar, 3–26, 45–46; Natalia, 257–58; Mosley, 37–39; Nikandrov, 102–6; Trotsky, "Stalin Seeks My Death," June 8, 1940, *Writings*, 12:233–35.

253 the guards were awakened by the gunfire: for the guards' perspectives on the raid, see Henry Schultz to Dobbs, May 25, 1940, TC 23:12; Harold Robins testimony, TEP 17193; Jake Cooper testimony, TEP 17025; Hansen, "The Attempted Assassination of Leon Trotsky," in *Leon Trotsky*, 5–12; Walter O'Rourke to Al Goldman, June 10, 1940, TEP 11412; Hansen, "Memorandum," June 30, 1940, Hansen papers, 70:3; Mark Harris, ed., *My Brother, My Comrade: Remembering Jake Cooper* (Walnut Publishing, 1994), 27–28.

254 famously Mephistophelian features: Salazar, 6.

255 Salazar grew suspicious . . . Trotsky himself had staged the raid: Salazar, 7–10; Natalia, 258–59.

255 twenty-five men on duty . . . They put the odds at fifty-fifty . . . "No, no, please don't" . . . tricky ignition switch: Henry Schultz to Dobbs, May 25, 1940, TC 23:12; also Hansen to Goldman, June 20, 1940, TEP 11402; Hansen to Reba Hansen, June 24, 1940, Hansen papers, 19:4.

255 Suspicion fell on Sergeant Casas . . . self-assault, *auto-asalto*: Hansen to Goldman, June 20, 1940, TEP 11402; Hansen to Reba Hansen, June 24, 1940, Hansen papers, 19:4.

256 Casas was compromised: *Writings*, 12:223–27.

256 Jesse Sheldon Harte: Mosley, 43–44.

256 a photograph of Stalin, warmly inscribed . . . Trotsky sent a telegram: Salazar, 93–94.

256 to bury the story for good: Jesse Harte to Trotsky, May 30, 1940, TEP 1841.

256 Trotsky's household servants: Salazar, 18–19.

256 Salazar arrested Charley and Otto: Salazar, 20; O'Rourke to Cannon, June 2, 1940, TC 24:11; Hansen, "The Attempted Assassination of Leon Trotsky," in *Leon Trotsky*, 5–12.

257 came to arrest Robins: Robins memoir, TC 30:1.

257 "We are always holding conferences": Salazar, 24.

257 Trotsky did not follow his usual routine: Salazar, 94.

257 the cook lied and ought to be fired: Salazar, 25; *Writings*, 12:316–22; Hansen to Reba Hansen, August 4, 1940, Hansen papers, box 19:4.

257 a put-up job, staged by Trotsky: Trotsky, "The Comintern and the GPU."

257 "moral preparation of the terrorist act" . . . "David Alfaro Siqueiros": Trotsky, "Letter to the Mexican Attorney General, May 27, 1940, *Writings*, 12:223–27.

257   Room 37 at the Hotel Europa . . . payoff money: Salazar, 49.
258   could not have been bought . . . why organize twenty to thirty raiders: Trotsky,
       "False Suspicions about Robert Sheldon Harte" (unpublished manuscript in
       Russian), July 15, 1940, Hansen papers, 69:57; Natalia, 258; Natalia, "Father and
       Son," in *Leon Trotsky*, 42.
258   Harte's peculiar behavior: Salazar, 24.
258   intestinal problems: Trotsky, "False Suspicions"; Trotsky to Jesse Harte, July 10,
       1940, TEP 8452.
258   Trotsky's Russian secretary . . . "Pure coincidence": Salazar, 94–95.
258   Salazar broke the case: Salazar, 34–42; Levine, 95–96.
258   confession of Néstor Sánchez Hernández: text in Hansen papers, 70:3; Salazar
       42–49.
259   arrests of some two dozen people: Levine, 95–96; Salazar, 65–66; "Trotsky case,"
       undated press release, TC 23:12.
259   rearrest of Casas: Schultz to Solow, June 18, 1940, TEP 15388.
259   a farmhouse in the village of Santa Rosa . . . Lucky Strike: Salazar, 70; Levine,
       96–97.
260   the corpse was exhumed: Salazar, 70–78.
260   Bob's kinky red-brown hair . . . "Poor Bob" . . . his face streaked with tears . . . the
       morgue in San Angel: Hansen to Goldman, June 25, 1940, TEP 11402; Hansen
       to Reba Hansen, June 26, 1940, Hansen papers, 19:4; Salazar, 77.
260   One of the conspirators told the police: Salazar, 82–86; Levine 98; Kolpakidi,
       162; *Ocherki*, 101.
261   "His memory is spotless" . . . pantheon of his fallen secretaries: Trotsky, Natalia,
       and staff statement [June 25, 1940], TEP 8451; Trotsky statement of June 25,
       1940, *Writings*, 12:293–94; Hansen to Goldman, June 25, 1940, TEP 11402.
261   a stone plaque: Trotsky to Jesse Harte, July 10, 1940, TEP 8452.

Chapter Eleven: Deadline

262   "It was a real attack": Cannon to Goldman, June 14, 1940, TC 24:11.
262   bombs, not bullets: Hansen to Dobbs, July 31, 1940, TC 24:12.
262   several thousand dollars: "Memorandum for JPC," June 18, 1940, TC 23:12;
       Cannon to Goldman, June 11, 1940, TC 24:11.
262   "made the supreme sacrifice": Trotsky Defense Appeal, July 11, 1940, TC 23:12.
262   "depistolization": Hansen to Dobbs, July [?], 1940, TC 24:12; Hansen to Dave
       Hansen, July 21, 1940, Hansen papers, 17:9.
262   President Cárdenas had refused to name a successor . . . Camacho won an over-
       whelming victory: *Cambridge History*, 55–63.
263   the Socialist Workers Party raised over $2,250: Sylvia Caldwell to Hansen, Au-
       gust 6, 1940, TEP 6671.
263   sale of his archives: Trotsky to Rose Karsner, September 28, 1939, TC 11:42;
       Goldman to Trotsky, May 10, 1940, TEP 1569.
264   The precious cargo: Trotsky to Goldman, July 17, 1940, TC 10:72.
264   Hank Schultz, a comrade from Minneapolis: Dobbs, *Teamster Rebellion*, 138,
       152–53; *Teamster Power* (Monad Press, 1973), 52, 82; *Teamster Bureaucracy* (Monad
       Press, 1977), 49, 143.

264   "all caved under the overload": Robins to Rose Karsner, June 1, TC 24:11; Trotsky to Cannon, May 28, 1940, TEP 7569.

264   "indefatigable, absolutely selfless": Trotsky to Rose Karsner, July 27, 1940, TC 11:43.

264   due back at his job: Hansen to Cannon, July 9, 1940, TC 24:12; Dobbs to Hansen, July 12, 1940, TC 24:12.

264   renovations made to the house on Avenida Viena: Adam [Hank Schultz] to Smith [Dobbs], June 24, 1940, TC 24:11; Hansen, "With Trotsky to the End," in *Leon Trotsky*, 16–26; Hansen to Dobbs, July [?], 1940, TC 24:12; Adam [Schultz] to Dobbs, July 13, 1940, TEP 11156; Hansen to Dobbs, July 31, 1940, TC 24:12.

265   delayed by the police investigation: Hansen to Dobbs, July [?], 1940, TC 24:12; Adam [Schultz] to Dobbs, July 13, 1940, TEP 11156.

265   the heavily guarded city courthouse: Hansen to Reba Hansen, June 24, 1940, Hansen papers, 19:4.

265   the judge and his associates took about five hours ... hands on their guns: Hansen to Reba Hansen, July 21, 1940, Hansen papers, 19:4.

265   the Arenal brothers behaving like innocent museum-goers: FBI, 4:34, 5:46.

265   not to install a photoelectric alarm system: Dobbs to Hansen, July 31, 1940, TEP 6298; Hansen to Dobbs, August 6, 1940, TC 24:13; William Bryan to Dobbs, July 25, 1940, TC 24:12.

265   "the next attack will most likely be bombs": Hansen to Dobbs, July 31, 1940, TC 24:12.

266   "permanent threat of a new 'blitzkrieg' assault": Trotsky to Mr. Kay, August 3, 1940, TC 11:44; Craig II, 670–71.

266   "in the next attack the GPU will use other methods": Hansen, "With Trotsky to the End," in *Leon Trotsky*, 22.

266   Jacson had met Trotsky for the first time: "Memorandum of talk with Rosmers about Frank Jacson," August 23, 1940, TC 24:6; Rosmers, "We never saw 'Jacson' in Paris," TC 24:6; Henry Schultz statement, September 10, 1940, TC 24:7 [hereafter: HS]; Hansen, "With Trotsky to the End," in *Leon Trotsky*, 17.

267   Marguerite Rosmer was very close to Natalia: Van, 146.

267   Jacson entered the patio on May 28, at 7:58 a.m.: Levine, 93–94.

268   Trotsky surprised everyone: "Joe's notes on Trotsky," Hansen papers, 40:7.

268   Natalia wondered about this: Levine, 94.

268   Natalia was returned to the house: Levine, 94.

268   he was introduced to Dorothy Schultz: Dorothy Schultz undated statement, TC 24:7 [hereafter: DS].

268   question him about his name, which did not seem French: HS.

268   came by the apartment at least a dozen times ... he was close to the Trotskyist circle in Paris: DS, HS.

269   One of the casualties was Mark Zborowski: Sara Weber to Trotsky, July 16, 1940, TEP 5916.

269   Jacson's bragging tales ... Jacson was introduced to Cannon and Dobbs ... Jacson bought Natalia a gift of sour cream ... dinner at the Hotel Geneva ... difficult to pin him down: DS, HS.

270   that relatively harmless creature: Mosley, 79.

270   Jacson's conversations with Dorothy: DS.

270   his Buick, which he arranged to leave at the house: HS; Hansen, "With Trotsky to the End," in *Leon Trotsky*, 24; Levine, 101–2.

270   Mercader-Jacson went to New York: FBI, 1:109, 2:46; telegrams sent between Sylvia and Jac, November 1939 to July 1940, TC 24:5.

270   surprised to learn that her "husband" had visited the house: "Memorandum of talk with Hilda and Ruth Ageloff," August 24, 1940, TC 24:2.

270   he phoned Evelyn . . . a diamond-cutting syndicate: DS.

271   he came by the house at 2:40: Levine, 104.

271   She grew impatient, and then desperate: Sylvia to Jac, July 29, 1940, in FBI, 5:42.

271   very ill in a small town near Puebla: "Memorandum of talk with Hilda and Ruth Ageloff," August 24, 1940, TC 24:2.

271   an expensive box of chocolates: Hansen, "With Trotsky to the End," in *Leon Trotsky*, 24.

271   "Everything is in order": *Ocherki*, 102.

271   he had not dropped in on the headquarters of the Socialist Workers Party: Hansen, "With Trotsky to the End," in *Leon Trotsky*, 23.

271   a separate Workers Party: Cannon and Dobbs to Trotsky, April 13, 1940, TEP 801.

271   Burnham's astonishingly candid resignation letter: May 21, 1940, TEP 13826.

272   a petty-bourgeois fraud: Hansen to Reba Hansen, July 21, 1940, Hansen papers, 19:4.

272   an event arranged by Professor Hubert Herring: Herring to Trotsky, July 15, 1940, TEP 1977.

272   Charles Orr: Charles A. Orr, "Trotsky comme je l'ai vu à Mexico," *Cahiers Léon Trotsky*, No. 51, October 1993.

272   "The OM ripped into the democracies": Hansen to Reba Hansen, July 21, 1940, Hansen papers, 19:4.

272   a formal debate with the guards: Hansen to Reba Hansen, July 24, 1940, Hansen papers, 19:4.

273   Sylvia Ageloff, who flew in from New York: FBI, 1:114, 117–18.

273   the discussion centered on the Majority and Minority views: Levine, 112.

273   he barely said a word: Natalia, 265.

273   "he certainly didn't act like his old dynamic self": Hansen to Dobbs, August 11, 1940, TC 24:13; Hansen to Reba Hansen, August 9, 1940, Hansen papers, 19:4; Hansen, "With Trotsky to the End," in *Leon Trotsky*, 17.

273   an hourlong siesta after lunch . . . "It bores him stiff": Hansen to Reba Hansen, June 24, 1940, Hansen papers, 19:4.

273   "I will do everything to observe this new 'deadline' ": Trotsky to Malamuth, March 19, 1940, TEP 8984.

274   Trotsky accused the monthly *Futuro* . . . mobilizing Goldman in New York: Hansen to Goldman, June 28, 1940, TC 24:11; Trotsky to Goldman, July 3, 1940, TC 10:70; *Writings*, 12:305–15; Trotsky, "The Comintern and the GPU."

274   the preliminary hearing on July 2: Robins to Rose Karsner, July 2, 1940, TC 24:10.

275   "working like a steam engine": Hansen to Reba Hansen, June 11, 1940, Hansen papers, 19:4.

275   "It is imperative not to lose a single hour": Trotsky to Goldman, July 3, 1940, TC, 10:70.

275   His blood pressure was running extremely high. His lower back was giving him trouble: Hansen to Dobbs, July 31, 1940, TC 24:12.

275   a deposition asserting that Siqueiros was a Trotskyist: Trotsky, "The Comintern and the GPU"; Trotsky to Curtiss, August 2, 1940, TEP 7639.

275   On August 6 Trotsky held a press conference: Hansen to Dobbs, August 6, 1940, TC 24:13; *Writings*, 12:330.

275   Another head of the Stalinist hydra: Mosley, 124.

275   "Indignation, anger, revulsion?": Trotsky to Angelica Balabanoff, Deutscher III, 295.

275   "We await the new intrigue calmly": Trotsky, "The Comintern and the GPU."

276   The picnic on August 9: Hansen to Reba Hansen, August 9, 1940, Hansen papers, 19:4.

276   he liked to joke to Natalia: Natalia, "How It Happened," in *Leon Trotsky*, 35.

276   "My death . . . may lighten Seryozha's situation" . . . her husband grieving in his study: Natalia to Sara Weber, September 25, 1941, TC 26:32.

276   a private moment between father and son: Natalia, "Father and Son," in *Leon Trotsky*, 40.

277   he slid down low in the seat . . . "we must have two of the best drivers in the car": Hansen, "With Trotsky to the End," in *Leon Trotsky*, 23.

277   The guard now numbered seven: Robins to Comrade [Dobbs], July 12, 1940, TC 24:12.

277   Natalia . . . was now pushing for a threefold guard: Hansen to Dobbs, July 31, 1940, TC 24:12.

277   a Sioux Indian known as the Rainman: Dobbs, *Teamster Rebellion*, 120, 155; *Teamster Power*, 125; *Teamster Politics*, 141–43.

278   "sufficient experience, prestige, and authority": Dobbs to Hansen, July 26, 1940, TC 24:12; Schultz to Dobbs, July 30, 1940, TC 24:12.

278   Trotsky doubted the value of the Rainman coming down: Hansen to Dobbs, July 31, 1940, TC 24:12.

278   Trotsky's reaction annoyed the comrades: Dobbs to Rainbolt, August 9, 1940, TC 24:13; Dobbs to Hansen, August 13, 1940, TC 24:13.

278   Trotsky was not always the most cooperative subject to guard . . . the indignity of a personal search: Hansen, "With Trotsky to the End," in *Leon Trotsky*, 20.

278   Robins proposed that Trotsky always be accompanied: Robins memoir, TC 30:1.

278   a bulletproof vest and a siren: Trotsky to Charles Curtiss, August 16, 1940, TEP 7640.

279   room 113 of the Hotel Montejo: FBI, 1:114; Levine, 115–16.

279   Sylvia was troubled by the changes she observed in Ramón's health: Sylvia to Hilda Ageloff, August 16, 1940, TC 24:12.

280   Jacson's haggard appearance and nervous twitching: Hansen, "With Trotsky to the End," in *Leon Trotsky*, 20.

280   Van sent a telegram: Feferman, 195.

280   his accent was not quite French . . . infatuated with Jacson: Van, 146–47; Feferman, 192–93; Walta Karsner statement, August 21, 1940, TC 24:4; statement of the secretary of the Belgian Legation, in *Excelsior*, August 28, 1940, copy in TC 24:1.

280   "It would be really too cruel": Trotsky to Van, August 2, 1940, TEP 10706.

280   On August 17 at 4:35 p.m.: Levine, 114–15.

280   his broad shoulders slightly stooped: Levine, 161.

280   "his clothes flop on him like a scarecrow": Sylvia to Hilda Ageloff, August 16, 1940, TC 24:2.

281   six feet in his shoes: Robins memoir, TC 30:1.

281   souvenir slugs from the Siqueiros raid: Dugrand, 50; Hansen to Dave Hansen, July 21, 1940, Hansen papers, 17:9; Levine, 124.

281   The entire visit took only eleven minutes: Levine, 114.

281   "I don't like him": Natalia, "How It Happened," in *Leon Trotsky*, 38.

281   dark clouds gathered in clusters: Levine, 115; Hansen, "With Trotsky to the End," in *Leon Trotsky*, 24.

282   He told Natalia he felt well: Natalia, "How It Happened," in *Leon Trotsky*, 38.

282   a telegram from Al Goldman in New York: TEP 1581.

282   "Disloyalty is always bad": Trotsky to Goldman, August 17, 1940, TEP 8340.

282   "a difference in comfort between various cars in a railway train": *Writings*, 12:221.

282   "civil liberties and other good things in America": Trotsky to Friends, August 13, 1940, TEP 7570.

282   an American brand of militarism: Hansen to Dobbs, August 6, 1940, TC 24:13.

283   "very pretentious, very muddled, and stupid": *Writings*, 12:341, 410–18.

283   Trotsky's Mexican attorney: Natalia, "How It Happened," in *Leon Trotsky*, 38.

283   punctuating the end of each sentence: "*Tochka!*": Hansen, "With Trotsky to the End," in *Leon Trotsky*, 24.

283   two congratulatory letters to comrades in Minneapolis: TEP 10971, 10529.

283   The last letter of the day: Trotsky to Schultz, August 20, 1940, TC 12:9.

283   Hansen was on the roof near the blockhouse: account of the murder drawn from Hansen, "With Trotsky to the End," in *Leon Trotsky*, 16–26; Natalia, "How It Happened," in *Leon Trotsky*, 35–39; Natalia, 266–70; for French-language original of Natalia's account: Victor Serge, *Vie et Mort de Trotsky* (Amiot Dumont, 1951); Levine, 111–32; Salazar, 140–42; Hansen, "On the 23rd Anniversary of the Russian Revolution," undated manuscript, Hansen papers, 40:26.

284   wet grass made their bellies swell: Salazar, 135.

287   Caridad Mercader and Leonid Eitingon: Sudoplatov, 78; Levine, 120, 131.

289   "Look, we found a barber": Natalia to Sara Weber, January 8, 1952, TC 26:31.

289   The doctors trepanned an area of the right parietal bone: Salazar, 103.

290   The direction of the pickax: Salazar, 135.

290   The first medical bulletin: medical reports and press releases are in TC 23:13.

290   the patient's chances were one in ten: *The New York Times*, August 21, 1940.

290   She sat beside him, dressed in a white hospital gown: *The New York Times*, August 21, 1940.

290   She was waiting for him to wake up: Natalia to Sara Weber, September 25, 1941, TC 26:32.

290   Colonel Salazar arrived to question Mercader: Salazar, 125–37.

290   the police found a dagger: Salazar, 105.

290   letter of confession: TC 24:5; Salazar, 128–31.

291   "It was a veritable maze": Salazar, 142.

291   Mercader's account of the details of his crime: *Excelsior*, August 26, 1940, copy in TC 24:5.

291   she rushed over to the house: "Declaration of Sylvia Ageloff," *Excelsior*, August 27, 1940, copy in TC 24:2; Salazar, 143–44.

291   "Kill him! Kill him!": Levine, 130–31.

292   "the flickering life of our Old Man": TC 24:14.

292   Van was out taking a walk: Van, 147.

292   Trotsky's breathing had become more rapid: Natalia, "How It Happened," in *Leon Trotsky*, 39.

292   "Gentlemen! Trotsky is dead!": Salazar, 108.

292   pressed her face against the soles of her husband's feet . . . "Everything is finished": Natalia to Sara Weber, September 25, 1941, TC 26:32.

Epilogue: Shipwreck

293   Natalia was increasingly skeptical . . . the Korean War: Natalia to the Executive Committee of the Fourth International, May 9, 1951, TC 26:13; *The New York Times*, June 8, 1951.

294   "neo-Trotskyist deviation": *Time*, April 10, 1964.

294   Natalia addressed a letter to the Soviet government: TC 26:30.

295   Soviet intelligence hatched an escape plan for Mercader: Salazar, 216–29; Sudoplatov, *Raznye dni tainoi voiny i diplomatii, 1941 god* (Olma-Press, 2001), 141–42; Kolpakidi, 170–85; *Venona*, 279; *Ocherki*, 106.

295   Caridad seems to have lost her bearings: Levine, 215–22.

295   private Kremlin ceremony on June 17, 1941: Sudoplatov, *Raznye dni*, 141–42; Nikandrov, 134.

295   Ramón's recognition would have to wait: Sudoplatov, *Raznye dni*, 141.

295   his true identity was discovered in 1950: Salazar, 231–35; Levine, 187–214.

295   Ramón never forgave her: Nikandrov, 133.

295   Hero of the Soviet Union, the Order of Lenin, and the Gold Star medal: Kolpakidi, 12.

296   Diego Rivera and Frida Kahlo became ardent supporters of Stalin's USSR: Herrera, 249, 341–42.

296   a small bust of Stalin: Van, 160.

296   Frida's funeral: Herrera, 436.

296   Colonel Salazar finally caught up with the fugitive painter: Salazar, 184–201.

296   At the trial, Siqueiros spoke passionately . . . moving to Chile: Stein, 121–30.

297   massive interior mural: Rochfort, 199, 207–11.

297   National Art Prize . . . hero's burial: *The New York Times*, January 7, 1974.

297   a team of FBI agents and U.S. marshals: Dobbs, *Teamster Bureaucracy*, 137, 145, 169–283; Myers, *The Prophet's Army*, 177–88.

298   Sylvia Caldwell: *Venona Secrets*, 359–61.

298   the law also caught up with Mark Zborowski: Poretsky, 271–74; *Venona*, 257–58; *Venona Secrets*, 368–74.

299   Alexander Orlov . . . surfaced in New York: *Deadly Illusions*, 339–48.

299   The FBI assumed . . . "Trotskyist invention": *Deadly Illusions*, 290.

299   Krivitsky . . . was found dead: Kern, *A Death in Washington*.

300   U.S. Senate's Internal Security Subcommittee: *Legacy*, 15–31.

300   Zborowski skillfully ducked and weaved: *Scope of Soviet Activity in the United States: Hearing before the Subcommittee to Investigate the Administration of the Internal Security Act and Other Internal Security Laws of the Committee on the Judiciary of the United States Senate, Eighty-Fourth Congress, Second Session* (United States Government Printing Office, 1956), 77–101, 103–35.

300   Zborowski appeared in court to testify . . . Zborowski was convicted: *The New York Times*, November 6, 1956; April 22 and 26, 1958; December 14, 1962.

300   Van had been out of the Trotskyist movement: Feferman, 215–18.

301   Max Eastman and Sidney Hook: Wald, 271–74, 290–94.

301   Max Shachtman: Peter Drucker, *Max Shachtman and His Left: A Socialist's Odyssey Through the "American Century"* (Humanities Press, 1994), 218–311.

301   James Burnham . . . *The Managerial Revolution*: Kelly, 97.

301   Burnham fell out with Cold War liberals . . . *National Review*: Kelly, 183–237.

301   President Reagan famously declared . . . Presidential Medal of Freedom: http://www.reagan.utexas.edu/archives/speeches; Kelly, 365.

301   Trotskyist sects endured: Wald, 295–310.

302   Columbia University . . . Yulia Akselrod . . . Seryozha was shot: Yulia Akselrod, "Why My Grandfather Leon Trotsky Must Be Turning in His Grave," *Commentary*, April 1989.

303   Van met a violent end: Feferman, 361–62.

303   from Gorbachev's point of view, there was no room for Trotsky: *The New York Times*, November 8, 1987.

304   the Soviet government granted Seva a visa . . . "people from a shipwreck": "Trotsky's Grandson in Moscow," *Workers Vanguard*, March 31, 1989, available online at http://www.ucc.ie/acad/appsoc/tmp_store/mia_2/Library/history/etol/document/family/volkov.htm.

304   Albert Glotzer . . . "simply bullshit": Glotzer to Bellow, August 30, 1991, Glotzer papers, box 48; on Bellow, see Wald, 246–47.

305   "Trotsky was lying dead with a bloody turban of bandages": Bellow to Glotzer, August 7, 1990, Glotzer papers, box 48; Saul Bellow, "Writers, Intellectuals, Politics," *The National Interest*, Spring 1993.

305   "The Soviet Union will live and develop" . . . "Optimism was all he really had": Glotzer, 314.

INDEX

Entries in *italics* refer to illustrations.